VOLUME TWO HUNDRED AND SIX

Advances in
IMAGING AND
ELECTRON PHYSICS

EDITOR-IN-CHIEF

Peter W. Hawkes
CEMES-CNRS
Toulouse, France

VOLUME TWO HUNDRED AND SIX

Advances in
IMAGING AND
ELECTRON PHYSICS

Edited by

PETER W. HAWKES
CEMES-CNRS
Toulouse, France

ACADEMIC PRESS
An imprint of Elsevier

Cover photo credit:
The cover picture is taken from Fig. 2.5 of the chapter by Axel Lubk (p. 36).

Academic Press is an imprint of Elsevier
125 London Wall, London EC2Y 5AS, United Kingdom
525 B Street, Suite 1650, San Diego, CA 92101-4495, United States
50 Hampshire Street, 5th Floor, Cambridge, MA 02139, United States
The Boulevard, Langford Lane, Kidlington, Oxford OX5 1GB, United Kingdom

Notices

Knowledge and best practice in this field are constantly changing. As new research and experience broaden our understanding, changes in research methods, professional practices, or medical treatment may become necessary.

Practitioners and researchers must always rely on their own experience and knowledge in evaluating and using any information, methods, compounds, or experiments described herein. In using such information or methods they should be mindful of their own safety and the safety of others, including parties for whom they have a professional responsibility.

To the fullest extent of the law, neither the Publisher nor the authors, contributors, or editors, assume any liability for any injury and/or damage to persons or property as a matter of products liability, negligence or otherwise, or from any use or operation of any methods, products, instructions, or ideas contained in the material herein.

ISBN: 978-0-12-815216-4
ISSN: 1076-5670

For information on all Academic Press publications
visit our website at https://www.elsevier.com/books-and-journals

  Working together
to grow libraries in
developing countries

www.elsevier.com • www.bookaid.org

Publisher: Zoe Kruze
Acquisition Editor: Jason Mitchell
Editorial Project Manager: Shellie Bryant
Production Project Manager: Divya KrishnaKumar
Designer: Matthew Limbert

Typeset by VTeX

CONTENTS

CONTRIBUTOR

Axel Lubk
Institute for Structure Physics, Physics Department, Faculty of Mathematics and Natural Sciences, Technical University of Dresden, Dresden, Germany

PREFACE

This study of electron tomography and holography by Axel Lubk makes a valuable contribution to thinking about the phase problem, which arises from the fact that detectors record amplitude and not phase. Many ways round this difficulty are known and here, Lubk reconsiders the problem by regarding electron holography as a quantum mechanical phase-space reconstruction procedure. The role of the Wigner function, with which much formerly disparate theory can be unified, is central to this approach. The theoretical part is complemented by several practical examples.

I am extremely pleased to include this very original work in these Advances, in which the first long publication on the phase problem was published by Owen Saxton when the subject was still young. Axel Lubk has gone to great trouble to make his work accessible to audiences from different fields.

Peter W. Hawkes

ACKNOWLEDGMENT

With the following lines, I would like to express my deep gratitude to those people who significantly contributed to this thesis in one or the other way. However, the time frame of several years for creating the work involves an interaction and stimulation in a variety of work and life situations so immense that adequately thanking all the loved ones, friends, colleagues, and discussion partner seems a hopeless endeavor. I apologize to anybody involved and not mentioned here. The most important support were Claudia, Anton, and Helene, without your love and joy, the study of the rather abstract natural sciences would make no sense. I want to thank you, mum, for your continued support in all walks of life, and you, papa, for your unfailing inspiration. The same goes for you my little brother, in particular for having an eye on my personal physical condition. Without your perpetual help in all situations, Waltraut and Rainer, life would be much more difficult. The "Triebenberger" and particularly Hannes Lichte have been an invaluable support, inspiration, and friends. I owe you all my knowledge about electron microscopy, holography, and tomography. Thank you Daniel, Falk, Jonas, Andreas, Bernd, John, Karin, Marianne, Petr, Jan, Felix, and Heide. It is a shame what happened to the lab. Time spent at the CEMES institute in Toulouse was a source of joy and inspiration very special for me because of you, Martin, Stephen, alter Schwede, Shay, Christophe, Florent, Patricia, and Elsa. I would also like to thank Johan Verbeeck, Jan Rusz, Giulio Pozzi, Juri Barthel, and Heiko Müller for a multitude of stimulating discussions on the different aspects of electron microscopy.

ABBREVIATIONS AND SYMBOLS

e	elementary charge
m_0	electron mass
c	velocity of light
μ_B	Bohr magneton
$\hbar$	reduced Planck constant
ϵ_0	vacuum permittivity
$\mathbf{r, s}$	spatial coordinates
$\mathbf{k, q}$	reciprocal coordinates
$\mathbf{R}$	second rank tensor/2-dimensional matrix
$\hat{H}, \hat{p}$	Hermitian operator
$\tilde{\Psi} = \mathcal{F}\{\Psi\}$	Fourier transformed function
$F = \mathcal{R}\{f\}$	Radon transformed function
$\langle \cdot, \cdot \rangle$	scalar product
(S)TEM	(Scanning) Transmission Electron Microscopy/Microscope
EH	Electron Holography
EHT	Electron Holographic Tomography
DFEH	Dark Field Electron Holography
HR/MR	High-Resolution/Medium-Resolution
EELS	Electron Energy Loss Spectroscopy
POA	Phase Object Approximation
MIP	Mean Inner Potential
MFPL	Mean Free Path Length

CHAPTER ONE

Holography and Tomography with Electrons
From Quantum States to Three-Dimensional Fields and Back

Axel Lubk

Institute for Structure Physics, Physics Department, Faculty of Mathematics and Natural Sciences, Technical University of Dresden, Dresden, Germany
e-mail address: axel.lubk@tu-dresden.de

Contents

> *"Science is not everything, but science is very beautiful."*
>
> ***(Robert Oppenheimer)***

The prefixes "holo" ($\cong$ whole) and "tomo" ($\cong$ section) combined with the suffix "graphy" ($\cong$ writing, drawing) denote two prevalent imaging techniques, which even affect our daily lives occasionally, for instance, when looking into the holographic face of Beethoven on a credit card, or, more unpleasantly, when being checked for fractures in a Computed Tomographic (CT) scanner at the hospital.

Like so many other great discoveries, the holographic principle was somewhat unexpectedly stumbled upon by the Hungarian–British physicist Gabor, while attempting to correct the large spherical aberration of Transmission Electron Microscopes in the late 1940s (Gabor, 1948). He employed the term "holo" to describe the most important outcome of his interferometric technique, the reconstruction of a complete wave function in amplitude and phase (Fig. 1.1). Holography proved such a general principle that he was honored with the Nobel Prize in 1971 for this invention.

The tomographic imaging of 3D volumes by 2D sections, which in many cases are computed from projections (Fig. 1.2), has various roots. Its mathematical foundations have been laid at the beginning of the 20th century, when Radon and Funk investigated, how to represent functions by their integrals on certain submanifolds (Funk, 1913;

Advances in Imaging and Electron Physics, Volume 206
ISSN 1076-5670
https://doi.org/10.1016/bs.aiep.2018.05.001
1

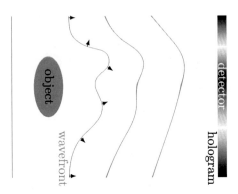

Figure 1.1 Gabor's holographic principle. A wave impinging on an object is modulated in its phase. The bended wave front generates an amplitude modulation further downstream in the hologram, because different partial waves of the transmitted wave superimpose constructively or destructively according to their relative phase shifts. By illuminating the hologram with a replica of the undisturbed wave the scattered wave may be reconstructed in amplitude and phase.

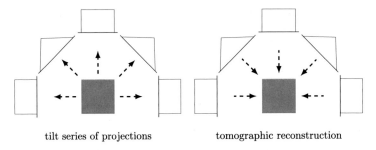

tilt series of projections   tomographic reconstruction

Figure 1.2 Tomographic reconstruction from projections. The set of projections recorded in a first step is backprojected (reconstructed) in a second step to recover the underlying object.

Radon, 1917). These studies provided the mathematical basis for computed X-ray tomography, supplanting earlier planographic methods (Pollak, 1953) suffering from shadowing effects. Cormack and Hounsfield have been honored with the Nobel prize for medicine in 1979 for their ground-breaking contributions (Cormack, 1963; Hounsfield, 1973) in the development of X-ray tomography to become one of the most important diagnostic tools in modern medicine.

By now both holography and tomography penetrate a large number of different scientific disciplines and found their ways in a large number of applications. For instance, the holographic recording and reconstruction of photon, electron, or sound waves is used for holographic data storage or security holograms, for measuring electric or magnetic fields, or for localizing sound sources. The pervasiveness of tomographic techniques is even larger. X-ray CT scanners can be found in hospitals around the world, often joined by magnetic resonance tomographic scanners (Lauterbur, 1973; Mansfield & Grannell, 1973) and positron emission tomographic scanners (Townsend, 2008). Seismic tomography of subterranean sound waves is used to study deep geologic structures (Nolet, 1987). Quantum state tomography is employed for the characterization of quantum particles such as atoms (Lvovsky & Raymer, 2009). Tomographic methods even play a fundamental role in pure mathematics such as the theory of partial differential equations (Helgason, 2011).

Thus, retrospectively, both terms, holography and tomography, have been chosen with great foresight, enabling later generalizations to equally fit into the given schemes. This is insofar remarkable because both techniques share a number of common aspects. Tomography seeks a reconstructing of the complete data from lower dimensional projections, hence has a holographic aspect. Holograms, on the other hand, often represent certain sections in the phase space representation of the wave function to be reconstructed. Indeed, tomography and holography could be conceived as antipodes, which attract each other in the sense that they can be combined in various ways to obtain more complete information about (quantum mechanical) wave fields and their interaction with matter.

In this volume we will explore the above outlined connections between electron holography and tomography; and exploit them for novel characterization techniques within the framework of Transmission Electron Microscopy (TEM). The two central themes are (A) the reconstruction of three-dimensional physical, i.e., electric, magnetic, and strain, fields by combining the holographic recording of projected fields with tomographic tilt series reconstruction, and (B) the extension of the holographic reconstruction of wave functions (so-called pure quantum states) to that of incoherent superpositions of wave functions (mixed quantum states) by recording multiple holograms (i.e., sections of that mixed state) under varying interference conditions. Both theoretic and experimental aspects will be discussed, albeit with a certain imbalance towards the former, owing to the bias of the author.

In the remainder of these introducing remarks we will embark on a very brief historical journey through Transmission Electron Microscopy, Electron Holography (EH), and Electron Tomography (ET), which serves us to provide a red ribbon of topics discussed in this volume. The Transmission Electron Microscope (also denoted by TEM in the following) is an indispensable tool in today's materials science. Its widespread use owes to the convergence of the high spatial resolution and the strong interaction of the electron beam with the sample. The capabilities of a TEM to analyze the scattered electrons is determined by its optical components. That includes the electron gun, the accelerator, the condenser lenses, the objective lens, the projector system, and the detector.

All these components have been improved and revolutionized throughout the decades following the invention of the TEM by Ruska in 1931. Milestones in that evolution have been the development of field emission electron sources (Crewe, Eggenberger, Wall, & Welter, 1968; Tonomura, Matsuda, Endo, Todokoro, & Komoda, 1979), hardware aberration correctors (Haider et al., 1998; Krivanek, Dellbya, & Lupinic, 1999), energy filters (Reimer, 1992), and CCD detectors (Daberkow, Herrmann, Liu, & Rau, 1991; Krivanek & Mooney, 1993) (see Fig. 1.3 for a state-of-the-art TEM). These instrumental developments have been paralleled by an ever improving understanding of the elastic and inelastic scattering processes occurring in the sample, resulting in the development of dynamic scattering theory (Bethe, 1928; Reimer, 1989; Kirkland, 1998; De Graef, 2003). Fundamentals of dynamic electron scattering and TEM image formation relevant for this work are provided in Chapters 2 and 4, respectively.

Historically, electron microscopists mainly studied structure properties such as the symmetry of crystal lattices and deviations from which, e.g., stacking faults or dislocations. Energy filters and spectrometers added the possibility to analyze the chemical composition of the sample through characteristic electron-energy-loss processes. A further degree of freedom represented the in-situ application of external stimuli, such as varying temperatures or strain. The sum of these developments nowadays permits the quantitative analysis of TEM micrographs in terms of physical properties of the sample much beyond simple morphological features. It should be noted, however, that the ever increasing complexity (in terms of optical degrees of freedom) of the TEM requires ever more experienced microscopists.[1] For instance, a hardware aberration corrector contains a multitude of lenses and multipole elements (see aberration corrector in Fig. 1.3), which need to be adjusted to eliminate or deliberately tailor specific aberrations of the

---

[1] The neat coffee-machine-like appearance of modern TEMs can easily feign the opposite.

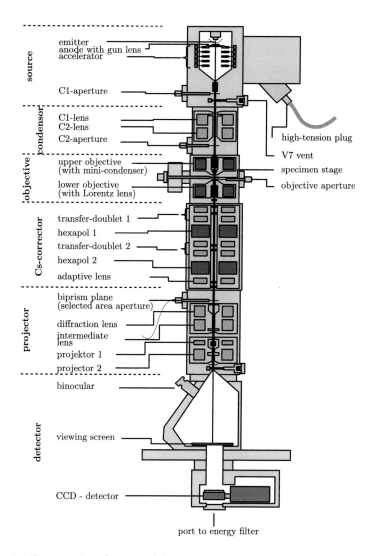

Figure 1.3 Cross-section of a state-of-the-art TEM.

TEM's electron optics. Large parts of this volume rely on the capabilities of expert microscopists to control these increased degrees of freedom in an almost artistic way. I cannot overemphasize this fact, in particular because the funding of TEM specialists seems to be frequently neglected in the funding process for expensive state-of-the-art TEM equipment. I hope the reader will pardon me this personal note and keeps it in the back of his mind, when reading the text.

Electron Holography extended the scope of TEM to the investigation of electric and magnetic fields as follows. Consider an electron wave transmitting a thin object of refractive index unequal one, which is proportional to electric and magnetic fields in the sample (Glaser, 1933). When leaving the sample the wavefront will be modulated by a corresponding phase shift (Fig. 1.1). The amplitude on the other hand remains constant, which is why such objects are classified as pure phase objects. This type of scattering is a limiting case to the general dynamic scattering of electrons elaborated on in Chapter 2. The intensity recorded under perfect imaging conditions (no aberrations), however, will remain unaffected by this phase shift, i.e., blind to the refractive index of the object. To overcome this obstacle in light microscopy, Zernike proposed his famous phase plate (Nobel Prize in Physics, 1953), which shifts the relative phase of the unscattered with respect to the scattered beam in order to reveal the phase distribution as an intensity modulation (Zernike, 1935). This phase contrast represents a huge benefit for studying biologic specimen under the light microscope because they are mostly phase objects.

Unfortunately, it remains technologically demanding to realize the Zernike phase plate in a TEM (Nagayama, 2005; Gamm et al., 2010), mainly due to contamination and stability issues. Moreover, wave aberrations (e.g., Scherzer focus) mimicking a Zernike phase-plate (Scherzer, 1949) over a wide band of spatial frequencies in the back focal plane (e.g., "Scherzer band"), proved inadequate, e.g., to detect large-area phase structures such as extended electromagnetic fields.

Gabor solved that problem by showing how to reconstruct a wave in amplitude and phase from a hologram, which he considered as a diffraction grating containing a replica of the complete wave (Gabor, 1948). From the start he noted two general obstacles marring his holographic setup (Gabor & Goss, 1966).

The first results from the extraction of a wave function from recorded intensities. Mathematically, holography can be considered as some sort of square root operator applying to the recorded image intensity, which results in multiple solutions as a matter of fact. The associated ambiguities manifest as twin image problems (Haine & Mulvey, 1952; Gabor & Goss, 1966; Collier, Burckhardt, & Lin, 1971; Völkl, Allard, & Joy, 1999), non-converging holographic reconstruction algorithms (Fienup & Wackerman, 1986), or non-unique wave reconstructions (Lubk, Guzzinati, Börrnert, & Verbeeck, 2013) depending on the particular holographic technique. To resolve these issues, the off-axial superposition of an undisturbed reference wave, separated from the initial wave by some sort of wave splitter, proved particularly powerful, because it permitted the complete removal of the twin image for band-limited wave functions. Möllenstedt and coworkers pioneered off-axis electron interferometry, which they used to measure electric potentials, e.g., occurring at contacts between different metals (Möllenstedt & Buhl, 1957; Möllenstedt & Keller, 1957; Krimmel, Möllenstedt, & Rothemund, 1964; Jönsson, Hoffmann, & Möllenstedt, 1965; Lichte, Möllenstedt, & Wahl, 1972), as well as magnetic fields in and around solids, including the Ehrenberg–Siday–Aharonov–Bohm effect (Bayh, 1962), which was later repeated by Tonomura (Tonomura, Osakabe, Matsuda, Kawasaki, & Endo, 1986). By now further advanced holographic methods (see Cowley, 1992 for an overview), including those using a focal series of inline holograms (Focal Series Inline Holography; Coene, Janssen, de Beeck, & Van Dyck, 1992; Thust, Coene, de Beeck, & Van Dyck, 1996) and two slightly defocused inline holograms (Transport of Intensity Holography; Teague, 1983), have been used in abundant measurements of electric and magnetic fields (Tonomura, 1987; Tonomura, Allard, Pozzi, Joy, & Ono, 1995; Völkl et al., 1999; Dunin-Borkowski et al., 2004; Lehmann & Lichte, 2005; Lichte et al., 2013; Pozzi, Beleggia, Kasama, & Dunin-Borkowski, 2014). The fundamentals of various holographic methods including Off-Axis Holography, Inline Holography, Differential Phase Contrast and Ptychography, which are important for this work, are provided in Chapter 5.

The second obstacle to holography is the presence of partial coherence dampening and eventually destroying the observed interference patterns. With the advent of the LASER this problem was experimentally resolved in light optics (Leith & Upatnieks, 1962; Denisyuk, 1962), whereas the wide-spread realization of electron holographic schemes was delayed until the introduction of highly coherent electron emitters, namely the cold field emission gun and field-assisted thermionic Schottky gun (Wahl, 1975; Tonomura et al., 1979). Partial coherence results from the incoherent superposition of wave functions, e.g., originating from different non-correlated points of the emitter surface or from inelastic scattering at the object. As a matter of fact it cannot be avoided in the TEM to a certain extent, which has important consequences not only for electron holography but also for the formation of conventional images, in particular at atomic resolution (Rose, 1976). Incoherent superpositions, also referred to as mixed electron states in contrast to pure wave functions, cannot be described by a single amplitude and phase. Thus, an extension of the wave function calculus to density operators (von Neumann, 1927) (mutual coherence function in optics; Born & Wolf, 1999) is required. Similar to other operators, the density operator may be represented in various bases, e.g., the position representation or the momentum representation, where the coefficients of the operator are referred to as density matrices. Contrary to the wave function, however, it is also possible to use a mixed position and momentum representation, referred to as quantum mechanical phase space (Schleich, 2001). This representation exhibits remarkable analogies to classical phase space permitting a concerted discussion of classical and quantum mechanical electron optics. Therefore, the theory of partial coherent imaging in the TEM, required within the scope of this volume, is developed in Chapter 4 by employing the density operator calculus in the phase space representation.

Since both, the twin image and coherence problem have their very origin in the sought-after reconstruction of a pure state from image intensities, a change of perspective seeking the reconstruction of the density operator or components of which instead of the wave, is suited to elegantly circumvent both issues. These so-called quantum state reconstructions or quantum state tomographies (Smithey, Beck, Raymer, & Faridani, 1993; Schleich & Raymer, 1997; Paris & Rehacek, 2004; Lvovsky & Raymer, 2009), have been very successfully employed under experimental conditions preventing sufficiently coherent wave fields, e.g., for the study of heavy quantum particles, such as atoms or molecules (Leibfried, Pfau, & Monroe, 1998; Kienle, 2000), or incoherent light sources (Breitenbach, Schiller, & Mlynek, 1997; Schleich, 2001; Cámara, 2015). Within the context of TEM, however, quantum state reconstructions have been rarely discussed (see Röder, Lubk, Wolf, & Niermann, 2014 and references therein). One main objective of this work is to fill this gap. The fundamental properties of electron quantum states including the equations of motion relevant for electron holographic experiments in the TEM are provided in Chapter 2. Chapter 5 elaborates on the extension of Electron Holography to Quantum State Reconstructions.

With the advent of the computer a widespread application of computed tomography for medical purposes became feasible. The first tomographic TEM studies also concerned biological specimen (de Rosier & Klug, 1968; Hoppe, Langer, Knesch, & Poppe, 1968; Hart, 1968; Hoppe, 1969) (see Hoppe & Hegerl, 1980; Baumeister, Grimm, & Walz, 1999; Frank, 2008; Stahlberg & Walz, 2008 for a more comprehensive account). Subsequently, electron tomography was also employed to other materials, such as catalysts (Koster, Ziese, Verkleij, Janssen, & de Jong, 2000; Kuebel et al., 2005) and polymers (Midgley, Ward, Hungría, & Thomas, 2007; Midgley & Dunin-Borkowski, 2009). In order to reconstruct electric and magnetic fields in three dimensions, Tonomura et al. combined Electron Holography and Tomography (EHT) (Lai, Hirayama, Fukuhara, et al., 1994; Lai, Hirayama, Ishizuka, & Tonomura, 1994). After this early proof of concept experimental issues delayed a quick development of this complex technique. With the development of automated holographic tilt series acquisition procedures (Wolf, Lubk, Lichte, & Friedrich, 2010) these experimental obstacles could be largely removed, facilitating a multitude of EHT studies, e.g., to reveal electric fields in depletion regions (Twitchett-Harrison, Yates, Newcomb, Dunin-Borkowski, & Midgley, 2007; Wolf, Lubk, Lenk, Sturm, & Lichte, 2013) or compositional variations (Wolf, Lichte, Pozzi, Prete, & Lovergine, 2011; Lubk, Wolf, Prete, et al., 2014). Current challenges pertain to improving the signal-to-noise ratio and spatial resolution in the reconstructed data. Moreover, an extension of the reconstructed quantities, e.g., to all three components of a magnetic vector field or strain field, is envisaged. Proposal and partial realization of novel electron holographic tomographies suited to overcome the above obstacles is the second focus of this volume. To this end the ramifications of (dynamic) electron scattering, provided in Chapter 2, need to be combined with the tomographic principle, laid out in Chapter 3, leading to novel experimental procedures and reconstruction principles realized in the final Chapter 6 of this volume. Since the material presented in this volume has been mainly gathered during my two PostDoc periods at the CEMES (CNRS) in Toulouse (France) and the Triebenberg Laboratory of the Technische Universität Dresden (Germany), it is also part of a number of publications concerned with Electron Holography and Tomography, most important those published from 2013 to 2016. Therefore, additional information may be drawn from the following list of references (Lubk et al., 2013; Lubk, Javon, et al., 2014; Lubk, Wolf, Kern, et al., 2014; Lubk, Wolf, Prete, et al., 2014; Lubk, Béché, & Verbeeck, 2015; Lubk & Röder, 2015a, 2015b; Lubk & Rusz, 2015; Lubk & Zweck, 2015; Lubk et al., 2016).

It is needless to say that the understanding of nanoscale materials and solid state physics phenomena greatly benefits from a comprehensive three-dimensional and nondestructive characterization method as provided by Electron Holography and Tomography. For instance, determining the structure and composition of small volumes is vital to control semiconductor device manufacturing. Similarly, the efficient and reliable syntheses of nanoparticles or nanowires for novel optoelectronic, photovoltaic, and magnetic memory devices, or strain engineering of transistor channels, improving the charge carrier mobility and thus the performance of processors, greatly benefit from a comprehensive understanding of the involved electric, magnetic and strain field distributions in three dimensions. This is because the functionality of the material or device often depends on minute changes in the local crystal structure and chemical composition; and various elementary processes, often depending on the fabrication parameters of the device, may significantly change its performance. Current nanoscale characterization methods only inadequately convey this information, e.g., because they probe surfaces (e.g., Scanning Tunneling Microscopy), record projections (e.g., conventional Transmission Electron Microscopy), or lack spatial resolution (e.g., X-ray Microscopy). Moreover, they often probe an insufficient set of the material's properties or destroy the sample while measuring (e.g., Atom Probe Tomography), rendering a repeatable measurement with different methods or under different conditions impossible. The main purpose of the holographic and tomographic methods discussed in this volume is to overcome this limitation by mapping electric and magnetic fields as well as crucial properties of the underlying atomic charge distribution, such as the chemical composition or mechanical strains in three dimensions.

REFERENCES

Baumeister, W., Grimm, R., & Walz, J. (1999). Electron tomography of molecules and cells. *Trends in Cell Biology, 9*, 81.

Bayh, W. (1962). Messung der kontinuierlichen Phasenschiebung von Elektronenwellen im kraftfeldfreien Raum durch das magnetische Vektorpotential einer Wolfram-Wendel. *Zeitschrift für Physik, 169*(4), 492–510.

Bethe, H. (1928). Theorie der Beugung von Elektronen an Kristallen. *Annalen der Physik, 392*(17), 55–129.

Born, M., & Wolf, E. (1999). *Principles of optics: Electromagnetic theory of propagation, interference and diffraction of light*. Cambridge: Cambridge University Press.

Breitenbach, G., Schiller, S., & Mlynek, J. (1997). Measurement of the quantum states of squeezed light. *Nature, 387*(6632), 471–475.

Cámara, A. (2015). *Optical beam characterization via phase-space tomography*. Switzerland: Springer International Publishing.

Coene, W., Janssen, G., de Beeck, M. O., & Van Dyck, D. (1992). Phase retrieval through focus variation for ultra-resolution in field-emission transmission electron microscopy. *Physical Review Letters, 69*(26), 3743–3746.

Collier, R. J., Burckhardt, C. B., & Lin, L. H. (1971). *Optical holography* (student edition). Academic Press.

Cormack, A. M. (1963). Representation of a function by its line integrals, with some radiological applications. *Journal of Applied Physics, 34*, 2722–2727.

Cowley, J. M. (1992). Twenty forms of electron holography. *Ultramicroscopy*, *41*(4), 335–348.

Crewe, A. V., Eggenberger, D. N., Wall, J., & Welter, L. M. (1968). Electron gun using a field emission source. *Review of Scientific Instruments*, *39*, 576.

Daberkow, I., Herrmann, K.-H., Liu, L., & Rau, W. D. (1991). Performance of electron image converters with YAG single-crystal screen and CCD sensor. *Ultramicroscopy*, *38*(3–4), 215–223.

De Graef, M. (2003). *Introduction to conventional transmission electron microscopy*. Cambridge: Cambridge University Press.

de Rosier, D. J., & Klug, A. (1968). Reconstruction of three dimensional structures from electron micrographs. *Nature*, *217*(5124), 130–134.

Denisyuk, Y. N. (1962). On the reflection of optical properties of an object in a wave field of light scattered by it. *Doklady Akademii Nauk SSSR*, *144*, 1275–1278.

Dunin-Borkowski, R. E., Twitchett, A. C., Barnard, J. S., Broom, R. F., Midgley, P. A., Robins, A. C., & Fischione, P. E. (2004). An ultra-high-tilt two-contact electrical biasing specimen holder for electron holography and electron tomography of semiconductor devices. *Microscopy and Microanalysis*, *10*(S02), 1012–1013.

Fienup, J. R., & Wackerman, C. C. (1986). Phase-retrieval stagnation problems and solutions. *Journal of the Optical Society of America A, Optics and Image Science*, *3*(11), 1897–1907.

Frank, J. (Ed.). (2008). *Electron tomography: Methods for three-dimensional visualization of structures in the cell*. Springer.

Funk, P. (1913). Über Flächen mit lauter geschlossenen geodätischen Linien. *Mathematische Annalen*, *74*.

Gabor, D. (1948). A new microscopic principle. *Nature*, *161*, 777.

Gabor, D., & Goss, W. P. (1966). Interference microscope with total wavefront reconstruction. *Journal of the Optical Society of America*, *56*(7), 849–858.

Gamm, B., Dries, M., Schultheiss, K., Blank, H., Rosenauer, A., Schröder, R., & Gerthsen, D. (2010). Object wave reconstruction by phase-plate transmission electron microscopy. *Ultramicroscopy*, *110*(7), 807–814.

Glaser, W. (1933). Über geometrisch-optische Abbildung durch Elektronenstrahlen. *Zeitschrift für Physik*, *80*, 451–464.

Haider, M., Uhlemann, S., Schwan, E., Rose, H., Kabius, B., & Urban, K. (1998). Electron microscopy image enhanced. *Nature*, *392*, 768.

Haine, M. E., & Mulvey, T. (1952). The formation of the diffraction image with electrons in the Gabor diffraction microscope. *Journal of the Optical Society of America*, *42*, 763–773.

Hart, R. G. (1968). Electron microscopy of unstained biological material: The polytropic montage. *Science*, *159*(3822), 1464–1467.

Helgason, S. (2011). *Integral geometry and Radon transforms*. New York: Springer.

Hoppe, W. (1969). Beugung im inhomogenen Primärstrahlwellenfeld. I. Prinzip einer Phasenmessung von Elektronenbeugungsinterferenzen. *Acta Crystallographica. Section A*, *25*(4), 495–501.

Hoppe, W., & Hegerl, R. (1980). Three-dimensional structure determination by electron microscopy (nonperiodic specimens). In P. W. Hawkes (Series Ed.), *Topics in current physics: Vol. 13. Computer processing of electron microscope images*. Berlin, Heidelberg: Springer.

Hoppe, W., Langer, R., Knesch, G., & Poppe, C. (1968). Protein-Kristallstrukturanalyse mit Elektronenstrahlen. *Naturwissenschaften*, *55*(7), 333–336.

Hounsfield, G. N. (1973). Computerized transverse axial scanning (tomography): Part 1. Description of system. *British Journal of Radiology, 46*(552), 1016–1022.

Jönsson, C., Hoffmann, H., & Möllenstedt, G. (1965). Messung des mittleren inneren Potentials von Be im Elektronen-Interferometer. *Physik der Kondensierten Materie, 3*, 193.

Kienle, S. H. (2000). *Matter wave reconstruction: Interferometric methods* (PhD thesis). University of Ulm.

Kirkland, E. J. (1998). *Advanced computing in electron microscopy.* New York: Plenum Press.

Koster, A. J., Ziese, U., Verkleij, A. J., Janssen, A. H., & de Jong, K. P. (2000). Three-dimensional transmission electron microscopy: A novel imaging and characterization technique with nanometer scale resolution for materials science. *The Journal of Physical Chemistry B, 104*(40), 9368–9370.

Krimmel, E., Möllenstedt, G., & Rothemund, W. (1964). Measurement of contact potential differences by electron interferometry. *Applied Physics Letters, 5*(10), 209–210.

Krivanek, O. L., Dellbya, N., & Lupinic, A. R. (1999). Towards sub-Å electron beams. *Ultramicroscopy, 78*, 1–11.

Krivanek, O. L., & Mooney, P. E. (1993). Applications of slow-scan CCD cameras in transmission electron microscopy. *Ultramicroscopy, 49*(1–4), 95–108.

Kuebel, C., Voigt, A., Schoenmakers, R., Otten, M., Su, D., Lee, T.-C., & Bradley, J. (2005). Recent advances in electron tomography: TEM and HAADF-STEM tomography for materials science and semiconductor applications. *Microscopy and Microanalysis, 11*(05), 378–400.

Lai, G., Hirayama, T., Fukuhara, A., Ishizuka, K., Tanji, T., & Tonomura, A. (1994). Three-dimensional reconstruction of magnetic vector fields using electron–holographic interferometry. *Journal of Applied Physics, 75*, 4593–4598.

Lai, G., Hirayama, T., Ishizuka, K., & Tonomura, A. (1994). Three-dimensional reconstruction of electric-potential distribution in electron-holographic interferometry. *Journal of Applied Optics, 33*, 829–833.

Lauterbur, P. C. (1973). Image formation by induced local interactions: Examples employing nuclear magnetic resonance. *Nature, 242*(5394), 190–191.

Lehmann, M., & Lichte, H. (2005). Electron holographic material analysis at atomic dimensions. *Crystal Research and Technology, 40*(1–2), 149–160.

Leibfried, D., Pfau, T., & Monroe, C. (1998). Shadows and mirrors: Reconstructing quantum states of atom motion. *Physics Today, 51*, 22–28.

Leith, E. N., & Upatnieks, J. (1962). Reconstructed wavefronts and communication theory. *Journal of the Optical Society of America, 52*, 1123–1130.

Lichte, H., Börrnert, F., Lenk, A., Lubk, A., Röder, F., Sickmann, J., & Wolf, D. (2013). Electron holography for fields in solids: Problems and progress. *Ultramicroscopy, 134*, 126–134.

Lichte, H., Möllenstedt, G., & Wahl, H. (1972). A Michelson interferometer using electron waves. *Zeitschrift für Physik, 249*(5), 456–461.

Lubk, A., Béché, A., & Verbeeck, J. (2015). Electron microscopy of probability currents at atomic resolution. *Physical Review Letters, 115*(17), 176101.

Lubk, A., Guzzinati, G., Börrnert, F., & Verbeeck, J. (2013). Transport of intensity phase retrieval of arbitrary wave fields including vortices. *Physical Review Letters, 111*(17), 173902.

Lubk, A., Javon, E., Cherkashin, N., Reboh, S., Gatel, C., & Hÿtch, M. (2014). Dynamic scattering theory for dark-field electron holography of 3D strain fields. *Ultramicroscopy, 136*, 42–49.

Lubk, A., & Röder, F. (2015a). Phase-space foundations of electron holography. *Physical Review A, 92*(3), 033844.

Lubk, A., & Röder, F. (2015b). Semiclassical TEM image formation in phase space. *Ultramicroscopy, 151,* 136–149.

Lubk, A., & Rusz, J. (2015). Jacob's ladder of approximations to paraxial dynamic electron scattering. *Physical Review B, 92*(23), 235114.

Lubk, A., Vogel, K., Röder, F., Wolf, D., Clark, L., & Verbeeck, J. (2016). Fundamentals of focal series inline electron holography. In *Advances in imaging and electron physics: Vol. 197* (pp. 105–147).

Lubk, A., Wolf, D., Kern, F., Röder, F., Prete, P., Lovergine, N., & Lichte, H. (2014). Nanoscale three-dimensional reconstruction of elastic and inelastic mean free path lengths by electron holographic tomography. *Applied Physics Letters, 105*(17), 173101.

Lubk, A., Wolf, D., Prete, P., Lovergine, N., Niermann, T., Sturm, S., & Lichte, H. (2014). Nanometer-scale tomographic reconstruction of three-dimensional electrostatic potentials in GaAs/AlGaAs core–shell nanowires. *Physical Review B, 90*(12), 125404.

Lubk, A., & Zweck, J. (2015). Differential phase contrast: An integral perspective. *Physical Review A, 91*(2), 023805.

Lvovsky, A. I., & Raymer, M. G. (2009). Continuous-variable optical quantum-state tomography. *Reviews of Modern Physics, 81*(1), 299–332.

Mansfield, P., & Grannell, P. K. (1973). NMR 'diffraction' in solids? *Journal of Physics. C. Solid State Physics, 6*(22), L422.

Midgley, P. A., & Dunin-Borkowski, R. E. (2009). Electron tomography and holography in materials science. *Nature Materials, 8*(4), 271–280.

Midgley, P. A., Ward, E. P. W., Hungría, A. B., & Thomas, J. M. (2007). Nanotomography in the chemical, biological and materials sciences. *Chemical Society Reviews, 36,* 1477–1494.

Möllenstedt, G., & Buhl, R. (1957). Ein Elektronen-Interferenz-Mikroskop. *Physikalische Blätter, 13,* 357–360.

Möllenstedt, G., & Keller, M. (1957). Elektroneninterferometrische Messung des inneren Potentials. *Zeitschrift für Physik, 148,* 34.

Nagayama, K. (2005). Phase contrast enhancement with phase plates in electron microscopy. In *Advances in imaging and electron physics: Vol. 138.* Elsevier (pp. 69–146).

Nolet, G. (1987). *Seismic tomography: With applications in global seismology and exploration geophysics. Modern approaches in geophysics.* Netherlands: Springer.

Paris, M., & Rehacek, J. (2004). *Quantum state estimation. Lecture notes in physics.* Springer.

Pollak, B. (1953). Experiences with planography. *Chest, 24*(6), 663–669.

Pozzi, G., Beleggia, M., Kasama, T., & Dunin-Borkowski, R. E. (2014). Interferometric methods for mapping static electric and magnetic fields. *Comptes Rendus. Physique, 15,* 126–139.

Radon, J. (1917). Über die Bestimmung von Funktionen durch ihre Integralwerte längs gewisser Mannigfaltigkeiten. *Sächsische Akademie der Wissenschaften, 69,* 262–277.

Reimer, L. (Ed.). (1989). *Transmission electron microscopy.* Springer Verlag.

Reimer, L. (1992). Energy-filtering transmission electron microscopy in materials science. *Microscopy Microanalysis Microstructures, 3,* 141.

Röder, F., Lubk, A., Wolf, D., & Niermann, T. (2014). Noise estimation for off-axis electron holography. *Ultramicroscopy, 144,* 32–42.

Rose, H. (1976). Image formation by inelastically scattered electrons in electron microscopy. I. *Optik, 45,* 139–158.

Scherzer, O. (1949). The theoretical resolution limit of the electron microscope. *Journal of Applied Physics, 20*, 20–29.

Schleich, W. P. (2001). *Quantum optics in phase space.* Berlin: Wiley VCH.

Schleich, W. P., & Raymer, M. G. (Eds.). (1997). Quantum state preparation and measurement. *Journal of Modern Optics.* Taylor & Francis.

Smithey, D. T., Beck, M., Raymer, M. G., & Faridani, A. (1993). Measurement of the Wigner distribution and the density matrix of a light mode using optical homodyne tomography: Application to squeezed states and the vacuum. *Physical Review Letters, 70*(9), 1244–1247.

Stahlberg, H., & Walz, T. (2008). Molecular electron microscopy: State of the art and current challenges. *ACS Chemical Biology, 3*(5), 268.

Teague, M. R. (1983). Deterministic phase retrieval: A Green's function solution. *Journal of the Optical Society of America, 73*, 1434–1441.

Thust, A., Coene, W., de Beeck, M. O., & Van Dyck, D. (1996). Focal-series reconstruction in HRTEM: Simulation studies on non-periodic objects. *Ultramicroscopy, 64*, 211–230.

Tonomura, A. (1987). Applications of electron holography. *Reviews of Modern Physics, 59*, 639.

Tonomura, A., Allard, L. F., Pozzi, G., Joy, D. C., & Ono, Y. A. (Eds.). (1995). *Electron holography. Proceedings of the international workshop on electron holography. North-Holland delta series* (pp. 267–276).

Tonomura, A., Matsuda, T., Endo, J., Todokoro, H., & Komoda, T. (1979). Development of a field emission electron microscope. *Journal of Electron Microscopy, 28*(1), 1–11.

Tonomura, A., Osakabe, N., Matsuda, T., Kawasaki, T., & Endo, J. (1986). Evidence for Aharonov–Bohm effect with magnetic field completely shielded from electron wave. *Physical Review Letters, 56*, 792–795.

Townsend, D. W. (2008). Combined positron emission tomography – computed tomography: The historical perspective. *Seminars in Ultrasound, CT and MRI, 29*(4), 232–235.

Twitchett-Harrison, A. C., Yates, T. J. V., Newcomb, S. B., Dunin-Borkowski, R. E., & Midgley, P. A. (2007). High-resolution three-dimensional mapping of semiconductor dopant potentials. *Nano Letters, 7*(7), 2020–2023.

Völkl, E., Allard, L. F., & Joy, D. C. (Eds.). (1999). *Introduction to electron holography.* Kluwer Academic/Plenum Publishers.

von Neumann, J. (1927). Thermodynamik quantenmechanischer Gesamtheiten. *Göttinger Nachrichten, 3*, 273–291.

Wahl, H. (1975). *Bildebenenholographie mit Elektronen* (Habilitationsschrift). Universität Tübingen.

Wolf, D., Lichte, H., Pozzi, G., Prete, P., & Lovergine, N. (2011). Electron holographic tomography for mapping the three-dimensional distribution of electrostatic potential in III–V semiconductor nanowires. *Applied Physics Letters, 98*(26), 264103.

Wolf, D., Lubk, A., Lenk, A., Sturm, S., & Lichte, H. (2013). Tomographic investigation of Fermi level pinning at focused ion beam milled semiconductor surfaces. *Applied Physics Letters, 103*(26), 264104.

Wolf, D., Lubk, A., Lichte, H., & Friedrich, H. (2010). Towards automated electron holographic tomography for 3D mapping of electrostatic potentials. *Ultramicroscopy, 110*(5), 390–399.

Zernike, F. (1935). Das Phasenkontrastverfahren bei der mikroskopischen Beobachtung. *Physikalische Zeitschrift, 36*, 848.

CHAPTER TWO

Paraxial Quantum Mechanics

Axel Lubk

Institute for Structure Physics, Physics Department, Faculty of Mathematics and Natural Sciences, Technical University of Dresden, Dresden, Germany
e-mail address: axel.lubk@tu-dresden.de

Contents

For me, the important thing about quantum mechanics is the equations, the mathematics. If you want to understand quantum mechanics, just do the math. All the words that are spun around it don't mean very much. It's like playing the violin. If violinists were judged on how they spoke, it wouldn't make much sense.

(Freeman Dyson)

Before starting to paint a picture on holography and tomography, a palette of colors, that is fundamental physical and mathematical concepts from which others can be "mixed" when necessary, is required. Our first base color is a restricted quantum mechanics for relativistic electrons traveling along the optical axis of a TEM. Two flavors of which are given below. First, the more familiar paraxial wave dynamics, which also serves as to define a large part of the symbols used in the course of this text. Second, the concepts of the density operator and the quantum mechanical phase space, including their paraxial dynamics, which provides the natural language for describing the various electron holographic imaging techniques including the ramifications of partial coherence. The phase space representation of the imaging process and the unified description of different holographic

15

schemes discussed in Chapters 4 and 5 are founded in these principles. The second base color is the tomographic principle discussed in Chapter 3.

The major incentive for the development of the TEM was to overcome the limited spatial resolution of light microscopes by using radiation of smaller wavelength. De Broglie (1924) was the first to realize that a wave with wave vector $\mathbf{k}$ can be assigned to massive particles with momentum $\mathbf{p}$ by dividing the latter with the reduced Planck constant, i.e., $\mathbf{k} = \mathbf{p}/\hbar$. Shortly before de Broglie's realization Cockcroft and Walton (1932a, 1932b) developed a particularly effective accelerator for electrons (and other charged particles) based on a cascade of electrodes, which is still found in modern TEMs in a modified form. Such devices can easily generate voltages U_A in the range of hundreds of kilovolts, which accelerate electrons to several ten percent of the speed of light with corresponding wavelengths in the order of picometers. Such wavelengths permit to resolve the atomic constituents of matter as typical distances between them are in the order of angstroms.

In Chapter 4 we will see which electron optical elements are required to focus such electron beams in a TEM. For the moment it is sufficient to note that particularly strong forces or electromagnetic fields are required to deviate these fast electrons from their path along the optical axis of the microscope. Indeed, the majority of fields within a TEM only scatter the electrons into small angles around the optical axis, which is referred to as paraxial scattering. Prominent examples are the magnetic fields of the coils typically employed as lenses and the atomic potentials of a thin TEM specimen. Under such conditions the electrons basically move along the optical axis and this axis may be identified with the time of flight via the electron's velocity. The following sections contain a detailed account of a restricted relativistic paraxial quantum mechanics, providing the principle laws governing the motion of these electrons.

2.1. STATE VECTOR AND REPRESENTATIONS

It is one of the main axioms of quantum mechanics that a vector from a Hilbert space – the state vector $|\Psi\rangle$ – completely describes a quantum system (Landau & Lifshitz, 1977). This state vector can be projected onto various bases, for instance the position basis $\langle x|\Psi\rangle = \Psi(x)$, the momentum basis $\langle q|\Psi\rangle = \Psi(q)$, or the energy basis $\langle m|\Psi\rangle = c_m$. Such basis functions are eigenfunctions of Hermitian operators, with the eigenvalues representing the observables of a quantum system. In the above noted examples, we have

$\hat{x}|x\rangle = x|x\rangle$, $\hat{q}|q\rangle = q|q\rangle$, and $\hat{H}|m\rangle = E_m|m\rangle$ with $\hat{x}$ the position operator, $\hat{q}$ the momentum operator and $\hat{H}$ the Hamiltonian of the system. All bases fulfill the completeness

$$\mathbb{I} = \int_{-\infty}^{\infty} dx |x\rangle\langle x| \tag{2.1}$$

and orthogonality relations

$$\delta\left(x - x'\right) = \langle x|x'\rangle. \tag{2.2}$$

The position representation of the momentum operator $\hat{q} = -i\partial_x$ corresponds to the following momentum eigenfunctions in position representation

$$q(x) = \langle x|q\rangle = \frac{1}{\sqrt{2\pi}} e^{iqx}. \tag{2.3}$$

Here, the normalization is fixed by the completeness and orthogonality relations (2.1), (2.2), i.e.,

$$\begin{aligned}
\langle q|q'\rangle &= \int_{-\infty}^{\infty} dx \langle q|x\rangle\langle x|q'\rangle \tag{2.4} \\
&= \frac{1}{2\pi} \int_{-\infty}^{\infty} dx e^{i(q'-q)x} \\
&= \delta\left(q' - q\right).
\end{aligned}$$

In the last line the following definition of the δ–distribution was inserted

$$\frac{1}{2\pi} \int_{-\infty}^{\infty} dq e^{iqx} = \delta(x). \tag{2.5}$$

With that normalization, the change from position to momentum representation of an arbitrary state vector is accomplished by a unitary Fourier transformation, i.e.,

$$\begin{aligned}
\Psi(q) &= \langle q|\Psi\rangle \tag{2.6} \\
&= \int_{-\infty}^{\infty} dx \langle q|x\rangle\langle x|\Psi\rangle \\
&= \frac{1}{\sqrt{2\pi}} \int_{-\infty}^{\infty} dx e^{-iqx} \Psi(x) \\
&= \mathcal{F}\{\Psi\}(q).
\end{aligned}$$

The latter convention for the Fourier transformation will be used throughout.

2.2. PARAXIAL WAVE DYNAMICS

In principle, one needs to employ fully relativistic quantum mechanics to describe the dynamics of relativistic electrons in a TEM. Fortunately various simplifications are possible. We first restrict our considerations to a sharp electron energy

$$E = eU_A + m_0c^2 \tag{2.7}$$

Here, e is the elementary charge, m_0 the electron rest mass, and c the velocity of light. In some cases, such as the scattering of beam electrons at thin TEM specimen, the small energy width of the beam may be completely neglected. When considering electron optical imaging on the other hand the correct result is obtained only after integrating of the beam's energy distribution.

Given a sharp electron energy, it is sufficient to consider the stationary Dirac equation

$$\left(\underbrace{E + e\Phi - c\boldsymbol{\alpha} \left(\hat{\mathbf{p}} + e\mathbf{A} \right) - \beta m_0 c^2}_{-\hat{H}_D} \right) \boldsymbol{\psi} = 0, \tag{2.8}$$

where Φ and $\mathbf{A}$ denote the electric and magnetic potential, respectively. In the following, we will always use the standard (Dirac) representation of the Dirac matrices

$$\boldsymbol{\alpha} = \begin{pmatrix} 0 & \boldsymbol{\sigma} \\ \boldsymbol{\sigma} & 0 \end{pmatrix}, \quad \beta = \begin{pmatrix} \mathbf{I}_2 & 0 \\ 0 & -\mathbf{I}_2 \end{pmatrix}. \tag{2.9}$$

Using the Dirac algebra

$$\{\alpha_i, \alpha_k\} = 0, \ i \neq k$$
$$\{\alpha_i, \beta\} = 0 \tag{2.10}$$
$$\alpha_i^2 = \beta^2 = \mathbf{I}_4$$

a squared version of the Dirac equation is obtained by multiplying the negative energy Dirac operator from the left to the original Dirac equation

$$\left(E - e\Phi + c\boldsymbol{\alpha}\left(\hat{\mathbf{p}} + e\mathbf{A}\right) + \beta m_0 c^2\right)\left(E + e\Phi - c\boldsymbol{\alpha}\left(\hat{\mathbf{p}} + e\mathbf{A}\right) - \beta m_0 c^2\right)\boldsymbol{\psi} = 0,$$

(2.11)

which gives four Klein–Gordon–Fock equations coupled by electric and magnetic fields terms (Fujiwara, 1961; Strange, 1998)

$$\left[(E + e\Phi)^2 - c^2\left(\hat{\mathbf{p}} + e\mathbf{A}\right)^2 - m_0^2 c^4 + e\hbar c\left(i\boldsymbol{\alpha}\mathbf{E} - c\boldsymbol{\Sigma}\mathbf{B}\right)\right]\boldsymbol{\psi} = 0.$$

(2.12)

Noting that the spin matrix

$$\boldsymbol{\Sigma} = \begin{pmatrix} \boldsymbol{\sigma} & 0 \\ 0 & \boldsymbol{\sigma} \end{pmatrix}$$

(2.13)

does not couple the large and the small spinor components, we may reduce (decouple) the squared Dirac equation to a relativistic Pauli equation for two-spinors

$$\left[(E + e\Phi)^2 - c^2\left(\hat{\mathbf{p}} + e\mathbf{A}\right)^2 - m_0^2 c^4 - e\hbar c^2 \boldsymbol{\sigma}\mathbf{B}\right]\boldsymbol{\psi} = 0,$$

(2.14)

because the electric coupling term is typically small and may be neglected therefore (Rother & Scheerschmidt, 2009). The remaining magnetic coupling between the two spinor components gives rise to weak scattering effects in magnetic materials (Edström, Lubk, & Rusz, 2016), which have not been observed to date of this work in the TEM. We may therefore dispense with spinor dynamics in favor of scalar relativistic dynamics (Rother & Scheerschmidt, 2009) based on the stationary Klein–Gordon–Fock equation

$$\left[E^2 + 2eE\Phi + e^2\Phi^2\right]\psi = \left[-c^2\hbar^2\Delta + m_0^2 c^4\right]\psi.$$

(2.15)

To simplify the following derivations, only the influence of an electrostatic potential is considered. The ramifications of magnetic fields will be discussed at a latter stage.

Due to large energy E of highly accelerated electrons in a TEM, the potential squared term can be neglected in the above stationary Klein–Gordon–Fock equation, which is referred to as high-energy approximation (Fujiwara, 1961)

$$\left[E^2 - m_0^2 c^4\right]\psi = \left[-2eE\Phi - c^2\hbar^2\Delta\right]\psi.$$

(2.16)

This **high-energy limit of the Klein–Gordon–Fock equation** then takes a form, which is mathematically equivalent to a stationary Schrödinger equation (Fujiwara, 1961; Rother & Scheerschmidt, 2009)

$$m_0 c^2 \frac{\gamma^2 - 1}{2\gamma} \psi = -\frac{\hbar^2}{2\gamma m_0} \Delta \psi - e\gamma \Phi \psi, \qquad (2.17)$$

where the energy and mass have been corrected according to the above prescription (γ is the Lorentz factor).

At this point the following observation is crucial. In spite of the high energy of the beam electrons, the high-energy limit of the Klein–Gordon–Fock equation generally does not admit an accurate first order Born type approximation as employed in high-energy particle physics or X-ray scattering because of the singular nature of the screened atomic Coulomb potentials at the core (Glauber & Schomaker, 1953; Zeitler & Olsen, 1964). This behavior is both a blessing and a curse for TEM investigations. The strong interaction of the electron beam with the atomic potentials allows imaging of atoms with manageable experimental effort. However, the interpretation of atomic resolution electron micrographs is typically complicated and requires elaborate electron scattering simulations.

This obstacle has been appreciated from the heydays of electron diffraction prompting the development of the dynamical scattering theory as opposed to the first order Born approximation (i.e., kinematic scattering). Important contributions have been given by Bethe (1928), Lamla (1938), Cowley and Moodie (1957), Fujimoto (1959), Howie and Whelan (1961), Fujiwara (1961), Kambe (1967), Berry (1971), and Gratias and Portier (1983) to name just a few (see, e.g., Cowley, 1995 for a more comprehensive account). In the following, basic elements of the dynamical scattering theory are provided, partly drawing from results published in Lubk and Rusz (2015).

From now on we shall use natural units $c = \hbar = 1$ to shorten the notation. As a consequence of the large kinetic energy of the electrons, typical scattering potentials brought into their path within a TEM (specimen or optical elements) do not deflect the electrons into large angles.[1] This observation motivates the following paraxial approximation. First, the wave is separated into a fast plane wave and a slowly varying envelope function in

---

[1] Notable exceptions are thick specimen or sector magnets used in energy filters.

propagation direction z

$$\psi(\mathbf{r}, z) = \Psi(\mathbf{r}, z)e^{ik_0 z}, \qquad (2.18)$$

where $\mathbf{r}$ denotes the 2D coordinates in the $x - y$ plane and

$$k_0 = \sqrt{m_0^2 (\gamma^2 - 1)}. \qquad (2.19)$$

Inserting this ansatz into the above high-energy limit of the Klein–Gordon–Fock equation

$$k_0^2 \Psi(\mathbf{r}, z) = \left(-\Delta - \partial_z^2 + k_0^2 - 2ik_0\partial_z - 2em\Phi(\mathbf{r}, z)\right)\Psi(\mathbf{r}, z), \qquad (2.20)$$

and neglecting the second-order derivative of the slowly varying envelope function along z, one obtains the **paraxial Klein–Gordon–Fock equation** (Kirkland, 1998)

$$i\partial_z \Psi(\mathbf{r}, z) = \underbrace{\left(-\frac{1}{2k_0}\Delta - \frac{e}{v}\Phi(\mathbf{r}, z)\right)}_{\hat{T}+\hat{U}}\Psi(\mathbf{r}, z). \qquad (2.21)$$

Here, $\hat{T}$ denotes the paraxial kinetic energy operator and $\hat{U}$ the paraxial potential, incorporating a scaling with the inverse of the electron's velocity v. This equation largely governs the dynamics of electrons within the TEM, for example the elastic electron–specimen interaction as well as the deflection of the electrons by optical elements such as magnetic lenses. Deviations from the paraxial behavior have to be considered when analyzing aberrations imposed by the optical elements or large angle scattering by the specimen.

The structure of (2.21) corresponds to a 2D time-dependent Schrödinger equation

$$i\partial_z \Psi(\mathbf{r}, z) = \underbrace{\left(\hat{T} + \hat{U}(\mathbf{r}, z)\right)}_{\hat{H}(\mathbf{r}, z)}\Psi(\mathbf{r}, z) \qquad (2.22)$$

with z and the paraxial Hamiltonian $\hat{H}$ taking the place of the time coordinate and the usual Hamilton operator, respectively (Berry, 1971). In the following we will dwell on this remarkable analogy, allowing for a substantial portion of the time-dependent Schrödinger equation's properties and solutions to be transferred to paraxial scattering and vice versa.

In full analogy to the time-dependent Schrödinger equation one can rewrite (2.21) as a coupled system of two differential equations for the wave's probability density $I = \Psi\Psi^*$ and phase

$$\frac{\partial I(\mathbf{r}, z)}{\partial z} = -\frac{1}{k_0} \underbrace{\nabla(I(\mathbf{r}, z)\nabla\varphi(\mathbf{r}, z))}_{\mathbf{j}} \tag{2.23a}$$

$$\frac{\partial\varphi(\mathbf{r}, z)}{\partial z} = \frac{e}{v}\Phi - \frac{1}{2k_0}(\nabla\varphi(\mathbf{r}, z))^2 + \underbrace{\frac{1}{8k_0}\frac{2I(\mathbf{r}, z)\Delta I(\mathbf{r}, z) - (\nabla I(\mathbf{r}, z))^2}{I(\mathbf{r}, z)^2}}_{\text{quantum potential}}$$

$$\tag{2.23b}$$

The first line equates the change of the probability density (intensity) along z with the lateral divergence of the paraxial probability current $\mathbf{j}$. It represents the continuity equation for the paraxial regime. In the context of the corresponding holographic technique discussed in Section 5.2, (2.23a) is also referred to as Transport of Intensity Equation (TIE). The second equation is the quantum version of the paraxial Hamilton–Jacobi equation differing from the latter by the so-called quantum potential. The numerical integration of the quantum Hamilton Jacobi equation is at the core of Focal Series Inline Holography discussed in Section 5.3.2.

An analytical solution of the quantum Hamilton–Jacobi equation may be obtained by employing the method of characteristics (Courant & Hilbert, 1962). Using this approach and neglecting the quantum potential, one obtains the paraxial equations of motion

$$\frac{\partial^2 \mathbf{r}_e(z)}{\partial z^2} = -\frac{e}{v}\nabla\Phi(\mathbf{r}_e(z), z) \tag{2.24}$$

$$= \frac{e}{v}\mathbf{E}(\mathbf{r}_e(z), z)$$

determining the electron trajectory $\mathbf{r}_e(z)$ within an electric field $\mathbf{E}$. A second differential equation determines the paraxial phase (classically referred to as point eikonal or action)

$$\frac{\partial\varphi(\mathbf{r}_e(z), z)}{\partial z} = \frac{e}{v}\Phi(\mathbf{r}_e(z), z). \tag{2.25}$$

When including the magnetic term, the paraxial equations of motion form the basis for describing stigmatic imaging within the TEM, that is Gaussian electron optics (Hawkes & Kasper, 1996). The paraxial eikonal is readily integrated to

$$\varphi(\mathbf{r}_e(z), z) = \frac{e}{v} \int_0^z \Phi\left(x_e(z'), y_e(z'), z'\right) dz' + \varphi_0 \tag{2.26}$$

within any z-interval containing no crossings of classical trajectories. Thus, the scope of the above semiclassical approximation (consisting of neglecting the quantum potential in the quantum Hamilton–Jacobi equation) is limited. Nevertheless, expression (2.26) provides a first, albeit complicated, example of a projection law linking the phase of the electron wave to the electrostatic potential integrated along classical paraxial trajectories. The line integral (2.26) prompted Glaser to interpret $e\Phi/v$ as the index of refraction for electrons (Glaser, 1933). We will come back to this expression, when discussing generalized tomographic schemes in Chapter 6.2.

The full solution to the paraxial equation (2.21) can be formally obtained as (Greiner, Reinhardt, 1984)

$$|\Psi(z)\rangle = \underbrace{\hat{Z} \exp\left(-i \int_{-\infty}^z \hat{H}(z') dz'\right)}_{\hat{K}(z, -\infty)} |\Psi(0)\rangle, \tag{2.27}$$

where $\hat{Z}$ denotes the z-ordering operator in analogy to Dyson's time ordering operator and $\hat{K}(z, -\infty)$ is the z-evolution operator corresponding to the time-evolution operator.[2] In full analogy to time-evolution the z-evolution is unitary, conserving the norm of the wave function along z, if the Hamiltonian is Hermitian.

Moreover, the z-propagation is a continuous transformation, which implies that the topology of the initial wave $\Psi(0)$ remains constant throughout propagation. For 2D wave functions considered in the paraxial regime, the topology is measured by the winding number w computed through anticlockwise loop integrals (directional path element $d\mathbf{s}$) around the outer boundary of the wave function according to (Lubk, Clark, Guzzinati, & Verbeeck, 2013)

$$w = \frac{1}{2\pi} \oint d\mathbf{s} \cdot \nabla\varphi = \sum_{j=1}^N w_j, \tag{2.28}$$

---

[2] Note that the frequently encountered practice of disregarding the z-ordering in the pertinent literature is problematic, e.g., when developing (numerical) approximations.

where the last equality says that w may be alternatively expressed as the sum over individual winding numbers around *all* individual phase vortices in the wave field. We will use that result when considering inline holographic reconstruction algorithms in Section 5.3.

To integrate the above equation (2.27), a multitude of numerical methods may be employed. A straight choice consists of a numerical integrator, i.e., a step-by-step propagation along z, which is well adapted to the problem (Rother & Scheerschmidt, 2009; Wacker & Schröder, 2015). The error pertaining to the numerical integration accumulates with propagation distance (specimen thickness t) and can be reduced by decreasing the integration step size. Current implementations of these methods still need considerable computation time, which can be overcome by fixing the step size and approximating the propagator by analytical expressions. This leads to the popular multislice method (Cowley & Moodie, 1957), which bears some similarities with the split operator method (Askar & Cakmak, 1978) used in the context of numerically solving the time-dependent Schrödinger equation. It consists of an alternate application of the free space propagator

$$\hat{K}_T(\delta z) := \exp\left(-i\underbrace{(z - z')}_{\delta z}\hat{T}\right) \tag{2.29}$$

and the transmission operator

$$\hat{K}_U(z, z') := \exp\left(\int_{z'}^{z} U(z'')\mathrm{d}z''\right) \tag{2.30}$$

according to

$$|\Psi(z_N)\rangle = \sum_{n=0}^{N-1} \hat{K}_T(\delta z_n)\hat{K}_U(z_{n+1}, z_n)|\Psi(z_0)\rangle. \tag{2.31}$$

Similar to the split operator method, the successive application of transmission operator and free space propagator can be performed very efficiently by switching between position and Fourier space. The transmission operator diagonalizes in position space

$$\langle\mathbf{r}|\hat{K}_U(z; z')|\mathbf{r}'\rangle = \exp\left(\frac{ie}{v}\int_{z'}^{z} \Phi(\mathbf{r}, z'')\mathrm{d}z''\right)\delta(\mathbf{r} - \mathbf{r}'), \tag{2.32}$$

where it is referred to as transmission function. The position space propagator reads

$$\langle \mathbf{r} | \hat{K}_T(z; z') | \mathbf{r}' \rangle = K_T(\mathbf{r}, z; \mathbf{r}', z') \tag{2.33}$$

$$= \frac{1}{i\lambda(z - z')} e^{i\frac{k_0}{2(z-z')}(\mathbf{r}-\mathbf{r}')^2},$$

which diagonalizes in Fourier space

$$K_T(\mathbf{k}, z; \mathbf{k}', z') = e^{-i\frac{z-z'}{2k_0}\mathbf{k}^2} \delta\left(\mathbf{k} - \mathbf{k}'\right). \tag{2.34}$$

This free space propagator is called Fresnel propagator as it can be alternatively derived from the near field approximation to the Kirchhoff integral.

In the limit $\delta z \to 0$ the multislice algorithm converges to the exact solution (Goodman & Moodie, 1974), however, in order to reduce computation time, one integration step δz is typically chosen such to comprise an equivalent of one atomic layer (i.e. $\delta z \approx 1$ Å) in a crystal (referred to as atomic multislice in the following[3]). The accuracy of this choice stems from the quasi-discrete nature of the atomic scattering potentials within a large potential-free atomic interspace, the atomic lattice. This particular potential structure allows diabatically transmitting the wave with the transmission operator only (neglecting free-space propagation) through thin sheets containing atoms and propagating the wave without influence of the potential in the atomic interspace (Fig. 2.1). Consequently, the atomic multislice algorithm basically consists of a particular, well-adapted, approximation of the paraxial equation (2.21) omitting either the kinetic or potential term within certain z-intervals. This renders the evaluation of the z-ordering trivial because the non-vanishing commutator between the kinetic and potential operator does not have to be considered. Its simplicity, accuracy, and speed makes the multislice algorithm one of the principal methods used to determine the outcome of electron scattering within the TEM (Stadelmann, 1987; Koch, 2002; Rosenauer & Schowalter, 2008; Allen, D'Alfonso, & Findlay, 2015).

The second class of popular algorithms seeks the solution to the paraxial equation in two distinct steps, rather than numerically propagating the wave through the potential. First a basis of stationary solutions is generated by diagonalizing a certain approximation of the paraxial Hamiltonian

---

[3] Note that this choice also permits a very efficient consideration of the thermally vibrating lattice (Rother, Gemming, & Lichte, 2009).

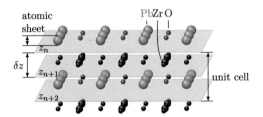

Figure 2.1 Working principle of the multislice algorithm with the example of a perovskite PbZrO$_3$ crystal lattice (space group Pm3m, $a_x = a_y = a_z = 4.18$ Å, orientation [1, 0, 0]). The strongly localized atomic potentials exert a dominant influence in thin sheets only, where the concomitant propagation may be neglected. In turn, the influence of the potential is negligible in the atomic interspace.

and second, the initial wave is expanded into this basis. To compute the basis function one typically employs periodic boundary conditions in all three spatial dimensions. Hence, the basis functions are Bloch waves and the algorithms are termed Bloch wave methods. Depending on whether including the z-direction (optical axis, propagation direction) in the expansion one speaks of 2D or 3D Bloch wave methods. The latter is founded in the principles of Floquet theory (Floquet, 1883) as detailed below.

Because of the numerical scaling of eigenvalue problems, Bloch wave methods are typically less favorable, if the basis is very large, i.e., if a large number of scattering directions has to be considered. A large class of problems, however, allows a considerable reduction of the basis, rendering Bloch wave methods a valuable alternative to direct integration schemes like multislice. In particular the basis truncation along the z-direction in 2D Bloch wave methods leads to a simple analytic z-dependency that can be favorably exploited under various circumstances. For instance, the prominent two-beam case (Howie & Whelan, 1961), where only one particular scattered beam additional to the unscattered one is considered, can be solved analytically, yielding the "Pendellösung". Similarly, the calculation of inelastic scattering matrix elements for electron energy loss spectra (EELS) can be efficiently organized by exploiting analytic expansions of the Bloch waves (Nelhiebel et al., 1999).

The 2D Bloch wave method may be derived from the multislice method in the following way. First, the Zassenhaus expansion (following from the Baker–Campbell–Haussdorf formula; Magnus, 1954) with some arbitrary operators $\hat{A}$ and $\hat{B}$

$$e^{(\hat{A}+\hat{B})t} = e^{\hat{A}t} e^{\hat{B}t} e^{-\frac{t^2}{2}[\hat{A},\hat{B}]} \cdots \tag{2.35}$$

can be truncated after the second factor, if the commutator

$$\frac{t^2}{2}[\hat{A}, \hat{B}] \ll 1 \tag{2.36}$$

is small, i.e.,

$$e^{\hat{A}t} e^{\hat{B}t} \approx e^{(\hat{A}+\hat{B})t} . \tag{2.37}$$

If substituting $\hat{A} = \hat{K}_U$, $\hat{B} = \hat{K}_T$, and $t = z_{n+1} - z_n = a_z$, where the slice thickness was chosen such to comprise one unit cell, one obtains

$$\Psi(\mathbf{r}, z_{n+1}) = \exp\left(i\left(\frac{1}{2k_0}\Delta + \frac{e}{v}\overline{\Phi}(\mathbf{r})\right) a_z\right) \Psi(\mathbf{r}, z_n) , \tag{2.38}$$

where the potential

$$\overline{\Phi}_n(\mathbf{r}) = \frac{1}{a_z} \int_{z_n}^{z_{n+1}} \Phi(\mathbf{r}, z) \, \mathrm{d}z \tag{2.39}$$

averaged along one unit cell in z-direction has been introduced. Because a slice thickness of one unit cell has been chosen, the exponential in (2.38) does not depend on z anymore (note the absence of the z-ordering) and (2.38) is the formal solution to the fundamental equation of the **2D Bloch wave formalism** (Bethe, 1928; Howie & Whelan, 1961; Kambe, 1967)

$$i\partial_z \Psi(\mathbf{r}, z) = \left[-\frac{1}{2k_0}\Delta - \frac{e}{v}\overline{\Phi}(\mathbf{r})\right] \Psi(\mathbf{r}, z) . \tag{2.40}$$

Here, the z-dependency of the paraxial Hamiltonian has vanished, since the potential of the individual atoms has been averaged along the propagation direction. Accordingly, the propagation diagonalizes in the eigenspace of the Hamiltonian. 2D Bloch wave methods exploit that property by diagonalizing the above Hamiltonian and expanding the 2D incoming wave at the entrance face of the crystal into the corresponding basis (2D Bloch waves). The wave at any z-coordinate is then readily obtained by the analytic behavior of the 2D Bloch waves without additional effort.

The above derivation contains two approximations, which lead to the following characteristic deviations of the 2D Bloch wave method compared to more accurate numerical integration schemes. First, the slice thickness was set to one unit cell, which may be significantly smaller in the numerical integration. As a consequence, the internal structure and symmetries of the unit cell determining the strength of systematic scattering directions into particular Laue zones may be absent from 2D Bloch wave calculations. Second, the commutator between transmission and propagation operator

$$-\frac{a_z^2}{2}[\hat{K}_U, \hat{K}_T] = -\frac{a_z^2 e}{4k_0 v}\left(\nabla\overline{\Phi}\,(\mathbf{r})\,\nabla + \Delta\overline{\Phi}\,(\mathbf{r})\right) \qquad (2.41)$$

$$= \frac{a_z^2 e}{4k_0 v}\left(\overline{\mathbf{E}}\,(\mathbf{r})\,\nabla + \overline{\rho}\,(\mathbf{r})\right)$$

is not negligible at the site of the atomic core, which, in spite of the smallness of the prefactor, leads to a violation of the approximation (2.38). Given the singular nature of this peak in position space this mainly translates into an erroneous behavior of the Bloch wave solution close to the atomic site. The large lateral spatial frequency components of that peak will quickly disperse upon propagation, dampening the discrepancy at larger distance to the atom.

Within the framework of Floquet theory (Floquet, 1883), the approximations leading to the 2D Bloch wave method may be dropped, if the potential is z-periodic, i.e.

$$\hat{H}\,(\mathbf{r}, z + a_z) = \hat{H}\,(\mathbf{r}, z)\ . \qquad (2.42)$$

For such z-periodic Hamiltonians the Floquet theorem states that a fundamental matrix $\boldsymbol{\Psi}\,(z + a)$ of linearly independent solution at a point $z + a$ can be computed from the fundamental matrix at $\boldsymbol{\Psi}\,(z)$ via

$$\boldsymbol{\Psi}(z + a_z) = \boldsymbol{\Psi}(z)\mathbf{B}\ \text{for}\ \forall z, \qquad (2.43)$$

where the monodromy matrix

$$\mathbf{B} = \boldsymbol{\Psi}^{-1}(0)\boldsymbol{\Psi}\,(a_z) \qquad (2.44)$$

does not depend on z. Consequently, a fundamental matrix at any z may be computed from its value on the first interval $z\epsilon\,[0, a_z]$ by successive application of the Floquet theorem. Furthermore, the eigenvalues $\rho^{(n)}$ of $\mathbf{B}$, called the characteristic multipliers, do not depend on the particular choice of the fundamental matrix. They define the (Floquet) exponents $\mu^{(n)}$ through the relations $\rho^{(n)} = \exp\left(\mu^{(n)}z\right)$. As a consequence of Floquet's theorem we then have a solution $\Psi^{(n)}(\mathbf{r}, z) = \exp\left(\mu^{(n)}z\right)u^{(n)}\,(\mathbf{r}, z)$ with z-periodic $u^{(n)}\,(\mathbf{r}, z)$. Consequently, the Floquet exponents play the role of the Bloch vector encountered when solving the stationary Schrödinger equation in periodic potentials.

The above considerations suggest the following numerical solution to dynamical scattering. First, one has to compute a fundamental matrix of a sufficiently large number of linearly independent solutions within the first unit cell at the crystal entrance face, e.g., by numerically integrating differently tilted plane waves through the first unit cell. Subsequently the fundamental matrix at any z may be obtained by virtue of (2.43). The latter permits the computation of any solution to a given incoming wave because we can expand that wave into the particular set of linearly independent solutions forming the fundamental matrix. A numerical implementation of the above paraxial scattering algorithm has not been realized to date of this work.

Instead, 3D Bloch wave methods approximate the Floquet solution by solving the associated eigenvalue problem

$$\mu^{(n)} u^{(n)}\left(\mathbf{r}, z\right) = -\left(\hat{H} + \partial_z\right) u^{(n)}\left(\mathbf{r}, z\right) \tag{2.45}$$

and decomposing the *unperturbed* incoming wave within the first unit cell into the z-periodic $u^{(n)}\left(\mathbf{r}, z\right)$ (e.g. Peng, Dudarev, & Whelan, 2004) defining the following expansion coefficients

$$C^{(n)} \approx \int_\Omega u^{(n)*}\left(\mathbf{r}, z\right) \Psi^{(0)}\left(\mathbf{r}, z\right) \mathrm{d}^2 r \mathrm{d}z. \tag{2.46}$$

Accordingly, the 3D Bloch wave solution reads

$$\Psi\left(\mathbf{r}, z\right) = \sum_n C^{(n)} e^{i\mu^{(n)}z} \sum_{\mathbf{g}} u_{\mathbf{g}}^{(n)} e^{i\mathbf{g}\mathbf{r}}. \tag{2.47}$$

Consequently, two approximations with respect to the Floquet solution are involved in the 3D Bloch wave approach. (I) The unperturbed instead of the true wave is expanded into (II) the periodic part of the Bloch waves only, thereby neglecting the phase factor containing the Bloch vector. These approximations seem to have been previously overlooked in the pertinent literature (Peng et al., 2004). The error of this 3D Bloch wave approximation grows with increasing atomic weight increasing the neglected modulations of the wave in the first unit cell. This error does not accumulate as the propagation distance grows. A second error stems from the deviations of the computed 3D Bloch wave basis the exact basis, which leads to an increasing mismatch of the 3D Bloch wave solution with growing propagation distance.

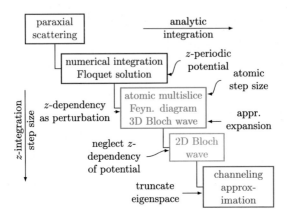

Figure 2.2 Hierarchy of approximations for dynamical scattering. The order roughly scales with the characteristic integration step size leading to simpler analytic expressions for the propagation along z.

The above analytical discussion of dynamical scattering can be summed up to a schematic Jacob's ladder of approximations depicted in Fig. 2.2. An "exact" numerical approximation is obtained by integrating the paraxial scattering equation with some numerically or semianalytical integration algorithm such as Runge–Kutta or multislice with sufficiently small step size. In the case of z-periodic potentials, the same level of approximation is obtained by the Floquet solution, incorporating direct integration schemes for the first unit cell. In practice, the accurate reproduction of the z-dependency is sacrificed to obtain acceptable computation times and analytical simplicity. The most radical approximation, i.e., the 2D Bloch wave method, completely neglects the z-dependent potential structure, which permits a full analytic description of the z-propagation within the eigenspace of the 2D Bloch wave Hamiltonian. Additionally, the separability of z and the lateral x, y-dimensions in the 2D Bloch wave method facilitates further approximations borrowed from band computations in solid state physics, notably the channeling approximation (Buxton, Loveluck, & Steeds, 1978; De Beeck & Van Dyck, 1995; Geuens & Van Dyck, 2002; Hovden, Xin, & Muller, 2012). The region in between the exact solutions and that of 2D Bloch waves is filled by "atomic" multislice integrating over atomic sheets, diagrammatic perturbation schemes (Gratias & Portier, 1983) and the approximate 3D Bloch wave method.

A vivid picture of the differences between the various schemes can be obtained with the help of scattering simulations based on numerically integrating (2.21) with very small step sizes (the numerical integrator (NI) solution) and atomic step size (the multislice (MS) solution), and computing the approximate 3D Bloch wave (3D-BW) and 2D Bloch wave (2D-BW) solution. In order to allow a maximal comparability (in terms of sampling, algorithms, etc.), we use the same numerical integration algorithm (Adams–Bashforth–Moulton PECE (Shampine & Gordon, 1975) as implemented in the ode113 routine of Matlab) for the NI and the 2D BW solution instead of performing the (expensive) diagonalization of the BW Hamiltonian with a large number of beams ($96 \times 96 = 9216$ in our case). The approximate 3D BW solution is obtained on a truncated basis consisting of $\mathbf{g}$-vectors, for which the dimensionless product $w_{\mathbf{g}}$ of excitation error $s_{\mathbf{g}}$ with extinction distance $\xi_{\mathbf{g}}$ fulfills $w_{\mathbf{g}} < 10^4$. This results in a basis comprising 3649 elements. In the final summation we include terms only, for which the product of Bloch coefficients was above 5×10^{-7}. The procedure follows the first steps of the MATS algorithm introduced by Rusz, Muto, and Tatsumi (2013) for inelastic scattering calculations.

As test target we use the high-temperature cubic perovskite phase of $PbZrO_3$ (Fig. 2.1), because it represents a relatively simple test structure combining light and heavy elements, thereby covering different scattering regimes. The maximal thickness was set to $t = 24a \approx 10$ nm, which covers the typical thickness range used in high-resolution TEM studies (Wei et al., 2014). The crystal potential was assembled from neutral independent atoms parametrized according to Weickenmeier and Kohl (1991) (see Kirkland, 1998; Lobato & Van Dyck, 2014 for alternative parametrizations). Further simulation parameters are $U_A = 200$ kV and a spatial sampling of 4.4 pm. Note that the ramifications of the thermal motion of the lattice have been completely ignored. They may be taken into account by performing an expensive frozen lattice summation of the results corresponding to the imaging conditions (Rother et al., 2009), which represents a further level of complexity and is beyond the scope of this study. For the same reasons aberrations of the microscope as well as the ramifications of partial coherence have been completely ignored.

Fig. 2.3 contains a compilation of various results of these computations. Accordingly, the full propagation from $z = 0$ nm to $z = t$, depicted in the left column for different $x - z$ cross-sections, exhibits the typical periodic channeling effect of the electron beam along atomic columns, with the periodicity depending on the atomic weight of the column. This effect is present in all four solutions. Both the $x - z$ cross-sections as well as the $x - y$ take outs furthermore reveal a substantial larger amount of high-spatial frequency data in the NI, MS, and 3D-BW data compared to the 2D-BW

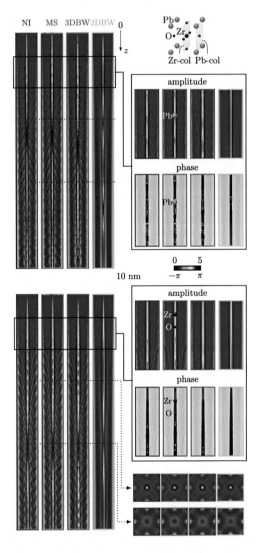

Figure 2.3 Results of the scattering simulations using the numerical forward integration (NI), the multislice (MS), and the 3D and 2D Bloch wave formalisms (3DBW, 2DBW). The prominent differences are mainly due to the absence of the higher order Laue zones in the 2D-BW calculations. The zoom-ins in the right column additionally reveal a beating structure of the NI, MS, and 3D-BW solution imprinted by the localized atomic potentials.

solution. This originates mainly from the missing higher order Laue zones in the 2D-BW solution (see also Fig. 2.4). We now zoom into the propagation along z to reveal the second characteristic feature of the 2D-BW

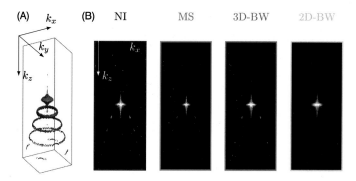

Figure 2.4 (A) 3D isosurface from the Fourier spectrum of the NI solution illustrating the appearance of higher order Laue zones. (B) Cross-sections of the 3D Fourier decomposition of the wave functions obtained from the NI, atomic MS, 3D-BW, and 2D-BW algorithms.

approximation (right column in Fig. 2.3). In the NI, MS, and 3D-BW solutions we observe a characteristic beating stemming from the pulse train of atomic potentials in z-direction. It consists of a sharp ramp in the phase (due to the projected potential of the atom) followed by a rapid interference effect in the amplitude. This effect is completely absent in the 2D-BW calculation. Note, however, that after each beat the NI, MS, and 3D-BW solutions converge to the 2D-BW solution as the localized effect behind the atoms quickly disperses. One could vividly rephrase this behavior as the difference between a staccato and legato interpretation of the same melody. The deviations of the 2D-BW solution agree well with the above theoretic predictions, where we identified a non-vanishing strongly peaked commutator as the main source of difference between the MS and 2D-BW approximation. Accordingly, the deviations are predicted to be more pronounced at the strongly scattering Pb column, which is confirmed by the numerical results. Moreover, we observe a significant difference between the 3D-BW solution on the one side and the NI and MS solution on the other side, when focusing on the first part of the crystal (in particular the Pb column). This deviation is the consequence of the approximate computation of the 3D-BW expansion based on an unperturbed incoming wave within the first unit cell discussed previously.

In addition to multislice and Bloch wave methods, a large number of additional approaches to compute the outcome of dynamical electron scattering has been developed in the past. Some of them merely differ in the type of basis used for the calculations, some apply further approximations to simplify the computations and some are mainly of theoretical value as they provide further insight into the nature of paraxial scattering. For instance, time-dependent approximation schemes (Feynman diagram techniques), well-known from quantum electron dynamics, have been applied to approximate the influence of the z-dependent potential terms in the paraxial Hamiltonian (Gratias & Portier, 1983; Houdellier, Altibelli, Roucau, & Casanove, 2008). We, however, also remark that the full scope of the analogy between paraxial scattering and the time-dependent Schrödinger equation has not been exploited yet. For instance, the prominent semiclassic propagator techniques, such as the van Fleck–Gutzwiller propagator (Van Vleck, 1928; Gutzwiller, 1967) or the Herman–Kluk propagator (Heller, 1981; Herman & Kluk, 1984), have not been elaborated on within the field of paraxial electron scattering so far (Lubk, Béché, & Verbeeck, 2015). Similarly, alternative direct integration schemes based on the Magnus (1954) and related expansions (Blanes, Casas, Oteo, & Ros, 2009) have not found their way into dynamical scattering computations. More comprehensive treatments of dynamical scattering are provided in the books of Peng et al. (2004), Kirkland (1998), and De Graef (2003).

For us the above considerations of dynamical scattering are particularly important when considering the holographic image formation at atomic resolution in high-resolution holographic tomography. For medium-resolution holography and tomography, it turns out that the most simple approximation consisting of neglecting the free space propagator completely, can be utilized under surprisingly general conditions. Here, the impact of dynamical scattering reduces to an orientation-dependent correction of the energy.

2.3. AXIAL SCATTERING

If the scattering angles within the wave function stay small, the action of the free space propagator reduces to the identity and (2.21) can be approximated to

$$i\partial_z \Psi(\mathbf{r}, z) = \underbrace{-\frac{e}{v}\Phi(\mathbf{r}, z)\Psi(\mathbf{r}, z)}_{H_{\mathrm{ax}}}, \qquad (2.48)$$

referred to as axial scattering in the following. Here, we may additionally incorporate an absorption term, approximating the interception of electrons scattered into angles exceeding the semi-angle α of some aperture stop, by a linear attenuation coefficient μ_{el} according to

$$i\partial_z \Psi(\mathbf{r}, z) = -\left(\frac{e}{v}\Phi(\mathbf{r}, z) + \frac{i}{2}\mu_{el}^{(\alpha)}(\mathbf{r}, z)\right)\Psi(\mathbf{r}, z). \tag{2.49}$$

This phenomenological approach is based on the observation that electrons, which have been scattered into large angles, possess a small probability to be back-scattered into forward direction again, hence may, to a good approximation, be incrementally removed from the axial wave field. For randomly distributed scatterers, $\mu_{el}^{(\alpha)}$ may be approximated by an incoherent superposition of n_i first order Born atomic scatterers of atomic species i with a total elastic cross section $\sigma_i^{(\alpha)}$ according to

$$\mu_{el}^{(\alpha)} = \sum_i \mu_{el,i}^{(\alpha)} = \sum_i n_i \sigma_i^{(\alpha)}. \tag{2.50}$$

Here, the total elastic cross section for electrons scattered beyond the aperture's acceptance angle α is given by the differential cross-section integrated over the solid angle blocked by the aperture:

$$\sigma_i^{(\alpha)} = \int_0^{2\pi} \int_\alpha^\pi \frac{\partial \sigma_i(\varphi, \theta)}{\partial \Omega} \sin\theta \, d\theta \, d\varphi. \tag{2.51}$$

The differential elastic scattering cross-section in the first order Born approximation is proportional to the absolute square of the Fourier transformed scattering potential, yielding

$$\frac{d\sigma_{el}}{d\Omega}(\varphi, \theta) = \left| \frac{C_E}{\lambda} \tilde{\Phi}\left(2k_0 \sin\frac{\theta}{2}\right) \right|^2 \tag{2.52}$$

in the case of spherical symmetric atomic scattering potentials. If the target is approximated as an assembly of $m = 1 \ldots M$ independent atomic scatterers (atomic number Z_m, density n_m), and Φ is modeled as a Wentzel potential, the total elastic attenuation coefficient reads (Reimer, 1989)

$$\mu_{el}(\alpha) = \sum_m \frac{\lambda^4 \gamma^2 Z_m^{4/3} n_m}{\pi \left(\lambda^2 + \left(2\pi a_B Z_m^{-1/3}\alpha\right)^2\right)}, \tag{2.53}$$

with a_B being the Bohr radius, respectively. This formula is inadequate to describe small angle scattering and scattering at heavy atoms, because chemical bonding, the short-comings of the Wentzel parameterization (see, e.g., Weickenmeier & Kohl, 1991; Kirkland, 1998; Lobato & Van Dyck, 2014 for improved parameterizations), and dynamical scattering are ignored. However, this approximation has been proved

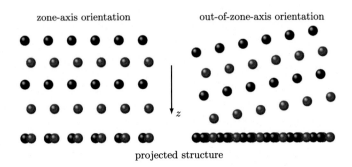

zone-axis orientation   out-of-zone-axis orientation

projected structure

Figure 2.5 Zone-axis and out-of-zone-axis orientation of a crystal and the corresponding structure projected into the electron's direction of flight.

valuable to describe total cross-sections into large angles for amorphous or out-of-zone axis oriented objects (Reimer, 1989), because high–angle scattering is rather accurately described by the corresponding Rutherford cross-section and some of the missing details favorably integrate out when considering total attenuation only.

It is one of the remarkable tricks of nature that axial scattering conditions do prevail for any type of scatterer, if the scattering centers are distributed randomly from the beam electron's perspective (Fig. 2.5) (Lubk, 2010). The latter condition can be adjusted even for a periodic crystal, if it is oriented out-of-zone-axis with respect to the optical axis, the typical orientation used for medium resolution (above one nanometer) investigations. The non-trivial proof of this assertion is given in Appendix A, which furthermore reveals that the influence of off-axial scattering may be approximated by a dynamical eigenenergy term Ξ_{dyn}.

The significance of the axial scattering approximation (2.49) for Electron Holography derives from its particularly simple analytic solution

$$\Psi(\mathbf{r}, z) = \underbrace{\exp\left(i\frac{e}{v}\int_{-\infty}^{z}\Phi(\mathbf{r}, z')\mathrm{d}z'\right)\exp\left(-i\int_{-\infty}^{z}A_z(\mathbf{r}, z')\mathrm{d}z'\right)}_{\text{electric and magnetic phase}} \quad (2.54)$$

$$\times\underbrace{\exp\left(i\int_{-\infty}^{z}\Xi_{\mathrm{dyn}}(\mathbf{r}, z')\mathrm{d}z'\right)}_{\text{dynamic phase}}$$

$$\times\underbrace{\exp\left(-\frac{1}{2}\int_{-\infty}^{z}\mu_{\mathrm{el}}(\mathbf{r}, z')\mathrm{d}z'\right)}_{\text{elastic damping}}\Psi(\mathbf{r}, 0).$$

Here, the Aharanov–Bohm phase shift, representing the dominant magnetic contribution,[4] and the additional phase shift due to the dynamical eigenenergy, derived in Appendix A, have been included for the sake of generality. The first line in the above equation is commonly referred to as phase object approximation (POA) or phase grating approximation (PGA) within the scope of electron scattering.

As it stands, the phase function in (2.54) has no particular meaning because the potentials are not gauge invariant. However, the various holographic setups discussed in Chapter 5 measure gauge invariant quantum mechanical observables such as electric or magnetic fields, permitting a straightforward interpretation of the phase term. In Sections 5.2 and 5.4, for instance, we will discuss a class of holographic techniques reconstructing the paraxial quantum probability current. Inserting (2.54) into the definition of the lateral paraxial probability current (2.23a) one obtains

$$\mathbf{j}(\mathbf{r}, z) = \frac{1}{k_0} I(\mathbf{r}, z) \nabla \varphi(\mathbf{r}, z)$$

$$= \mathbf{j}^{(\mathrm{in})}(\mathbf{r}, z) + \frac{1}{k_0} I(\mathbf{r}, z) \nabla \left(\frac{e}{v} \int_{-\infty}^{z} \Phi(\mathbf{r}, z') \mathrm{d}z' - \int_{-\infty}^{z} A_z(\mathbf{r}, z') \mathrm{d}z' \right)$$

(2.55)

$$= \mathbf{j}^{(\mathrm{in})}(\mathbf{r}, z)$$

$$- \frac{1}{k_0} I(\mathbf{r}, z) \left(\frac{e}{v} \int_{-\infty}^{z} \begin{pmatrix} E_x(\mathbf{r}, z') \\ E_y(\mathbf{r}, z') \end{pmatrix} \mathrm{d}z' - \int_{-\infty}^{z} \begin{pmatrix} B_y(\mathbf{r}, z') \\ -B_x(\mathbf{r}, z') \end{pmatrix} \mathrm{d}z' \right),$$

which is directly proportional to the projected electric and magnetic fields. Notably, the last result does not rely on the notion of a wave function or a phase, which permits a slight generalization of the POA to situations, where the electron beam cannot be described by a single wave function anymore (see below). Accordingly, the POA generally refers to a situation, where the scattering does not affect the intensity I in the object plane, rendering the probability current proportional to the projected electric and magnetic field in the sense defined above.

The axial scattering approximation leads to simple projection laws between the electron wave's phase and the static electromagnetic potentials on the one hand and the logarithm of the amplitude and linear (in)elastic attenuation coefficients on the other hand. By reconstructing the wave with

---

[4] The Zeeman term resulting from the electron spin in a magnetic field is very small (Edström et al., 2016) and can be neglected therefore.

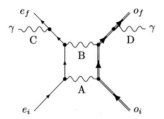

Figure 2.6 Feynman diagram representation of various inelastic interactions between a beam electron *e* and the object *o*: (A) first order object–electron interaction, (B) second order interaction, (C) bremsstrahlung, and (D) cathodoluminescence.

Electron Holography (cf. Chapter 5) one gains direct access to these projected fields, which renders Electron Holography a valuable technique for investigating material properties at the nanoscale (cf. Section 5.1). Moreover, the simple line integrals allow a reconstruction of local quantities resolved in three dimensions with the help of tomographic methods, which will be extensively discussed in Chapter 6. Since the dynamical eigenenergy depends on the crystal orientation relative to the beam direction, it gives rise to small tilt-angle-dependent fluctuations in the measured phase shifts, which need to be considered, when recording a holographic tilt series.

So far, only the elastic interaction between beam electrons and static electric or magnetic fields have been considered. However, these scattering events constitute only a subset of all possible electron–specimen or electron–TEM interactions. An electron traversing an object is not only deflected by the object's static fields but also exchanges virtual photons and hence energy with the latter, thereby changing the state of the object (see Fig. 2.6). Due to conservation of energy, the object (including eventually emitted photons) finds itself in a higher energetic state if the electron looses the corresponding amount of energy and vice versa after such a process. This one-to-one relationship between electron energy loss and object state change is the basis for electron energy loss spectroscopy (EELS), one of the main disciplines of TEM. In the following we will shortly sketch the main ramifications of inelastic scattering for Electron Holography. That implies a certain disregard to the details of scattering amplitude computations (i.e., the computation of Feynman diagrams). The theory of such excitations is generally very involved as it requires to take into account both elastic and inelastic scattering. Excellent overviews are to be found in Schattschneider (1986), Egerton (1996), García de Abajo (2010).

We continue by considering the special case of axial inelastic scattering in close analogy to axial elastic scattering. In particularly, the previously

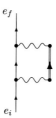

Figure 2.7 Feynman diagram of the eigenenergy correction. The initial and final object states have been omitted because they coincide.

obtained result that the axial state may be described by a diagonal Hamiltonian in position space, containing the effect of high-angle scattering as a damping term, may be generalized to approximate the outflow of intensity into inelastically scattered electrons, i.e.,

$$\Psi_{el}(\mathbf{r}, z) \to \Psi_{el}(\mathbf{r}, z) \exp\left(-\int \mu_{inel}(\mathbf{r}, z)\,dz\right), \qquad (2.56)$$

with the inelastic attenuation coefficient denoted by μ_{inel}. Eigenenergy corrections similar to that obtained for the elastic scattering, i.e.,

$$\Psi_{el}(\mathbf{r}, z) \to \Psi_{el}(\mathbf{r}, z) \exp\left(-i\Xi_{inel}z\right), \qquad (2.57)$$

correspond to loop corrections or eigenenergy terms in the presence of a polarizable medium (Fig. 2.7). Although, there are several approximations predicting different magnitudes of these loop corrections (Yoshioka, 1957; Ichikawa & Ohtsuki, 1969; Rez, 1978), second order perturbation theory, similar to that employed in calculating the related Lamb shift of atomic energy levels, predicts magnitudes well below 0.1 eV, which may be detectable with extraordinary instrumentation only.

In the following μ_{inel} is estimated from the dominant plasmonic attenuation (Egerton, 1996). Here, above some critical angle θ_c no significant inelastic plasmon scattering occurs, because the momentum transfer becomes large enough for single electron excitation (electron Compton scattering) (Nolting, 2005). Consequently, the critical angle may be approximated by the Bethe ridge

$$\theta_c = \sqrt{\frac{E_p m_0 c^2}{eU(E - E_p)}} \qquad (2.58)$$

derived from the energy and momentum conservation in this free particle collision. Using this assumption in combination with a simple Drude model for the differential cross-section pertaining to plasmon scattering (plasmon energy E_p, atom density n) leads to the following Lorentzian differential cross-section for bulk plasmon losses (Ritchie, 1959; Egerton, 1996)

$$\frac{d\sigma_p}{d\Omega} = \frac{E_p}{2\pi a_B m_0 v^2 n} \left(\frac{1}{\theta^2 + \theta_E^2} \right), \tag{2.59}$$

and the total inelastic attenuation reads

$$\mu_{\text{inel}}^{(\alpha)} = \int_\alpha^{\theta_c} \frac{d\sigma_p}{d\Omega} d\Omega \tag{2.60}$$

$$= \frac{E_p}{2 a_B m_0 v^2} \ln \left(\frac{\theta_c^2 + \theta_E^2}{\alpha^2 + \theta_E^2} \right),$$

where $\theta_E = E_p/\gamma m_0 v^2$ is the so-called characteristic scattering angle. Accordingly, the maximal inelastic attenuation

$$\mu_{\text{inel}} = \frac{E_p}{2 a_B m_e v^2} \ln \left(\frac{\theta_c^2 + \theta_E^2}{\theta_E^2} \right), \tag{2.61}$$

mainly depending on the plasmon energy E_p, is attained at $\alpha = 0$.

It is important to emphasize that the log–linear relationships, (2.54) and (2.56), are only valid up to a certain thickness, because multiple scattering eventually increases the angular distribution of the beam, thereby increasing the probability of scattering absorption in a nonlinear manner. The linear thickness range increases for materials with low scattering power, large acceleration voltages and large aperture angles (Smith & Burge, 1963; Wang, Zhang, Cao, Nishi, & Takaoka, 2010; Zhang, Egerton, & Malac, 2010).

We close this section with a note of caution. The scope of the axial approximation is limited (Vulovic, Voortman, van Vliet, & Rieger, 2014) under dynamical scattering conditions, that is low–index zone-axis orientation (Fig. 2.5) and/or high-resolution imaging. Here, the systematic interference of partial waves scattered from symmetric atomic positions in a crystal lattice leads to an amplification of large angle scattering, which requires dynamical scattering methods from the previous section to compute the transfer of an electron wave through a crystal.

2.4. DENSITY OPERATOR

Above, inelastic effects could be incorporated into a wave description of the scattering process, because we were solely interested in the shape of the elastic wave function axially scattered on an object. If we want to describe the total outcome, i.e., the intensity distribution, of *all* axially scattered electrons, these considerations are inadequate and the coherence of inelastically scattered electrons has to be included. A significant extension of the wave vector formalism is necessary in this case, because the beam electrons become entangled to a multitude of object states by virtue of the inelastic interaction. A similar situation occurs if some degrees of freedom of the system are unknown and averaged over in the experiment. Typical examples are quantum systems with unknown preparation history (e.g., beam electrons in a TEM) or systems in thermal equilibrium.

One can model such systems by state vectors, who live in a product of two Hilbert spaces $\mathbb{H}_A \otimes \mathbb{H}_B$ with B representing the unknown degrees of freedom (e.g., of the target in the TEM). An observable of an operator acting solely on Hilbert space $\mathbb{H}_A$ is now obtained by

$$\langle O_A \rangle = \sum_B \langle \Psi_{AB} | \hat{O}_A | \Psi_{AB} \rangle \qquad (2.62)$$

where the trace over B stands for the summation over the unknown degrees of freedom. The calculation of this sum can be complicated if not impossible as the Hilbert space $\mathbb{H}_B$ is usually not sufficiently defined. In order to treat such systems, the last expression is therefore reformulated by inserting complete basis sets in $\mathbb{H}_A$ and $\mathbb{H}_B$

$$\sum_B \langle \Psi_{AB} | \hat{O}_A | \Psi_{AB} \rangle = \sum_{mn\,m'n'} c_{mn} c^*_{m'n'} \langle A_{m'} | \hat{O}_A | A_m \rangle \langle B_{n'} | B_n \rangle \qquad (2.63)$$

$$= \sum_{mm'} \left(\sum_n c_{mn} c^*_{m'n} \right) \langle A_{m'} | \hat{O}_A | A_m \rangle$$

$$= \sum_m \langle A_m | \hat{\rho} \hat{O}_A | A_m \rangle .$$

Here, a new operator has been introduced in the last line, the (reduced) density operator

$$\hat{\rho} = \sum_{m'm''} \underbrace{\sum_n c_{m''n} c^*_{m'n}}_{\rho_{m''m'}} | A_{m''} \rangle \langle A_{m'} | \qquad (2.64)$$

with coefficients denoted by $\rho_{m''m'}$. Consequently, the expectation value in the reduced system may be obtained from

$$\langle O_A \rangle = \mathrm{Tr}\hat{\rho}\hat{O}_A, \qquad (2.65)$$

if the density operator is known, even in the complete absence of information about the structure of $\mathbb{H}_B$ (Von Neumann, 1927; Landau, 1927; Schlosshauer, 2004).

By definition the density operator is Hermitian and positive semidefinite, which implies that it may be diagonalized in some basis

$$\hat{\rho} = \sum_m |c_m|^2 |\Psi_m\rangle \langle \Psi_m| . \qquad (2.66)$$

Moreover, a pure and normalized quantum state, which can be completely described by a state vector $|\Psi_A\rangle$ in $\mathbb{H}_A$, is described by a density operator obeying $\mathrm{Tr}\hat{\rho}^2 = 1$, because the state vector can always be made a basis vector of the Hilbert space. Normalized ($\mathrm{Tr}\hat{\rho} = 1$) but otherwise arbitrary quantum states obey $\mathrm{Tr}\hat{\rho}^2 \leq 1$. They are called mixed states if $\mathrm{Tr}\hat{\rho}^2 < 1$. Therefore, the value of

$$\zeta = \mathrm{Tr}\hat{\rho}^2 \qquad (2.67)$$

(or $\zeta = \mathrm{Tr}\hat{\rho}^2/\mathrm{Tr}\hat{\rho}$ in the not normalized case) can be interpreted as a measure of the purity of the quantum state.

The most important basis representation for the density operator in the context of this work will be the spatial

$$\hat{\rho} = \iint_{-\infty}^{\infty} dx dx' \, |x\rangle \langle x |\hat{\rho}| x'\rangle \langle x'| \qquad (2.68)$$

$$= \iint_{-\infty}^{\infty} dx dx' \, |x\rangle \, \rho\left(x, x'\right) \langle x'|$$

and momentum

$$\hat{\rho} = \iint_{-\infty}^{\infty} dq dq' |q\rangle \langle q |\hat{\rho}| q'\rangle \langle q'| \qquad (2.69)$$

$$= \iint_{-\infty}^{\infty} dq dq' |q\rangle \rho\left(q, q'\right) \langle q'|$$

representation. The matrix elements in position $\rho(x, x')$ and momentum representation $\rho(q, q')$ are usually referred to as density matrix. As a consequence of the definition of the density operator, they are symmetric in their arguments, e.g.,

$$\rho\left(x, x'\right) = \langle x | \hat{\rho} | x' \rangle \tag{2.70}$$

$$= \sum_m |c_m|^2 \langle x | \Psi_m \rangle \langle \Psi_m | x' \rangle$$

$$= \sum_m |c_m|^2 \langle x' | \Psi_m \rangle^* \langle \Psi_m | x \rangle^*$$

$$= \rho^*\left(x', x\right).$$

Consequently, the main diagonal is positive semidefinite, and by comparison with (2.65) one sees that an intensity measurement performed in position (reciprocal) space corresponds to the main diagonal of the density matrix in position (momentum) representation. The information on the antidiagonals, on the other hand, is related to the correlation or degree of coherence of the quantum state in the respective representation, which is measured by the following density matrix normalized with the intensities on the main diagonals

$$\mu\left(x, x'\right) = \frac{\rho(x, x')}{\sqrt{\rho\left(x, x\right)}\sqrt{\rho(x', x')}}. \tag{2.71}$$

Moreover, these density matrices can be mapped on Wigner functions, which turn out to be the phase space representation of the density operator (cf. next section).

Finally, the paraxial dynamics of the density operator is governed by the paraxial von Neumann equation (in the Heisenberg picture)

$$\frac{\partial \hat{\rho}}{\partial z} = \sum_m |c_m|^2 \left(\frac{\partial |\Psi_m\rangle}{\partial z} \langle \Psi_m | + |\Psi_m\rangle \frac{\partial \langle \Psi_m |}{\partial z} \right) \tag{2.72}$$

$$= \sum_m |c_m|^2 \left(-i\hat{H} |\Psi_m\rangle \langle \Psi_m | + i |\Psi_m\rangle \langle \Psi_m | \hat{H} \right)$$

$$= -i\left[\hat{H}, \hat{\rho}\right].$$

The formal solution for the free-space propagation reads

$$\hat{\rho}\left(z\right) = e^{-i(z-z')\hat{H}} \hat{\rho}\left(z'\right) e^{i(z-z')\hat{H}} = \hat{K}_T(z; z')\hat{\rho}\left(z'\right)\hat{K}_T^*(z; z'), \tag{2.73}$$

where the Fresnel propagator introduced previously has been used.

To solve the complete problem of (dynamical) scattering of a mixed quantum state, encompassing in particular inelastic scattering, one typically has to resort to perturbation schemes facilitating approximate solutions to the von Neumann equation (Dudarev, Peng, & Whelan, 1993; Peng et al., 2004). Employing such techniques, quantum state transfer in the parlance of density matrices has been successfully used to describe inelastic plasmon scattering (Verbeeck, Bertoni, & Schattschneider, 2008; Schattschneider & Verbeeck, 2008; Lee, Rose, Hambach, Wachsmuth, & Kaiser, 2013), core-loss scattering (Schattschneider, Nelhiebel, & Jouffrey, 1999; Nelhiebel et al., 1999; Rusz, Rubino, & Schattschneider, 2007), and frozen-lattice scattering (Dinges, Berger, & Rose, 1995; Müller, Rose, & Schorsch, 1998; Rother et al., 2009) in the TEM.

In the following, we again restrict ourselves to axial scattering, where the von Neumann equation in position space representation taking into account a variable mixing due to (inelastic) scattering with a possible energy transfer, reads

$$\frac{\partial \rho \left(\mathbf{r}, \mathbf{r}', z \right)}{\partial z} = -i \sum_m |c_m|^2 \tag{2.74}$$
$$\times \left(\Psi_m^* \left(\mathbf{r}', z \right) H_m \left(\mathbf{r}, z \right) \Psi_m \left(\mathbf{r}, z \right) - \Psi_m \left(\mathbf{r}, z \right) H_m \left(\mathbf{r}', z \right) \Psi_m^* \left(\mathbf{r}', z \right) \right) .$$

In the case of elastic axial scattering of a pure state, the Hamiltonian does not depend on m and we can write

$$\frac{\partial \rho \left(\mathbf{r}, \mathbf{r}', z \right)}{\partial z} = -i \left(H_{\mathrm{ax}} \left(\mathbf{r}, z \right) - H_{\mathrm{ax}} \left(\mathbf{r}', z \right) \right) \rho \left(\mathbf{r}, \mathbf{r}', z \right) , \tag{2.75}$$

which has the solution

$$\rho \left(\mathbf{r}, \mathbf{r}', z \right) = T_{\mathrm{el}} \left(\mathbf{r}, \mathbf{r}', z \right) \rho \left(\mathbf{r}, \mathbf{r}', 0 \right) . \tag{2.76}$$

The function T, depending on two spatial coordinates, is referred to as mutual object transparency (Kohl & Rose, 1985). In the case of purely elastic axial scattering, the transparency

$$T_{\mathrm{el}} \left(\mathbf{r}, \mathbf{r}' \right) = \exp \left(\frac{ie}{\nu} \int_{-\infty}^z \left(\Phi(\mathbf{r}, z') - \Phi(\mathbf{r}', z') \right) \mathrm{d}z' \right) \tag{2.77}$$
$$\times \exp \left(-i \int_{-\infty}^z \left(A_z(\mathbf{r}, z') - A_z(\mathbf{r}', z') \right) \mathrm{d}z' \right)$$
$$\times \exp \left(-\frac{1}{2} \int_{-\infty}^z \left(\mu_{\mathrm{el}}(\mathbf{r}, z') + \mu_{\mathrm{el}}(\mathbf{r}', z') \right) \mathrm{d}z' \right)$$

is separable into two wave transmission functions containing the projected electrostatic and magnetostatic potentials as phase and the projected elastic damping coefficient as amplitude argument.

Similar to the phenomenological treatment of scattering absorption within elastic axial scattering, we may now approximate the redistribution of the electrons among the various incoherent channels due to inelastic scattering by incorporating a linear "mixing" coefficient into the axial von Neumann equation

$$\frac{\partial \rho\left(\mathbf{r}, \mathbf{r}', z\right)}{\partial z} = -\left(iH_{\mathrm{ax}}\left(\mathbf{r}, z\right) - iH_{\mathrm{ax}}(\mathbf{r}', z) + \mu_{\mathrm{inel}}(\mathbf{r}, \mathbf{r}', z)\right) \rho\left(\mathbf{r}, \mathbf{r}', z\right).$$

(2.78)

By comparison with the "exact" expression (2.74) we see that this mixing term can be approximated as the average of m-dependent Hamiltonians (stripped from the m-independent elastic part)

$$\mu_{\mathrm{inel}}(\mathbf{r}, \mathbf{r}', z) \approx \sum_{m} \left(H_m\left(\mathbf{r}, z\right) - H_{\mathrm{ax}}\left(\mathbf{r}, z\right)\right) - \left(H_m\left(\mathbf{r}', z\right) - H_{\mathrm{ax}}\left(\mathbf{r}', z\right)\right).$$

(2.79)

The general mutual object transparency then reads

$$T\left(\mathbf{r}, \mathbf{r}', z\right) = T_{\mathrm{el}}\left(\mathbf{r}, \mathbf{r}', z\right)$$

$$\times \underbrace{\exp\left(-\int_0^z \mu_{\mathrm{inel}}(\mathbf{r}, \mathbf{r}', z') \mathrm{d}z'\right)}_{T_{\mathrm{inel}}(\mathbf{r}, \mathbf{r}', z)},$$

(2.80)

and one readily observes that the "elastic" separability is lost under inelastic interaction because the beam electrons become entangled to the object's degrees of freedom (Zurek, 2003; Schlosshauer, 2004).

Here, the dominant influence of bulk plasmon excitations

$$\mu_{\mathrm{inel}}(\mathbf{r}, \mathbf{r}', z) = \mu_{\mathrm{inel}}(|\mathbf{r} - \mathbf{r}'|, z),$$

(2.81)

only depends on the distance between two spatial coordinates, because the electron plasma is isotropic and homogeneous to a good approximation (Schattschneider & Lichte, 2005). This expression converges against the previously considered maximum μ_{inel} (2.61) at very large distances, where any inelastic plasmon excitation becomes incoherent. Moreover, the mutual inelastic damping may depend additionally on the integration limit αk_0 in Fourier space integral defining the attenuation

$$\mu_{\text{inel}}^{(\alpha)}(\mathbf{r} - \mathbf{r}', z) = \int_{>\alpha k_0} \tilde{\mu}_{\text{inel}}(\mathbf{k}, z) e^{i\mathbf{k}(\mathbf{r} - \mathbf{r}')} \mathrm{d}^2 k. \qquad (2.82)$$

In the TEM, this limit is physically defined by an aperture of semiangle α blocking (in)elastically scattered electrons with lateral momenta larger than αk_0. Thus reducing the aperture semiangle increases the scattering absorption due to inelastic scattering as obtained from the main diagonal of the mixing coefficient

$$\mu_{\text{inel}}^{(\alpha)}(\mathbf{r}, z) = \mu_{\text{inel}}^{(\alpha)}(\mathbf{r}, \mathbf{r}' = \mathbf{r}, z). \qquad (2.83)$$

2.5. WIGNER FUNCTION

In this section we introduce one particular representation of quantum mechanical phase space – the Wigner function. The main goal is to review all its necessary properties without giving an exhaustive overview over phase space quantum mechanics. The **Wigner function** of a one-dimensional electron wave function is defined as (Schleich, 2001)

$$W(r, k) := \frac{1}{2\pi} \int_{-\infty}^{\infty} \mathrm{d}r' \Psi^* \left(r - \frac{1}{2}r' \right) \Psi \left(r + \frac{1}{2}r' \right) e^{-ikr'}, \qquad (2.84)$$

where the normalization factor $(2\pi)^{-1}$ guarantees

$$\iint_{-\infty}^{\infty} \mathrm{d}r \mathrm{d}k \, W(r, k) = 1 \qquad (2.85)$$

for normalized wave vectors Ψ.[5] The two-dimensional version required for electron optics follows in a straightforward manner and will be introduced later.

The Fourier space coordinate k can be interpreted as the momentum coordinate of phase space, because changing the representation of the electron wave function to momentum space, i.e.,

$$\langle q | \Psi \rangle = \Psi(q) = \frac{1}{\sqrt{2\pi}} \int_{-\infty}^{\infty} \Psi(r) e^{-iqr} \mathrm{d}r, \qquad (2.86)$$

---

[5] The momentum coordinate is denoted by k (instead of the more common p). Both are related by $\hbar$, which is one in the natural units adopted here.

yields a symmetric version of the original definition

$$W(r, k) = \frac{1}{2\pi} \int_{-\infty}^{\infty} \Psi^* \left(k - \frac{1}{2}k' \right) \Psi \left(k + \frac{1}{2}k' \right) e^{ik'r} dk', \qquad (2.87)$$

with the role of the phase space coordinates r and k exchanged.

The full scope of the Wigner function approach becomes evident when identifying the product $\Psi^*(r) \Psi(r')$ as a pure state density matrix and generalizing the **Wigner function to mixed states** according to

$$W(r, k) = \frac{1}{2\pi} \int_{-\infty}^{\infty} \sum_{m} |c_m|^2 \Psi_m^* \left(r - \frac{1}{2}r' \right) \Psi_m \left(r + \frac{1}{2}r' \right) e^{-ikr'} dr' \qquad (2.88)$$

$$= \frac{1}{2\pi} \int_{-\infty}^{\infty} \rho \left(r + \frac{1}{2}r', r - \frac{1}{2}r' \right) e^{-ikr'} dr'.$$

Consequently, the Wigner function is the Fourier transform of the density matrix along the antidiagonal as indicated in Fig. 2.8. When discussing Electron Holography in Chapter 5 we will frequently make use of the (symplectic) Fourier transform of the Wigner function along both phase space coordinates

$$\tilde{W}(k', r') = \frac{1}{2\pi} \iint_{-\infty}^{\infty} W(r, k) e^{-ik'r} e^{ikr'} dr dk, \qquad (2.89)$$

which is referred to as Moyal's characteristic function or **ambiguity function** in literature (Fig. 2.8). Inserting the definition of the Wigner function one readily obtains that Moyal's characteristic function corresponds to the Fourier transform of the density matrix along the main diagonal

$$\tilde{W}(k', r') = \frac{1}{2\pi} \int_{-\infty}^{\infty} \rho \left(r + \frac{1}{2}r', r - \frac{1}{2}r' \right) e^{-ik'r} dr, \qquad (2.90)$$

instead of the antidiagonal as in the case of the Wigner function. The density matrix in position and momentum state, and the Wigner function and its Fourier transform (ambiguity function) form a "magical square" of density operator representations (Fig. 2.8) facilitating a complete description of practically all relevant optical setups and measurements performed in the TEM (cf. Chapter 4).

2.5.1 Basic Properties

The definition (2.88) implies that the Wigner function is strictly real. Projecting the Wigner function along k yields a strictly positive quantity, which

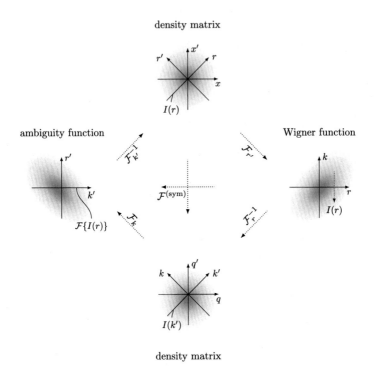

Figure 2.8 Relation between the Wigner function, the density matrix in position and momentum representation, and the ambiguity function.

can be identified as the quantum mechanical density in position space

$$\int_{-\infty}^{\infty} W(x,k)\, \mathrm{d}k = \frac{1}{2\pi} \iint_{-\infty}^{\infty} \left\langle r + \frac{1}{2}r' \middle| \hat{\rho} \middle| r - \frac{1}{2}r' \right\rangle e^{-ikr'}\, \mathrm{d}r'\mathrm{d}k \qquad (2.91)$$

$$= \frac{1}{2\pi} \int_{-\infty}^{\infty} \left\langle r + r' \middle| \hat{\rho} \middle| r - r' \right\rangle e^{-ikr'} \delta\left(2r'\right) \mathrm{d}r'$$

$$= I(r)\,.$$

The same holds for Fourier space, i.e.,

$$I(k) = \int_{-\infty}^{\infty} W(x,k)\, \mathrm{d}r\,. \qquad (2.92)$$

The latter projection along r could be alternatively understood as a rotation of phase space around 90° with a subsequent projection along the vertical axis (Fig. 2.9). The generalization of this concept to arbitrary tilt angles leads to the notion of fractional Fourier transforms (Almeida, 1994),

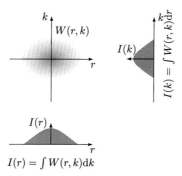

Figure 2.9 A Wigner function and its marginal densities.

which is closely related to the free space propagation in the paraxial regime (Fresnel propagation) discussed previously. In Section 5.3 we will discuss a particular holographic technique based on defocus variations, namely Focal Series Inline Holography, in terms of rotating phase space distributions.

The above lines illustrate that the Wigner function is indeed not only sufficient to compute the outcome of a quantum measurement, i.e., the quantum mechanical density, in various planes by simple projections, but also provides a rather straightforward image of the quantum state as the underlying function of these projections.

Furthermore, the quantum mechanical **probability current**

$$j(r) = \frac{1}{2i}\left(\Psi^*(r)\frac{\partial}{\partial r}\Psi(r) - \Psi(r)\frac{\partial}{\partial r}\Psi^*(r)\right) \qquad (2.93)$$

corresponds to the first moment of the Wigner function along the momentum coordinate

$$j(r) = \int_{-\infty}^{\infty} kW(r,k)\,dk, \qquad (2.94)$$

which may be verified by the following computations

$$\int kW(r,k)\,dk = \frac{i}{2\pi}\iint_{-\infty}^{\infty}\left(\Psi^*\left(r-\frac{1}{2}r'\right)\Psi\left(r+\frac{1}{2}r'\right)\right)\frac{\partial}{\partial r'}e^{-ikr'}\,dr'\,dk$$

$$= -i\int\delta(r')\frac{\partial}{\partial r'}\left(\Psi^*\left(r-\frac{1}{2}r'\right)\Psi\left(r+\frac{1}{2}r'\right)\right)dr' \qquad (2.95)$$

$$= \frac{1}{2i}\left(\Psi^*(r)\frac{\partial}{\partial r}\Psi(r) - \Psi(r)\frac{\partial}{\partial r}\Psi^*(r)\right)$$

$$= j(r).$$

Higher-order moments may be defined in a similar fashion

$$j^{(n)}(r) = \int_{-\infty}^{\infty} k^n W(r, k) \, dk \tag{2.96}$$

and one can show that their Fourier transforms correspond to derivatives of increasing order of the ambiguity function at $r' = 0$

$$\mathcal{F}\left\{ \int_{-\infty}^{\infty} k^n W(r, k) \, dk \right\} = i^n \frac{\partial^n}{\partial r'^n} \tilde{W}(k', r') \Big|_{r'=0}. \tag{2.97}$$

This includes that the diffractogram (i.e., the Fourier transform of the image intensity) is given by the cross section of the ambiguity function

$$\mathcal{F}\{I(r)\} = \tilde{W}(k', r' = 0). \tag{2.98}$$

In Section 5.2 we will discuss a holographic setup determining spatially resolved generalized currents (moments) of the Wigner function.

2.5.2 Quantum Mechanics in Phase Space

The above expression for the moments of the Wigner function can be considered as special cases of arbitrary operators acting on the wave function and the corresponding Wigner function. Indeed, the Wigner function forms the basis for a fully flexed formulation of quantum mechanics referred to as phase space quantification (Groenewold, 1946; Moyal, 1949). Within that formulation a quantum mechanical operator $\hat{A}$ is transformed to a scalar function by means of the Wigner map

$$A(r, k) = \int_{-\infty}^{\infty} \left\langle r + \frac{1}{2}r' \left| \hat{A} \right| r - \frac{1}{2}r' \right\rangle e^{-ikr'} \, dr' \tag{2.99}$$

Quantum mechanical expectation values are then obtained by a phase space average of this function weighted with the Wigner function of the system

$$\text{Tr}\left(\hat{A}\hat{\rho}\right) = \iint_{-\infty}^{\infty} dr \, dk \, A(r, k) W_{\hat{\rho}}(r, k) \tag{2.100}$$

in close analogy to phase space averages in classical mechanics. One particular instant of the above transformation yields the overlap between two arbitrary quantum states

$$\text{Tr}\left(\hat{\rho}_1 \hat{\rho}_2\right) = 2\pi \iint_{-\infty}^{\infty} dr \, dk \, W_{\hat{\rho}_1}(r, k) W_{\hat{\rho}_2}(r, k) \tag{2.101}$$

and two pure states

$$|\langle \Psi_1 | \Psi_2 \rangle|^2 = 2\pi \iint_{-\infty}^{\infty} drdk \, W_{|\Psi_1\rangle}(r, k) W_{|\Psi_2\rangle}(r, k). \qquad (2.102)$$

These expressions can be used to derive certain properties pertaining to the shape of the Wigner function. Inserting the upper bound of the normalized squared density operator into (2.101) yields

$$2\pi \iint_{-\infty}^{\infty} drdk \, W_{\hat{\rho}}^2(r, k) \leq 1 \qquad (2.103)$$

or equivalently

$$\zeta = \frac{2\pi}{\Omega} = 2\pi \iint_{-\infty}^{\infty} drdk \, W^2(r, k) \leq 1. \qquad (2.104)$$

Since the right hand side is a measure for the phase space volume covered by the quantum state (Schleich, 2001) one can deduce that a quantum state cannot occupy less than 2π in phase space. The fraction of the occupied volume is the purity measure introduced in Section 2.4, i.e., $\zeta = 1$ for a pure state and $\zeta < 1$ for mixed states. If, on the other hand, two quantum states with zero overlap are considered, one obtains

$$0 = \mathrm{Tr}\left(\hat{\rho}_1 \hat{\rho}_2\right) = 2\pi \iint_{-\infty}^{\infty} drdk \, W_{\hat{\rho}_1}(r, k) W_{\hat{\rho}_2}(r, k), \qquad (2.105)$$

which can only be true if the Wigner function takes on negative values. These negative values prevent an interpretation of the Wigner function as positive probability density in phase space, which is why the Wigner function is referred to as a quasi-probability function. Indeed, the appearance of negative values hallmarks the quantum nature of the system, as will be shown below (e.g., Table 2.1).

We finally discuss the relationship between the Wigner function and other phase space representations. Indeed, there is a whole set of quantum mechanical phase space representations $f(r, k)$ (Mandel & Wolf, 1995), which all facilitate the calculation of expectation values of an operator through quantum phase space averages according to the optical equivalence theorem (Agarwal & Wolf, 1970). Here, the ordering of creation and annihilation operators (or position and momentum operators) fix the particular choice of f. The most important phase space representations named after Wigner ($f \cong W$), Glauber–Sudarshan ($f \cong P$), and Husimi ($f \cong Q$)

Table 2.1 Wigner function of several pure quantum states (wave functions). The linear chirp is represented by a single line in phase space with the slope determined by the prefactor of the quadratic term. Coherent states are Gaussians in phase space and the weak phase object is represented by the imaginary part of its Fourier transform multiplied with a phase factor

Quantum state	Wave function	Wigner function
Linear chirp	$e^{i(ax^2+bx+c)}$	$\delta(k - 2ar - b)$
Coherent state	$\sqrt[4]{\frac{1}{\pi\sigma^2}}e^{-\frac{(x-x_\alpha)^2}{2\sigma^2}+ik_\alpha x}$	$\frac{1}{2\pi\sigma^2}e^{-\frac{(x-x_\alpha)^2}{2\sigma^2}-\frac{(k-k_\alpha)^2}{2\sigma^{-2}}}$
Weak phase object	$1 + i\varphi(x) \,\|\,\varphi \ll 1$	$\delta(k) + \sqrt{\frac{2}{\pi}}\mathrm{Im}\left(\tilde{\varphi}(2k)e^{i2kr}\right)$

correspond to a symmetric, normal, and antisymmetric operator ordering, respectively. They may be directly transformed into each other through successive convolutions (coarsening) of phase space with Gaussians (coherent states) in the order $P \to W \to Q$. All these different choices happen to coincide in the classical limit, where the operator ordering becomes meaningless. For instance, the Husimi representation is obtained from the following convolution

$$H(r, k) = \frac{1}{2\pi} \iint_{-\infty}^{\infty} dr' dk' e^{-\frac{(r'^2+k'^2)}{2}} W\left(r - r', k - k'\right).$$ (2.106)

It can be shown that this convolution leads to positive semidefinite values of the Husimi distribution.

This subsection is closed by providing a very short list of phase space representations, important in the course of this work. Moreover, these examples illustrate some of the previously derived properties of the Wigner function. In particular they are real and assume negative values.

2.5.3 Paraxial Dynamics

In order to consider the paraxial dynamics of the Wigner function, we first generalize the above definition to two dimensional wave functions, yielding a 4D Wigner function

$$W(\mathbf{r}, \mathbf{k}) = \frac{1}{4\pi^2} \int d^2r' \Psi^*\left(\mathbf{r} - \frac{1}{2}\mathbf{r}'\right) \Psi\left(\mathbf{r} + \frac{1}{2}\mathbf{r}'\right) e^{-i\mathbf{k}\mathbf{r}'}$$ (2.107)

$$= \frac{1}{4\pi^2} \int d^2r' \left\langle \mathbf{r} - \frac{1}{2}\mathbf{r}' \,\middle|\, \hat{\rho} \,\middle|\, \mathbf{r} + \frac{1}{2}\mathbf{r}' \right\rangle e^{-i\mathbf{k}\mathbf{r}'}$$

for the pure (first line) and mixed (second line) state. The paraxial dynamics of the 4D Wigner function may now be obtained by performing the

z-derivative in the above definition

$$\partial_z W(\mathbf{r}, \mathbf{k}, z) = \frac{1}{4\pi^2} \int d^2 r' \left\langle \mathbf{r} - \frac{1}{2}\mathbf{r}' \left| \partial_z \hat{\rho} \right| \mathbf{r} + \frac{1}{2}\mathbf{r}' \right\rangle e^{-i\mathbf{k}\mathbf{r}'} \tag{2.108}$$

$$= -\frac{i}{4\pi^2} \int d^2 r' \left\langle \mathbf{r} - \frac{1}{2}\mathbf{r}' \left| \hat{H}\hat{\rho} - \hat{\rho}\hat{H} \right| \mathbf{r} + \frac{1}{2}\mathbf{r}' \right\rangle e^{-i\mathbf{k}\mathbf{r}'},$$

where the paraxial Hamiltonian from (2.72) has been inserted in the last line. This equation is referred to as **quantum Liouville equation** for reasons, which will become clear shortly. First, the last expression is transformed to

$$\boxed{\partial_z W(\mathbf{r}, \mathbf{k}, z) = -2 W(\mathbf{r}, \mathbf{k}, z) \underbrace{\sin\left(\frac{1}{2}(\overleftarrow{\nabla}_r \overrightarrow{\nabla}_k - \overleftarrow{\nabla}_k \overrightarrow{\nabla}_r)\right) H(\mathbf{r}, \mathbf{k}, z)}_{\{\{W(\mathbf{r},\mathbf{k},z), H(\mathbf{r},\mathbf{k},z)\}\}}}$$

$$\tag{2.109}$$

by inserting the Wigner transform of the Hamilton operator. The compact notation on the second line, referred to as Moyal (or sine) bracket, bears a remarkable similarity to the Poisson bracket determining the time evolution of phase space functions in classical mechanics. Indeed, if the above expressions are simplified by collecting all higher order derivatives of the potential on the right hand side of the quantum Liouville equation

$$\left(\partial_z + \frac{1}{k_0}\mathbf{k}\nabla - \frac{e}{v}\nabla\Phi\nabla_k\right) W = \sum_{k=1}^{\infty}\sum_{l=1}^{\infty} (\ldots), \tag{2.110}$$

one immediately identifies the classical Liouville equation on the left hand side.

The solution of this equation for arbitrary potentials is complicated. Here, we only note the solution to free space propagation, which is frequently required in the remainder of this work. It may be obtained from solving the classical Liouville equation with the method of characteristics and reads (Lee, 1995)

$$W(\mathbf{r}, \mathbf{k}, z) = W\left(\mathbf{r} + \frac{z - z'}{k}\mathbf{k}, \mathbf{k}, z'\right). \tag{2.111}$$

This result exhibits a remarkable simplicity. The paraxially propagated Wigner function is merely a sheared version of its initial state. This relationship will be exploited at various instances in this work as it establishes a direct link to tomographic methods discussed in the next chapter. Moreover, similar transformations as the one above occur when considering the influence of aberrations of the optical system in a TEM on the quantum state of the beam electrons in Section 4.2.

REFERENCES

Agarwal, G. S., & Wolf, E. (1970). Calculus for functions of noncommuting operators and general phase-space methods in quantum mechanics. I. Mapping theorems and ordering of functions of noncommuting operators. *Physical Review D, 2,* 2161–2186.

Allen, L. J., D'Alfonso, A. J., & Findlay, S. D. (2015). Modelling the inelastic scattering of fast electrons. *Ultramicroscopy, 151,* 11–22.

Almeida, L. B. (1994). The fractional Fourier transform and time-frequency representations. *IEEE Transactions on Signal Processing, 42*(11), 3084–3091.

Askar, A., & Cakmak, A. S. (1978). Explicit integration method for the time-dependent Schrödinger equation for collision problems. *Journal of Chemical Physics, 68*(6), 2794–2798.

Berry, M. V. (1971). Diffraction in crystals at high energies. *Journal of Physics C, 4,* 697–722.

Bethe, H. (1928). Theorie der Beugung von Elektronen an Kristallen. *Annalen der Physik, 392*(17), 55–129.

Blanes, S., Casas, F., Oteo, J., & Ros, J. (2009). The Magnus expansion and some of its applications. *Physics Reports, 470*(5–6), 151–238.

Buxton, B. F., Loveluck, J. E., & Steeds, J. W. (1978). Bloch waves and their corresponding atomic and molecular orbitals in high energy electron diffraction. *Philosophical Magazine A, 38*(3), 259–278.

Cockcroft, J. D., & Walton, E. T. S. (1932a). Experiments with high velocity positive ions. (I) Further developments in the method of obtaining high velocity positive ions. *Proceedings of the Royal Society of London A: Mathematical, Physical and Engineering Sciences, 136*(830), 619–630.

Cockcroft, J. D., & Walton, E. T. S. (1932b). Experiments with high velocity positive ions. II. The disintegration of elements by high velocity protons. *Proceedings of the Royal Society of London A: Mathematical, Physical and Engineering Sciences, 137*(831), 229–242.

Courant, R., & Hilbert, D. (1962). *Methods of mathematical physics: Partial differential equations. Methods of mathematical physics.* Interscience Publishers.

Cowley, J. M. (1995). *Diffraction physics. North-Holland personal library.* Amsterdam: Elsevier Science.

Cowley, J. M., & Moodie, A. F. (1957). The scattering of electrons by atoms and crystals. I. A new theoretical approach. *Acta Crystallographica, 10,* 609–619.

De Beeck, M. O., & Van Dyck, D. (1995). An analytical approach for the fast calculation of dynamical scattering in HRTEM. *Physica Status Solidi A, 150*(2), 587–602.

De Broglie, L. (1924). *Recherches sur la theorie des quanta* (PhD thesis).

De Graef, M. (2003). *Introduction to conventional transmission electron microscopy.* Cambridge University Press.

Dinges, C., Berger, A., & Rose, H. (1995). Simulation of TEM images considering phonon and electronic excitations. *Ultramicroscopy, 60,* 49–70.

Dudarev, S. L., Peng, L.-M., & Whelan, M. J. (1993). Correlations in space and time and dynamical diffraction of high-energy electrons by crystals. *Physical Review B, 48*(18), 13408–13429.

Edström, A., Lubk, A., & Rusz, J. (2016). Elastic scattering of electron vortex beams in magnetic matter. *Physical Review Letters, 116*(12), 127203.

Egerton, R. F. (1996). *Electron energy-loss spectroscopy in the electron microscope.* Plenum Press.

Floquet, G. (1883). Sur les équations différentielles linéaires à coefficients périodiques. *Annales de l'École Normale Supérieure, 12*, 47–88.

Fujimoto, F. (1959). Dynamical theory of electron diffraction in Laue—case I. General theory. *Journal of the Physical Society of Japan, 14*, 1558–1568.

Fujiwara, K. (1961). Relativistic theory of electron diffraction. *Journal of the Physical Society of Japan, 16*, 2226–2238.

García de Abajo, F. J. (2010). Optical excitations in electron microscopy. *Reviews of Modern Physics, 82*, 209–275.

Geuens, P., & Van Dyck, D. (2002). The S-state model: A work horse for HRTEM. *Ultramicroscopy, 93*(3), 179–198.

Glaser, W. (1933). Über geometrisch-optische Abbildung durch Elektronenstrahlen. *Zeitschrift für Physik, 80*, 451–464.

Glauber, R., & Schomaker, V. (1953). The theory of electron diffraction. *Physical Review, 89*, 667–671.

Goodman, P., & Moodie, A. F. (1974). Numerical evaluations of N-beam wave functions in electron scattering by the multi-slice method. *Acta Crystallographica Section A, 30*(2), 280–290.

Gratias, D., & Portier, R. (1983). Time-like perturbation method in high-energy electron diffraction. *Acta Crystallographica Section A, 39*(4), 576–584.

Greiner, W., & Reinhardt, J. (Eds.). (1984). *Theoretische Physik: Vol. 7. Quantenelektrodynamik.* Thun/Frankfurt am Main: Verlag Harri Deutsch.

Groenewold, H. J. (1946). On the principles of elementary quantum mechanics. *Physica, 12*(7), 405–460.

Gutzwiller, M. C. (1967). The phase integral approximation in momentum space and the bound states of an atom. *Journal of Mathematical Physics, 8*, 1979–2000.

Hawkes, P. W., & Kasper, E. (1996). *Principles of electron optics: Vol. 1. Basic geometrical optics.* Academic Press.

Heller, E. J. (1981). Frozen Gaussians: A very simple semiclassical approximation. *Journal of Chemical Physics, 75*(6), 2923–2931.

Herman, M. F., & Kluk, E. (1984). A semiclassical justification for the use of non-spreading wavepackets in dynamics calculations. *Chemical Physics, 91*(1), 27–34.

Houdellier, F., Altibelli, A., Roucau, C., & Casanove, M.-J. (2008). New approach for the dynamical simulation of CBED patterns in heavily strained specimens. *Ultramicroscopy, 108*, 426–432.

Hovden, R., Xin, H. L., & Muller, D. A. (2012). Channeling of a subangstrom electron beam in a crystal mapped to two-dimensional molecular orbitals. *Physical Review B, 86*(19), 195415.

Howie, A., & Whelan, M. J. (1961). Dynamical theory of crystal lattice defects. II. The development of a dynamical theory. *Proceedings of the Royal Society of London A: Mathematical and Physical Sciences, 263*, 217–237.

Ichikawa, M., & Ohtsuki, Y.-H. (1969). The correction of the mean inner potential in electron diffraction. *Journal of the Physical Society of Japan, 27*(4), 953–956.

Kambe, K. (1967). Theory of electron diffraction by crystals. I. Green's function and integral equation. *Zeitschrift für Naturforschung, 22a*, 422–431.

Kirkland, E. J. (1998). *Advanced computing in electron microscopy.* New York: Plenum Press.

Koch, C. T. (2002). *Determination of core structure periodicity and point defect density along dislocations* (PhD thesis). Arizona State University.

Kohl, H., & Rose, H. (1985). Theory of image formation by inelastically scattered electrons in the electron microscope. *Advances in Electronics and Electron Physics*, *65*, 173–227.

Lamla, E. (1938). Zur Theorie der Elektronenbeugung bei Berücksichtigung von mehr als 2 Strahlen und zur Erklärung der Kikuchi-Enveloppen. I. *Annalen der Physik*, *424*(1–2), 178–189.

Landau, L. D. (1927). The damping problem in wave mechanics. *Zeitschrift für Physik*, *45*, 430.

Landau, L. D., & Lifshitz, E. M. (1977). *Quantum mechanics: Non-relativistic theory*. Butterworth-Heinemann.

Lee, H.-W. (1995). Theory and application of the quantum phase-space distribution functions. *Physics Reports*, *259*(3), 147–211.

Lee, Z., Rose, H., Hambach, R., Wachsmuth, P., & Kaiser, U. (2013). The influence of inelastic scattering on EFTEM images exemplified at 20 kV for graphene and silicon. *Ultramicroscopy*, *134*, 102–112.

Lobato, I., & Van Dyck, D. (2014). An accurate parameterization for scattering factors, electron densities and electrostatic potentials for neutral atoms that obey all physical constraints. *Acta Crystallographica Section A*, *70*(6), 636–649.

Lubk, A. (2010). *Quantitative off-axis electron holography and (multi-)ferroic interfaces* (PhD thesis). Technische Universität Dresden.

Lubk, A., Béché, A., & Verbeeck, J. (2015). Electron microscopy of probability currents at atomic resolution. *Physical Review Letters*, *115*(17), 176101.

Lubk, A., Clark, L., Guzzinati, G., & Verbeeck, J. (2013). Topological analysis of paraxially scattered electron vortex beams. *Physical Review A*, *87*, 033834.

Lubk, A., & Rusz, J. (2015). Jacob's ladder of approximations to paraxial dynamic electron scattering. *Physical Review B*, *92*, 235114.

Magnus, W. (1954). On the exponential solution of differential equations for a linear operator. *Communications on Pure and Applied Mathematics*, *7*(4), 649–673.

Mandel, L., & Wolf, E. (1995). *Optical coherence and quantum optics*. Cambridge: Cambridge University Press.

Moyal, J. E. (1949). Quantum mechanics as a statistical theory. *Mathematical Proceedings of the Cambridge Philosophical Society*, *45*(1), 99–124.

Müller, H., Rose, H., & Schorsch, P. (1998). A coherence function approach to image simulation. *Journal of Microscopy*, *190*(1–2), 73–88.

Nelhiebel, M., Louf, P.-H., Schattschneider, P., Blaha, P., Schwarz, K., & Jouffrey, B. (1999). Theory of orientation-sensitive near-edge fine-structure core-level spectroscopy. *Physical Review B*, *59*(20), 12807–12814.

Nolting, W. (2005). *Vielteilchentheorie*. Berlin: Springer.

Peng, L. M., Dudarev, S. L., & Whelan, J. (2004). *High energy electron diffraction and microscopy. Monographs on the physics and chemistry of materials*. Oxford University Press.

Reimer, L. (Ed.). (1989). *Transmission electron microscopy*. Springer Verlag.

Rez, P. (1978). Virtual inelastic scattering in high-energy electron diffraction. *Acta Crystallographica Section A*, *34*(1), 48–51.

Ritchie, R. H. (1959). Interaction of charged particles with a degenerate Fermi–Dirac electron gas. *Physical Review*, *114*(3), 644–654.

Rosenauer, A., & Schowalter, M. (2008). STEMSIM – a new software tool for simulation of STEM HAADF Z-contrast imaging. In A. G. Cullis, & P. A. Midgley (Eds.), *Springer proceedings in physics: Vol. 120. Microscopy of semiconducting materials 2007* (pp. 170–172). Netherlands: Springer.

Rother, A., Gemming, T., & Lichte, H. (2009). The statistics of the thermal motion of the atoms during imaging process in transmission electron microscopy and related techniques. *Ultramicroscopy, 109*(2), 139–146.

Rother, A., & Scheerschmidt, K. (2009). Relativistic effects in elastic scattering of electrons in TEM. *Ultramicroscopy, 109*(2), 154–160.

Rusz, J., Muto, S., & Tatsumi, K. (2013). New algorithm for efficient Bloch-waves calculations of orientation-sensitive ELNES. *Ultramicroscopy, 125*, 81–88.

Rusz, J., Rubino, S., & Schattschneider, P. (2007). First-principles theory of chiral dichroism in electron microscopy applied to 3D ferromagnets. *Physical Review. B, Condensed Matter and Materials Physics, 75*(21), 214425.

Schattschneider, P. (1986). *Fundamentals of inelastic electron scattering*. Springer.

Schattschneider, P., & Lichte, H. (2005). Correlation and the density-matrix approach to inelastic electron holography in solid state plasmas. *Physical Review B, 71*(4), 045130.

Schattschneider, P., Nelhiebel, M., & Jouffrey, B. (1999). Density matrix of inelastically scattered fast electrons. *Physical Review B, 59*, 10959.

Schattschneider, P., & Verbeeck, J. (2008). Fringe contrast in inelastic LACBED holography. *Ultramicroscopy, 108*(5), 407–414.

Schleich, W. P. (2001). *Quantum optics in phase space*. Berlin: Wiley VCH.

Schlosshauer, M. (2004). Decoherence, the measurement problem, and interpretations of quantum mechanics. *Reviews of Modern Physics, 76*, 1267–1305.

Shampine, L. F., & Gordon, M. K. (1975). *Computer solution of ordinary differential equations: The initial value problem*. San Francisco: W. H. Freeman.

Smith, G. H., & Burge, R. E. (1963). A theoretical investigation of plural and multiple scattering of electrons by amorphous films, with special reference to image contrast in the electron microscope. *Proceedings of the Physical Society, 81*(4), 612.

Stadelmann, P. (1987). EMS – a software package for electron diffraction analysis and (HREM) image simulation in material science. *Ultramicroscopy, 21*, 131–146.

Strange, P. (1998). *Relativistic quantum mechanics*. Cambridge University Press.

Van Vleck, J. H. (1928). The correspondence principle in the statistical interpretation of quantum mechanics. *Proceedings of the National Academy of Sciences of the United States of America, 14*, 178.

Verbeeck, J., Bertoni, G., & Schattschneider, P. (2008). The Fresnel effect of a defocused biprism on the fringes in inelastic holography. *Ultramicroscopy, 108*(3), 263–269.

Von Neumann, J. (1927). Thermodynamik quantenmechanischer Gesamtheiten. *Göttinger Nachrichten, 3*, 273–291.

Vulovic, M., Voortman, L. M., van Vliet, L. J., & Rieger, B. (2014). When to use the projection assumption and the weak-phase object approximation in phase contrast cryo-EM. *Ultramicroscopy, 136*, 61–66.

Wacker, C., & Schröder, R. R. (2015). Multislice algorithms revisited: Solving the Schrödinger equation numerically for imaging with electrons. *Ultramicroscopy, 151*, 211–223.

Wang, F., Zhang, H.-B., Cao, M., Nishi, R., & Takaoka, A. (2010). Determination of the linear attenuation range of electron transmission through film specimens. *Micron, 41*(7), 769–774.

Wei, X.-K., Tagantsev, A. K., Kvasov, A., Roleder, K., Jia, C.-L., & Setter, N. (2014). Ferro-electric translational antiphase boundaries in nonpolar materials. *Nature Communications*, *5*, 3031.

Weickenmeier, A., & Kohl, H. (1991). Computation of absorptive form factors for high energy electron diffraction. *Acta Crystallographica Section A*, *47*, 590–597.

Yoshioka, H. (1957). Effect of inelastic waves on electron diffraction. *Journal of the Physical Society of Japan*, *12*, 618–628.

Zeitler, E., & Olsen, H. (1964). Screening effects in elastic electron scattering. *Physical Review*, *136*, A1546–A1552.

Zhang, H., Egerton, R. F., & Malac, M. (2010). Local thickness measurement in TEM. *Microscopy and Microanalysis*, *16*(Suppl. S2), 344–345.

Zurek, W. H. (2003). Decoherence, einselection, and the quantum origins of the classical. *Reviews of Modern Physics*, *75*(3), 715–775.

CHAPTER THREE

Tomography

Axel Lubk

Institute for Structure Physics, Physics Department, Faculty of Mathematics and Natural Sciences, Technical University of Dresden, Dresden, Germany
e-mail address: axel.lubk@tu-dresden.de

Contents

"Oft liegen Dinge so, daß mathematische Theorien in abstrakter Form bereits vorliegen, vielleicht als unfruchtbare Spielerei betrachtet, die sich plötzlich als wertvolles Werkzeug für physikalische Erkenntnisse entpuppen und ihre latente Kraft in ungeahnter Weise offenbaren."

(Johann Radon, 1954)

Tomography is concerned with the reconstruction of a function from its lower dimensional projections. Its foundations have been developed when mathematicians started to look into how a certain multidimensional function can be retrieved from its integrals over certain submanifold (Funk,

Advances in Imaging and Electron Physics, Volume 206
ISSN 1076-5670
https://doi.org/10.1016/bs.aiep.2018.05.003

1913; Radon, 1917). A widespread utilization of the tomographic principle only occurred more than half a century later (Cormack, 1963, 1964; Hounsfield, 1973), when the advent of the computer facilitated sufficiently fast numerical implementations of the previously discovered reconstruction principles. Nowadays, the most prominent application is X-ray computed tomography, where the logarithmic attenuation of X-ray beams, traversing biological tissue from different angles, is used to compute the 3D distribution of the radiation attenuation characteristic for the tissue distribution. The typical tomographic problem encountered in TEM is closely related to this archetypal case.

Similar to the previous chapter, the preliminaries on tomography, required in the course of this work, are summarized subsequently. First, a brief introduction to general projection transformations, including the geometrical setting typically used when combining Electron Holography and Tomography, is given. Subsequently, the two-dimensional Radon transformation, providing the mathematical theory behind the tomographic reconstructions used in this work, is introduced. Based on these foundations, the discrete Radon transformation, providing the framework for tomographic reconstruction algorithms, is discussed. Two separate subsections are devoted to sampling and regularization, which are substantial for a thorough understanding of tomographic reconstructions from experimental data. Finally, three reconstruction algorithms, heavily used in the remainder of this work, are presented in detail.

Excellent and more comprehensive treatments on the foundations of tomography can be found in Deans' (1983), Natterer's (2001), and Helgason's (2011) books. Further recommendable monographs on the topic with varying focus on the mathematical foundations, algorithmic implementations or applications in various fields are those of Kak and Slaney (1988), Hsieh (2003), and Herman (2009).

3.1. PROJECTION TRANSFORMATIONS

The original works of Radon and Funk dealt with the problem of recovering functions from their projections along (curved) lines on certain 2D manifolds (surfaces). These initial examples can be considered as specific cases of the more general problem of reconstructing a function f on arbitrary dimensional homogeneous spaces from its projections F along certain submanifolds (Helgason, 2011), which amounts to asking for a bijection

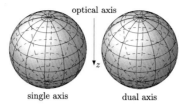

Figure 3.1 Single and dual axes projection geometries realized within the TEM. The red circles indicate the trajectory of a surface point on the sphere upon rotation. Accordingly, when tilting around one axis, the tomographic problem may be restricted to 2D planes perpendicular to the tilt axis, running through the midpoint of the circle. In the dual axes setup two rotation series around two orthogonal rotation axes are recorded.

between

$$f\,(\mathbb{R}^n) \overset{\text{bijection}}{\Longleftrightarrow} F\,(\mathbb{P}^n) \quad \text{or shortly} \quad \mathbb{R}^n \Longleftrightarrow \mathbb{P}^n. \qquad (3.1)$$

Here, $\mathbb{P}^n$ denotes the parameter space of the projections. The particular case of projections along $n-1$ dimensional submanifolds is referred to as Radon transformation in the literature. The study of these transformations revealed an extremely rich structure with manifold connections to the theory of differential equations, integral and differential geometry and algebraic topology. The following considerations heavily draw on these findings, in particular those pertaining to the projection geometries realized within the TEM, which have been excellently summarized in Helgason's (2011) book.

The most important projection geometry (cf. Sections 6.1, 6.2, and 6.4) is the simple projection from a two-dimensional plane to a two-dimensional projection space ($\mathbb{R}^2 \Longleftrightarrow \mathbb{P}^2$), which arises, when the line projections of a 3D object tilted around a single tilt axis are recorded (Fig. 3.1). This geometry allows for a separation of the 3D tomographic problem into 2D ones confined to planes perpendicular to the tilt axis. It is furthermore assumed that the projection lines are parallel, which corresponds to a parallel broad beam illumination in the TEM (cf. Chapter 4). Additionally, a dual axes projection $\mathbb{R}^2 \oplus \mathbb{R}^2 \leftrightarrow \mathbb{P}^2 \oplus \mathbb{P}^2$ is briefly discussed in Section 6.3 and projections from a 4D phase space to 2D images play a role in Section 5.3.

3.2. RADON TRANSFORMATION

As announced previously, this chapter focuses on parallel beam single tilt axis projections in 2D, which covers the majority of projection geome-

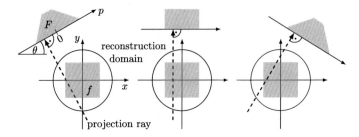

Figure 3.2 2D Radon transformation principle including some notation used in this context. The scalar function f ($= 1$ for $|(x, y)^T| < 1$ in this case) is projected under angles θ ranging from 0 to π to yield its Radon transform $F(p, \theta)$.

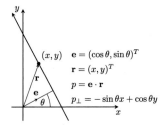

Figure 3.3 Coordinate system, normal form and geometrical parameters used in combination with the 2D Radon transformation.

tries encountered in this work. The other ones are discussed as extensions to this archetypal case. The corresponding integral transformation mapping a real function f on 2D Cartesian space equipped with coordinates x and y to a function F on a 2D projection space parametrized by the coordinates p (the detector coordinate) and θ (the tilt angle) is referred to as 2D Radon transformation (Fig. 3.2).

Further definitions, in particular the normal form of a line in two dimensions parametrized by its normal orientation θ and length p

$$p := \left\langle \underbrace{\begin{pmatrix} \cos\theta \\ \sin\theta \end{pmatrix}}_{\mathbf{e}}, \underbrace{\begin{pmatrix} x \\ y \end{pmatrix}}_{\mathbf{r}} \right\rangle, \tag{3.2}$$

are summarized in Fig. 3.3. It follows a general definition of the 2D Radon transformation before discussing its properties and inversion formulas.

3.2.1 Definition

The general **Radon transformation**

$$F := \mathcal{R}\{f\} \qquad (3.3)$$

in 2D Cartesian coordinates reads

$$F(p, \theta) := \int_{-\infty}^{\infty} f(p\cos\theta - p\sin\theta, p\sin\theta + p\cos\theta)\, dp \qquad (3.4a)$$

$$= \iint_{-\infty}^{\infty} f(x, y)\, \delta(p - x\cos\theta - y\sin\theta)\, dxdy \qquad (3.4b)$$

$$= \iint_{-\infty}^{\infty} f(x, y)\, \delta(p - \mathbf{e} \cdot \mathbf{r})\, dxdy \qquad (3.4c)$$

$$= \iint_{\mathbf{e}\cdot\mathbf{r}=p} f(x, y)\, dxdy \qquad (3.4d)$$

employing the coordinate system given in Fig. 3.3. The first definition uses the detector coordinate system to perform the integration, while in the second line the object's coordinate system is employed. Because of the third and fourth line of the above definition, the Radon transform is frequently parametrized by $F(p, \mathbf{e}(\theta))$ or shortly $F(p, \mathbf{e})$ in the following.

The corresponding formulation in 2D polar coordinates with $x = r\cos\varphi$, $y = r\sin\varphi$ reads

$$F(p, \theta) = \int_{0}^{\infty}\int_{0}^{2\pi} f(r, \varphi)\, \delta(p - r\cos(\theta - \varphi))\, rdrd\varphi \qquad (3.5)$$

$$= \int_{0}^{\infty} f(r, \theta) * \delta(p - r\cos\theta)\, rdr.$$

Note the convolution in azimuthal space in the last line, which can be diagonalized in the space of circular harmonics.

The above definition equips the Radon transformation with a number of specific properties. Some of them merely simplify computations, some lead to more fundamental aspects such as the existence, support and uniqueness theorems. In the next subsection an incomplete account of some simple properties, motivated by our needs in later chapters, is given. The important Fourier slice theorem and the support theorem are discussed in separate subsections as they directly lead to the inversion of the Radon transformation, i.e., the inverse Radon transformation. The latter *is* the mathematical terminus for the tomographic reconstruction carried out in Chapter 6.

3.2.2 Basic Properties

Obviously, the Radon transform is linear in its argument

$$\mathcal{R}\{af + bg\} = a\mathcal{R}\{f\} + b\mathcal{R}\{g\} \quad (a, b \in \mathbb{R}), \tag{3.6}$$

which allows us to transform different constituents of the underlying function separately. Linear affine transformations of the coordinate space $\mathbb{R}^2$ directly translate into well defined transformations of the projection space coordinates. Here, only the special case of a coordinate dependent scaling referred to as similarity

$$\mathcal{R}\{f(ax, by)\} = \frac{1}{|ab|} F\left(p, \frac{e_1}{a}, \frac{e_2}{b}\right) \tag{3.7}$$

and the shifting property

$$\mathcal{R}\{f(x - a, y - b)\} = F(p - ae_1 - be_2, \mathbf{e}) \tag{3.8}$$

are noted (Poularikas, 2010). These two relations are very useful for deriving the explicit Radon transforms of geometric objects such as a collection of ellipsoids from simple prototypes such as a circle in this case. For instance the Radon transform pair of an off-center δ-function

$$f(x, y) = \delta(x - a, y - b) \leftrightarrow F(p, \theta) = \delta(p - (a\cos\theta + b\sin\theta)) \tag{3.9}$$

is readily computed from the obvious transformation pair $f(x, y) = \delta(x, y) \leftrightarrow F(p, \theta) = \delta(p)$ using the shifting property (3.8). Fig. 3.4 shows the projections of a point (3.9) following a sinusoidal motion in the Cartesian p, θ plane

$$p(\theta) = a\cos\theta + b\sin\theta, \tag{3.10}$$

which lends the name *sinogram* to the graphical representation of the Radon transformed function.[1] One typically arranges the p, θ coordinates of the sinogram in a Cartesian setting, although other choices are also possible and preferred in this work (see below).

Furthermore, particular symmetry relations hold (Poularikas, 2010)

$$F(ap, a\mathbf{e}) = \frac{1}{|a|} F(p, \mathbf{e}) \tag{3.11}$$

---

[1] No Chinese character is involved here.

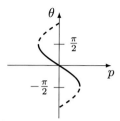

Figure 3.4 Sinogram of Radon transformed point (δ-function). The π interval of projection space is indicated and the extended 2π-trajectory is drawn dashed.

$$F\left(p, a\mathbf{e}\right) = \frac{1}{|a|} F\left(\frac{p}{a}, \mathbf{e}\right) \tag{3.12}$$

connecting warped functions in projection space with scaled counterparts. With the particular choice $a = -1$ the first relation yields

$$F\left(-p, -\mathbf{e}\right) = F\left(p, \mathbf{e}\right), \tag{3.13}$$

which says that, after tilting the projections around 180°, mirrored projections appear. This property defines the topology of the Radon space $\mathbb{P}^2$, which is important for analyzing or manipulating sinograms. It is not as trivial as that of the underlying space $\mathbb{R}^2$ and, therefore, requires some further explanation.

As a consequence of (3.13), the topology of the sinogram is that of a Möbius band, i.e., a strip of paper, half-twisted and glued together at the end to form a loop. Mathematically, one can define this as the square $[0, 1] \times [0, 1]$ with its top and bottom sides identified as indicated in Fig. 3.5A.

We now introduce an alternative representation of the Möbius band, which simplifies the following discussion and is therefore preferred throughout. When cutting the Möbius band in the middle one obtains an orientable manifold (in contrast to the non–orientable Möbius strip). That motivates the representation indicated in Fig. 3.5. Here, p and θ form a punctuated plane, where the opposite points at the puncture are identified. This representation corresponds to the polar coordinates introduced in Fig. 3.5 and one immediately observes that, contrary to polar coordinates in $\mathbb{R}^2$, a one-to-one correspondence between Cartesian and polar representations also holds at $p = 0$.

Fig. 3.6 exhibits the Radon transform of the point in the polar p, θ representation. The harmonic motion of the sinogram transforms into a

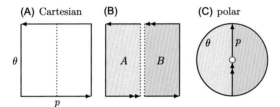

Figure 3.5 Sinogram representations and topology: (A) Cartesian representation with $(p, \theta) \mapsto (x, y)$. According to (3.13) bottom $(p, \theta = 0)$ and top $(-p, \theta = \pi)$ can be identified (indicated by arrows) endowing $\mathbb{P}^2$ with the topology of a Möbius band. (B) Cutting (indicated by dotted line) of the Cartesian representation into two halves. (C) Polar representation with $(p, \theta) \mapsto (r, \varphi)$ obtained from (B) by horizontally flipping B, gluing together A and B along the double arrowed edges, 360°-bending of the obtained stripe around the cut, and finally gluing along the single arrowed edges. Note that opposite sides on the central, now circular, cutting line have to be identified. Consequently, when contracting the hole to a single point the function value at the origin is obtained from $F(p = 0, \theta) = \lim_{p \to \pm 0} F(p, \theta)$ in the polar representation. Accordingly, the sinusoidal trajectory of the point projection in the Cartesian plane transforms to a circle in the polar representation.

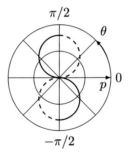

Figure 3.6 Rotating point projected into the polar representation of the sinogram. The extended 2π-trajectory of the point is drawn dashed.

circle

$$p = (a\cos\theta + b\sin\theta) \leftrightarrow \left(p_x - \frac{a}{2}\right)^2 + \left(p_y - \frac{b}{2}\right)^2 = \frac{a^2 + b^2}{4}, \qquad (3.14)$$

which is not only visually closer to the circular motion but also easier to analyze. For instance in Section 6.1, the center coordinates of the circle of various point features are detected to localize and align the rotation axis of experimental tilt series. Moreover, rotational symmetries of the specimen are readily identified in the polar representation of the sinogram.

We finally consider some basic analytic transformations namely convolution and derivation. The convolution of two functions translates into a convolution of their respective Radon transforms along the detector coordinate according to

$$\mathcal{R}\{g * h\} = \int G(s, \mathbf{e}) H(p - s, \mathbf{e}) \, ds. \tag{3.15}$$

Similarly, partial derivatives translate into partial derivatives along the detector coordinate weighted with the cosine or sine of the tilt angle

$$\mathcal{R}\left\{\frac{\partial f(x, y)}{\partial x}\right\} = e_1 \frac{\partial F(p, \theta)}{\partial p} \tag{3.16}$$

$$\mathcal{R}\left\{\frac{\partial f(x, y)}{\partial y}\right\} = e_2 \frac{\partial F(p, \theta)}{\partial p}. \tag{3.17}$$

These relations are important when discussing the reconstruction of magnetic fields by means of holographic tomography in Section 6.3. From the transformation properties of the two directional derivatives one may directly deduce that of the Laplace operator resulting in the second derivative of the Radon transform with respect to the detector coordinate

$$\mathcal{R}\{\Delta f(x, y)\} = \frac{\partial^2}{\partial p^2} F(p, \theta). \tag{3.18}$$

This property allows, e.g., the direct computation of charges from projected holographic potentials in Section 5.1.

We finally note a particularly interesting formula, connecting the moments along the detector coordinate of some Radon transformed function with corresponding integrals over homogeneous polynomials in the underlying Cartesian space (Helgason, 2011)

$$\int_{-\infty}^{\infty} F(p, \theta) p^n dp = \int_{\langle \mathbf{e}, \mathbf{r} \rangle = p} f(\mathbf{r}) p^n d^2 r dp \tag{3.19}$$

$$= \int_{\langle \mathbf{e}, \mathbf{r} \rangle = p} f(\mathbf{r}) \langle \mathbf{e}, \mathbf{r} \rangle^n d^2 r dp$$

$$= \int_{-\infty}^{\infty} f(\mathbf{r}) \langle \mathbf{e}, \mathbf{r} \rangle^n d^2 r.$$

These integral relations are not only important for identifying function spaces allowing a well-defined Radon transformation (Helgason, 2011),

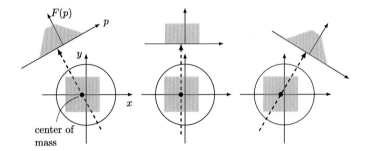

Figure 3.7 Relation between center of mass of the 2D function f density projected in 2D and centroid of the projected function F at particular projection angles θ.

they also prove very useful for the alignment of tilt series discussed in Section 6.1. In particularly the first moment relation

$$\int_{-\infty}^{\infty} F(p,\theta)\, p\, dp = \mathbf{e} \int_{-\infty}^{\infty} f(\mathbf{r})\, \mathbf{r}\, d^2 r \qquad (3.20)$$

$$= \mathbf{e}\mathbf{m}_1 \,,$$

stating that the first moment of the projected data corresponds to the projected center of mass of the underlying function, which may be used as a point feature, whose circular trajectory can be determined from its projected positions. The parameters of this trajectory then determine the relative position of center of mass and tilt axis (Fig. 3.7), which is used to align various tomographic tilt series later in this work. It is interesting to note that similar relations hold for higher order moments. For instance, the second moment relation explicitly reads

$$\int_{-\infty}^{\infty} F(p,\theta)\, p^2\, dp = \mathbf{e}^T \int_{-\infty}^{\infty} f(r) \begin{pmatrix} x^2 & xy \\ xy & y^2 \end{pmatrix} d^2 r\, \mathbf{e} \qquad (3.21)$$

$$= \mathbf{e}^T \mathbf{m}_2 \mathbf{e} \,,$$

which may be equally well exploited for tilt series alignment. To date of this work, however, higher-order alignment methods remain to be explored within the scope of Electron Tomography.

3.2.3 Fourier Slice Theorem

The **Fourier slice theorem** provides a very useful relation between the 2D Fourier transform of the original function and the 1D Fourier transform of its Radon transform along the detector coordinate, i.e.,

$$\tilde{f}(q,\theta) = \frac{1}{2\pi} \int_{-\infty}^{\infty} F(p,\theta) e^{-iqp} dp. \qquad (3.22)$$

Proof. Its significance for the understanding of the Radon transformation including reconstruction artifacts as well as the development of efficient reconstruction algorithms cannot be overestimated, which is why a short derivation of this theorem is provided. We start by expanding the 2D Fourier transform of f with a δ-integral in direction $\mathbf{e}$

$$\tilde{f}(\mathbf{q}) = \frac{1}{2\pi} \int_{-\infty}^{\infty} f(\mathbf{r}) e^{-i\mathbf{q}\cdot\mathbf{r}} d^2 r \qquad (3.23)$$

$$= \frac{1}{2\pi} \iint_{-\infty}^{\infty} f(\mathbf{r}) e^{-it} \delta(t - \mathbf{q}\cdot\mathbf{r}) dt d^2 r.$$

Collapsing the Fourier integral in a second step by substituting $t = qp$ and using $\delta(ax) = |a|^{-1}\delta(x)$, i.e.,

$$\tilde{f}(q\mathbf{e}) = \frac{|\mathbf{q}|}{2\pi} \iint_{-\infty}^{\infty} f(\mathbf{r}) e^{-iqp} \delta(qp - q\mathbf{e}\cdot\mathbf{r}) dp d^2 r \qquad (3.24)$$

we obtain the above Fourier slice theorem. $\square$

Accordingly, the set of projections inhomogeneously covers the complete Fourier space of the original function with a radially decreasing weight corresponding to the factor $1/|\mathbf{q}|$. A visual representation of the Fourier slice theorem is given in Fig. 3.8.

Several important properties of the (inverse) Radon transform can be directly derived from the Fourier space representation. First of all, an inverse Radon transform, denoted by $\mathcal{R}^{-1}$, exists and one possible analytic formula consists of Fourier transforming the projection data, removing the weight by multiplying with $|\mathbf{q}|$, and performing an inverse 2D Fourier transformation to obtain $f(x,y)$. The corresponding reconstruction algorithm, referred to as weighted or filtered back-projection, is discussed below. Second, the $|\mathbf{q}|$-filter applied in the reconstruction renders the inverse Radon transform non-local in the sense that line integrals, not passing a certain subvolume in the reconstruction domain, still contribute to the reconstructed value in that subvolume. We will see in Sections 6.1 and 6.3 that this behavior leads to a propagation of artifacts pertaining to inconsis-

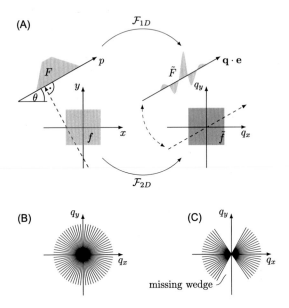

Figure 3.8 Sketch of the Fourier slice theorem (A) and corresponding sampling of Fourier space of a 180° tilt series (B). An incomplete tilt series containing a missing wedge of spatial frequencies is shown in (C).

tent projection data (e.g., due to misalignment) to the whole reconstruction domain. Last but not least we observe that the inverse Radon transform is not well-defined if the tilt interval of the projections does not encompass 180°, because spatial frequencies of the projected function laying in the so-called missing wedge (Fig. 3.8) are not contained in the projection data. Due to technical limitations imposed by the delicate sample holding mechanism in a TEM, however, Electron Tomography typically suffers from incomplete tilt intervals and the pertaining non-local reconstruction errors, referred to as missing wedge artifacts, represent one of the biggest obstacles to the technique.

3.2.4 Support Theorem

The support theorem defines conditions, the projected function has to fulfill in order to allow a tomographic reconstruction. Its says that, if a continuous function tends faster to zero than any power of $\mathbf{r}$ and its projections vanish outside some radial distance R, the function is zero outside of the support defined by a circle $|\mathbf{r}| \leq R$ with radius R. More formally, we have:

Theorem 1. *Let $f \in C(\mathbf{R}^2)$ (with C denoting the space of the continuous functions) satisfy the following conditions:*

For $\forall k > 0$, $|\mathbf{r}|^k f(\mathbf{r}) < c$ with some positive constant c, and $F(p, \theta) = 0$ for $p > R$ with some positive constant R. Then $f(\mathbf{r}) = 0$ for $|\mathbf{r}| > R$.

Proof. Subsequently, a short sketch for the proof of the support theorem is provided. First, one writes the complete function

$$f = f_{\text{ex}} + f_{\text{in}} \tag{3.25}$$

as the sum of its restrictions to the convex domain (denoted by f_{in}) and its exterior (denoted by f_{ex}). This sum can be equated to the sum of the inverse Radon transformation of projections containing (denoted by F_{in}) and excluding (denoted by F_{ex}) the convex domain by exploiting the linearity of the (inverse) Radon transformation:

$$f_{\text{ex}} + f_{\text{in}} = \mathcal{R}^{-1}\{F_{\text{ex}} + F_{\text{in}}\} = \mathcal{R}^{-1}\{F_{\text{ex}}\} + \mathcal{R}^{-1}\{F_{\text{in}}\} . \tag{3.26}$$

In a second step, the function on the convex domain f_{in} is computed from the inverse Radon transformation of the projections containing the domain minus the projections of the exterior part truncated (truncation operator $\mathcal{T}$) to the convex domain's projection range

$$f_{\text{in}} = \mathcal{R}^{-1}\left\{F_{\text{in}} - \mathcal{T}\mathcal{R}\{f_{\text{ex}}\}\right\} = \mathcal{R}^{-1}\{F_{\text{in}}\} - \mathcal{R}^{-1}\mathcal{T}\mathcal{R}\{f_{\text{ex}}\} . \tag{3.27}$$

Inserting (3.27) into (3.26) eliminates f_{in} and leads to an integral equation relating the function on the exterior of the convex domain with the projections outside of the convex domain

$$f_{\text{ex}} = \mathcal{R}^{-1}\{F_{\text{ex}}\} + \mathcal{R}^{-1}\mathcal{T}\mathcal{R}\{f_{\text{ex}}\} = \left(I - \mathcal{R}^{-1}\mathcal{T}\mathcal{R}\right)^{-1}\mathcal{R}^{-1}\{F_{\text{ex}}\} . \tag{3.28}$$

It remains to be shown that the above formula corresponds to the inverse Radon transformation restricted to the projections outside of the convex domain, which follows in the next lines:

$$f_{\text{ex}} = \underbrace{\left(I - \mathcal{R}^{-1}\mathcal{T}\mathcal{R}\right)^{-1}}_{\sum_{n=0}^{\infty}(\mathcal{R}^{-1}\mathcal{T}\mathcal{R})^n}\mathcal{R}^{-1}\{F_{\text{ex}}\} \tag{3.29}$$

$$= \mathcal{R}^{-1}\{F_{\text{ex}}\} .$$

In the second line the term in brackets has been reformulated as a geometric series, which collapses to the zeroth order term by using $\mathcal{T}F_{\text{ex}} = 0$. The

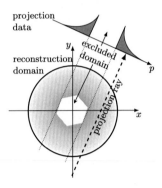

Figure 3.9 Schematics of the reconstruction of an exterior field from projections outside of a convex region. The support theorem ensures the unique reconstruction of that field.

last line, however, implies that $f_{ex} = 0$, if $F_{ex} = 0$, which completes the proof.   □

It follows that the exact shape of a function may be determined by virtue of the inverse Radon transformation, if the object is well contained in all projections. An immediate consequence for Electron Tomography in the TEM is that the object or field to be reconstructed must not escape the field of view during the tilt series. Therefore, needle type geometries tilted around the needle axis are particularly well adapted to tomography (see Chapter 6). Nevertheless, violations of the above condition cannot be avoided occasionally in the TEM (Twitchett-Harrison, Yates, Newcomb, Dunin-Borkowski, & Midgley, 2007), in particular if long range electromagnetic fields reach out of the field of view to a certain extent (Matteucci et al., 1991). Particular care is required in those cases, and it is necessary to assess the error due to the violation of the support constraint.

An interesting consequence of the above proof is that f_{ex} can be reconstructed from its projections F_{ex} outside of some convex domain (Fig. 3.9). In Sections 6.1 and 6.3 this non-obvious result is used to reconstruct electric and magnetic stray fields around nanowires with a significantly increased accuracy compared to the standard reconstruction suffering from non-local artifacts, e.g., due to misalignments.

3.2.5 Inverse Radon Transformation

Under the limitations imposed by the support theorem one can now derive a number of equivalent analytic formulas for the inverse Radon transformation, which form the basis of the tomographic reconstruction algorithms discussed in Section 3.6. It is convenient to start with a definition of the simple backprojection (Fig. 3.3)

$$B(x, y) := \int_0^\pi F(x\cos\theta + y\sin\theta, \theta)\, d\theta \qquad (3.30)$$

$$= \int_0^\pi F(p, \theta)\, d\theta ,$$

which consists of back-projecting the projected data along the respective projection paths into the reconstruction domain and summing over all tilt angles. Inserting the Fourier transform $\tilde{f}(q, \theta)$ of the Radon transform along p (3.22) into the above definition (3.30) then yields

$$B(x, y) = \int_0^\pi \int_{-\infty}^\infty \tilde{f}(q, \theta)\, e^{iqp}\, dq d\theta \qquad (3.31)$$

$$= \int_0^\pi \int_{-\infty}^\infty \tilde{f}(q, \theta)\, e^{iq(x\cos\theta + y\sin\theta)}\, dq d\theta ,$$

which, after introducing $x = r\cos\varphi$ and $y = r\sin\varphi$, gives

$$B(x, y) = \int_0^\pi \int_{-\infty}^\infty \tilde{f}(q, \theta)\, e^{iqr\cos(\varphi-\theta)}\, dq d\theta \qquad (3.32)$$

$$= \int_0^\pi \int_{-\infty}^\infty \frac{1}{q}\tilde{f}(q, \theta)\, e^{iqr\cos(\varphi-\theta)}\, q dq d\theta .$$

Finally, after going back to Cartesian coordinates, we obtain the following formula for the simple backprojection

$$B(x, y) = \int_{-\infty}^\infty \frac{1}{q}\tilde{f}(\mathbf{q})\, e^{i\mathbf{q}\mathbf{r}} d^2 q . \qquad (3.33)$$

Consequently, the simple backprojection corresponds to the original object weighted by a $1/|q|$ filter.

Inverting that filter leads to the Weighted Back Projection (WBP) or Filtered Back Projection (FBP)

$$f(x, y) = \mathcal{F}^{-1}\{|q|\}\, B(x, y) \qquad (3.34)$$

inversion formula of the Radon transform, which can be reformulated in various alternative ways, depending on how the filter is applied. For instance, in Radon's original formula (Radon, 1917)

$$f(x, y) = \frac{1}{4\pi^2} \int_0^\pi \int_{-\infty}^\infty \frac{\partial_p F(p, \mathbf{e})}{\mathbf{r} \cdot \mathbf{e} - p} \, dp d\theta \tag{3.35}$$

the filter is expressed as a derivative of the projection data with respect to the detector coordinate p. Both inversion formulas reveal a decrease of smoothness in the course of the inverse Radon transform due to the derivation with respect to p or the multiplication with the $|q|$-filter. The algorithmic implementation of the filtered backprojection and related forms provide one principle reconstruction algorithm used in the context of holographic tomography (see Section 3.6).

3.2.6 Circular Harmonic Decomposition

Further insight into the properties of the inverse Radon transformation is gained by separation of variables in polar coordinates (Cormack, 1963, 1964; Perry, 1975; Chapman & Cary, 1986). To derive the corresponding expressions, one first expands f into circular harmonics

$$f(r, \varphi) = \sum_{n=-\infty}^\infty f_n(r) \, e^{in\varphi} \tag{3.36}$$

with expansion coefficients

$$f_n(r) = \frac{1}{2\pi} \int_0^{2\pi} f(r, \varphi) \, e^{-in\varphi} \, d\varphi . \tag{3.37}$$

Next, one inserts the harmonic expansion in the original definition of the Radon transformation written in polar coordinates (3.5)

$$F(p, \theta) = \int_0^\infty \int_0^{2\pi} \sum_{n=-\infty}^\infty f_n(r) \, e^{in\varphi} \delta(p - r\cos(\theta - \varphi)) \, r dr d\varphi \tag{3.38}$$

$$= \sum_{n=-\infty}^\infty \int_0^\infty \int_0^{2\pi} f_n(r) \, e^{in\varphi} \delta(p - r\cos(\theta - \varphi)) \, r dr d\varphi ,$$

where summation and integration have been exchanged on the second line. In the next step both sides are Fourier transformed along the p coordinate, the integration order is exchanged, and the p-integration is carried out

$$\tilde{f}(q, \theta) = \int_{-\infty}^\infty F(p, \theta) \, e^{-iqp} dp \tag{3.39}$$

$$= \sum_{n=-\infty}^{\infty} \int_0^{\infty} \int_0^{2\pi} \int_{-\infty}^{\infty} f_n(r) e^{in\varphi} \delta\left(p - r\cos(\theta - \varphi)\right) e^{-iqp} r \, dr \, d\varphi \, dp$$

$$= \sum_{n=-\infty}^{\infty} \int_0^{\infty} \int_0^{2\pi} f_n(r) e^{in\varphi} e^{-iqr\cos(\theta-\varphi)} r \, dr \, d\varphi \, .$$

Inserting the following identity for the Bessel function

$$J_n(qr) = \frac{1}{2\pi i^n} \int_0^{2\pi} e^{in\varphi} e^{iqr\cos\varphi} d\varphi \tag{3.40}$$

and introducing true polar coordinates for q and θ, i.e.,

$$q, \theta = \begin{cases} q, \theta & q \geq 0, \\ -q, \theta + \pi & q < 0, \end{cases} \tag{3.41}$$

one finally obtains

$$\tilde{f}(q, \theta) = \sum_{n=-\infty}^{\infty} e^{in\theta} \tilde{f}_n(q) \tag{3.42}$$

with

$$\boxed{\tilde{f}_n(q) = 2\pi i^n \int_0^{\infty} f_n(r) J_n(qr) r \, dr.} \tag{3.43}$$

Thus, the harmonic expansion coefficients of the original function and the Fourier transform of the projected data are **Hankel transformation pairs** (Klug, Crick, & Wyckoff, 1958; Cormack, 1964; Crowther, DeRosier, & Klug, 1970; Chapman & Cary, 1986); and the inverse Hankel transform

$$\boxed{f_n(r) = 2\pi i^{-n} \int_0^{\infty} \tilde{f}_n(q) J_n(qr) q \, dq} \tag{3.44}$$

yields yet another inversion formula for the Radon transformation.

Note that this particular formulation of the inverse Radon transformation is separated into polar and azimuthal coordinates, i.e., each harmonic expansion coefficient is linked to its counterpart of the same order only. Using this property we can immediately write down the (inverse) Radon transformation of a radially symmetric function $f(r)$ as the (inverse) Hankel transformation of a single Fourier transformed projection $F(p)$, i.e.,

$$f(r, \varphi) = f_0(r) = 2\pi \int_0^{\infty} \tilde{f}_0(q) J_0(qr) q \, dq. \tag{3.45}$$

Similar to the weighted backprojection one may reformulate that result by expanding the Bessel function into some orthogonal basis functions. For instance, in his two Nobel prize honored papers, Cormack (1963, 1964) revealed that the circular harmonic expansion coefficients of the Radon transform defined by

$$F(p, \theta) = \sum_{n=-\infty}^{\infty} F_n(p) e^{in\theta} \tag{3.46}$$

are connected to those of the underlying function via

$$F_n(p) = 2 \int_p^{\infty} dr \frac{f_n(r) U_n(p/r) r}{\sqrt{r^2 - p^2}}, \tag{3.47}$$

where U_n are the Chebyshev polynomials of the first kind. The last expression may be inverted by

$$f_n(r) = -\frac{1}{\pi} \int_r^{\infty} dp \frac{\partial_p F_n(p) U_n(p/r)}{\sqrt{r^2 - p^2}}. \tag{3.48}$$

Considering again the special case of reconstructing a radially symmetric function $f(r)$ from one projection $F(p)$ only, the reconstruction corresponds to an inverse Abel transformation of the derivative of the projection along the detector coordinate p

$$f(r) = -\frac{1}{\pi} \int_r^{\infty} dr \frac{\partial_p F(p)}{\sqrt{r^2 - p^2}}. \tag{3.49}$$

3.3. DISCRETE RADON TRANSFORMATION

In practice, the projection data are collected on a pixel detector under a discrete set of tilt angles, i.e., the finite set of experimental values (serial index m) is given as the integral of the continuous Radon transform with some basis function (e.g., pixel support) B_m according to

$$F_m = \int F(p, \theta) B_m(p, \theta) \, dp d\theta. \tag{3.50}$$

Similarly, the reconstructed data f is computationally represented on some discrete basis (e.g., Cartesian pixels b_n) according to

$$f(\mathbf{r}) = \sum_n f_n b_n(\mathbf{r}). \tag{3.51}$$

Therefore, the tomographic reconstruction is also performed in a discrete setting, which implies to consider the discrete (inverse) Radon transformation, when dealing with practical tomographic reconstruction problems.

Fortunately, most of the properties of the continuous Radon transformation carry over to the discrete one, permitting the design of efficient reconstruction algorithms based on analytic formulas for the inverse Radon transformation (e.g., (3.34) or (3.44)). Basically, only two additional problems appear in the discrete setting. First, one has to discuss, how many sampling points (tilt angles and detector pixels) are required to ensure an accurate tomographic reconstruction of a given function. Second, it is paramount to consider the influence of ubiquitous experimental errors, such as misalignment of the tilt series and noise introduced by the detector.

The variety of reconstruction techniques (Deans, 1983; Natterer & Wübbeling, 2001; Kuba & Herman, 2008), i.e., implementations of the discrete inverse Radon transformation, can be distinguished along different lines, such as the underlying analytic inversion formula or the number of computation steps required to reach the solution. For instance, one of the most prominent inversion algorithms, referred to as Weighted or Filtered Back Projection (WBP/FBP) (Bracewell & Riddle, 1967; Vainshtein, 1970; Harauz & van Heel, 1986; Radermacher, 2006), is a direct implementation of (3.34). Closely related are Fourier methods based on the Fourier slice theorem (3.22) (Crowther, Amos, Finch, De Rosier, & Klug, 1970; Smith, Peters, & Bates, 1973; Miao, Förster, & Levi, 2005). In the two previous classes of algorithms, the solution is obtained in one step; hence they may be classified as direct reconstruction methods.

Beside direct inversions there are numerous reconstruction algorithms, which iteratively converge to a solution (Gordon, Bender, & Herman, 1970; Gilbert, 1972; Carazo, 1992; Fernandez et al., 2002). They are typically derivatives of numerical solvers of large and sparse linear equation systems, in particular Krylov subspace methods such as the Conjugate Gradient algorithm (Kaczmarz, 1937; Landweber, 1951; Stiefel, 1952). The iterative approach has some advantages for controlling the error of the reconstruction and facilitates the incorporation of additional boundary conditions for the reconstructed data, such as a known support, for instance. These advantages come with a reduction in speed in practice. Because projection data obtained in the TEM typically suffer from large noise levels, incomplete tilt ranges and artifacts, constraining the reconstructed data is often necessary; this is why iterative algorithms are considered in the following.

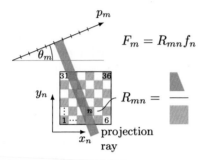

Figure 3.10 Schematic of the discrete (algebraic) Radon transformation. The projection weight of a particular Cartesian pixel in position space is given by the ratio between the projected area of a Cartesian pixel and its total area.

By virtue of the linearity of the Radon transformation, the discrete version of the Radon transformation can be formulated as a large system of linear equations

$$F_m = \int_{-\infty}^{\infty} \int_0^{\pi} \mathcal{R}(f) \, B_m \mathrm{d}p \mathrm{d}\theta = \sum_n \underbrace{\int_{-\infty}^{\infty} \int_0^{\pi} \mathcal{R}(b_n) \, B_m \mathrm{d}p \mathrm{d}\theta f_n}_{\text{Radon matrix } R_{mn}}, \qquad (3.52)$$

where the indices m, n denote discrete (pixel) coordinates in projection and position space (Fig. 3.10).

Consequently, the Radon matrix $\mathbf{R}$ is a discrete approximation to the Radon operator $\mathcal{R}$, where the entries are determined by the projections of the used basis functions. For instance, in Fig. 3.10 the function f is represented by Cartesian pixels, which are projected into the detector pixels with a weight given by the ratio between projected and total pixel area. In Fig. 3.11 the corresponding "filling" of the Radon matrix is exemplified for two different tilt angles.

In principle, any spatial basis $b_i(\mathbf{r})$ facilitating a faithful representation

$$f(\mathbf{r}) = \sum_n f_n b_n(\mathbf{r}) \qquad (3.53)$$

with

$$\sum_n b_n(\mathbf{r}) = 1 \qquad (3.54)$$

is admitted in the above sampling procedure. However, the existence of such a basis (for instance $b_n(\mathbf{r}) = \mathrm{sinc}(x - x_n) \times \mathrm{sinc}(y - y_n)$) depends on the

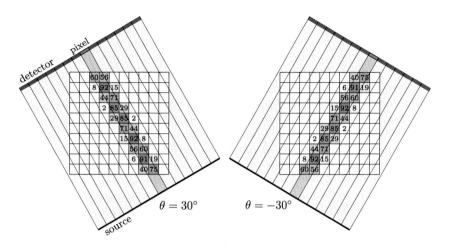

Figure 3.11 Explicit example for the construction of the Radon matrix for two different tilt angles. The numerical entries (in %) of the Radon matrix (i.e., projected pixel
weights) for one particular detector pixel are given at the position of the corresponding
Cartesian pixels. Additionally, the weight is indicated by the gray level saturation of the
corresponding pixel.

validity of sampling conditions, i.e., a band-limit in some function space
(cf. Section 3.4). Moreover, it can be computationally very convenient to
use spatially confined basis functions, because they yield sparse Radon matrices. Therefore Cartesian pixels, annular pixels, or even radially symmetric
"blobs" are typically used in practice. Keep in mind, however, that the last
three representations are typically not faithful, i.e.,

$$f(\mathbf{r}) \approx \sum_n f_n b_n(\mathbf{r}), \tag{3.55}$$

which means that the sampling introduces some error, which has to be
taken into account when discussing the reconstruction error (cf. Section 3.5).

In practice the basis functions, sampling the reconstruction domain, may
be used for optimizing different aspects of the reconstruction such as the
sampling error or the sparsity of the Radon matrix. For instance, the projection data acquired in the TEM typically suffer from experimental noise
and other artifacts such that the sampling error introduced by some spatially well-localized non-faithful basis functions (e.g., Cartesian pixels) may
be neglected in comparison. Thus, one can choose a basis minimizing the
total number of basis functions, thereby speeding up and stabilizing the

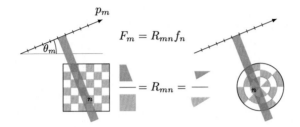

Figure 3.12 Cartesian and polar pixel basis.

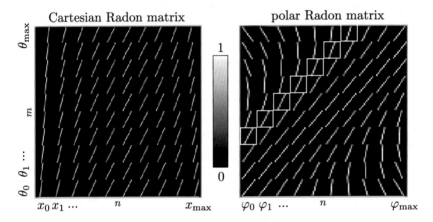

Figure 3.13 Radon matrices for Cartesian and polar pixel basis. Both matrices are sparse due to the well localized pixel basis. Additionally, the polar Radon matrix exhibits a Toeplitz (i.e., convolution matrix) structure (indicated by shifted white squares) owing to the azimuthal convolution structure of the analytic Radon transformation (see (3.5)).

reconstruction. We show below that the use of a polar grid often offers a valid strategy to reduce the number of reconstructed pixels without sacrificing resolution under the conditions prevalent in holographic tomography. Consequently, such grids are frequently utilized in the following.

Fig. 3.12 illustrates a polar pixel grid as basis in the discrete Radon transformation. Exemplary Radon matrices based on a Cartesian and a polar pixel basis are depicted in Fig. 3.13.

The first advantage of the polar grid is the decoupling of the azimuthal and radial sampling owing to the separability of the 2D Radon transformation in polar coordinates (cf. Section 3.2.6). In particular, it is shown in Section 3.4 that a tomographic reconstruction from a finite set of tilt angles requires azimuthally band-limited projection data. Here, the band-limit is the minimum from the set of band-limits associated to different choices of the origin of the polar grid. For example, a radially symmetric object (e.g., a homogeneous disk) can be reconstructed from a single projection only (via an inverse Abel transform, cf. (3.48)), i.e., the relevant azimuthal band-limit is the one around the symmetry axis. The analysis in Section 3.4 furthermore reveals that the rotation center with the smallest azimuthal band limit not only determines the number of required tilt angles in the experiment but also the number of azimuthal grid points of the polar reconstruction grid. Accordingly, some large annular pixels with radial dimensions determined by the radial resolution are sufficient for reconstructing the homogeneous circle from above on a polar grid. Obviously the number of Cartesian pixels would have been much larger in this case. We will see in Chapter 6 that the requirements on the azimuthal sampling in tomographic TEM investigations are often more relaxed compared to the radial one, for instance because of an approximate cylindrical geometry of the objects to be reconstructed. Consequently, memory requirements and computation time for the reconstruction can be greatly reduced by employing polar grids in this case.

The second advantage of the polar grid is that a certain class of spatial constraints (cf. Section 3.5), improving the reconstruction, can be implemented in a straightforward manner. For instance n-fold rotationally symmetric objects allow an identification of corresponding polar pixels, thereby reducing the number of reconstructed pixels and the size of the Radon matrix. Beside these two main advantages, it is noted that the radial grid exploits the whole circular reconstruction area more efficiently than Cartesian ones.

The tomographic reconstruction (or discrete inverse Radon transformation) is formally obtained by inverting the Radon matrix. The latter requires the matrix to be square and non-singular (i.e., $\det \mathbf{R} \neq 0$). For reason detailed below (Section 3.5), a direct inversion, such as obtained by the FBP algorithm, is, however, problematic in the presence of inconsistent projection data affected by artifacts and noise. Instead of inverting the Radon matrix directly one therefore solves the related problem of minimizing the residual Euclidean norm between projected solutions and experimental projections

$$f^+ = \arg\min \left\| \mathbf{R}f - F \right\|^2 , \qquad (3.56)$$

where f^+ is referred to as pseudosolution in the following. That formulation of the reconstruction problem is applicable to Radon matrices $\mathbf{R}$ of every shape and, even more importantly, admits a mitigation to the problem of inconsistent projection data through so-called regularization methods (cf. Section 3.5).

Minimizing the projection error can be transformed into solving the so-called normal equation

$$\nabla_{f^+} \left\| \mathbf{R}f^+ - F \right\|^2 = 2\mathbf{R}^T \mathbf{R}f^+ - 2\mathbf{R}^T F = 0, \tag{3.57}$$

obtained by equating the gradient of the residual norm (3.56) to zero. Bringing the pseudosolution to one side, one finally obtains

$$f^+ = \underbrace{\left(\mathbf{R}^T \mathbf{R}\right)^{-1} \mathbf{R}^T}_{\mathbf{R}^+} F, \tag{3.58}$$

defining the Moore–Penrose pseudoinverse $\mathbf{R}^+$. When considering reconstruction algorithms in Section 3.6, we will focus on the solution of the normal equation.

3.4. SAMPLING

Prior to performing a particular tomographic reconstruction on the computer, we have to settle the following important question regarding the discrete representation of the experimental projection data: How many tilt angles and how many detector pixels are necessary to faithfully represent the projection data of some underlying function? The answer is given by the pertinent theorems of Shannon–Whittacker sampling theory. Accordingly, a function, band-limited in Fourier space (band limit B_k), may be faithfully represented by a discrete set of samples taken at a distance $d = 1/(2B_k)$ (corresponding to a Nyquist sampling rate $q_s = 2B_k$) by virtue of the famous Shannon–Whittacker interpolation formula

$$F(p) = \sum_n F(nd) \operatorname{sinc}\left(\frac{p - nd}{d}\right). \tag{3.59}$$

If the sampling rate is smaller than the Nyquist rate, i.e., $q_s < 2B$, aliasing ($\cong$ indistinguishable copies) at "under-sampled" frequency values occur.

We may directly conclude that a sampling rate of $2B_k$ is sufficient to faithfully represent the projection data along the detector coordinate. To derive the azimuthal sampling we have to revert to the harmonic expansion (3.36)

$$f(r, \varphi) = \sum_{n=-n_{\max}}^{n_{\max}} f_n(r) e^{in\varphi}, \tag{3.60}$$

where the azimuthal band width $B_\varphi = n_{\max}/(2\pi)$ determines the maximal (minimal) n with $f_n \neq 0$. By virtue of the sampling theorem, the required azimuthal sampling rate then amounts to $2B_\varphi$ yielding an azimuthal sampling distance

$$d_\varphi = \frac{\pi}{n_{\max}}. \tag{3.61}$$

Note that the circular harmonic expansion is not shift invariant. As a consequence, the above azimuthal sampling criterion only applies for a particular choice of origin and we may identify one particular origin possessing a minimal B_φ. This smallest azimuthal band width then defines the required number of tilt angles in a tomographic tilt series because of the separability of the Radon transformation in polar coordinates (cf. Section 3.2.6). As a limiting case, we obtain the previously noted result that a radially symmetric function may be reconstructed from one projection only.

It is now natural to ask, whether there exists a relationship between the azimuthal band limit B_φ and the overall band limit in Fourier space B_k. By considering two limiting cases it can be shown that this is generally not the case. The first example pertains to radially symmetric functions, represented by $f(r) = 2\pi \int_0^\infty \tilde{f}(q) J_0(qr) q dq$. Obviously, the overall band limit may assume any value in this case, whereas the azimuthal band limit remains minimal, i.e., only one projection suffices for a complete reconstruction. The second example pertains to functions with very large azimuthal band limit, possessing harmonic expansion coefficients

$$\lim_{n \to \infty} f_n(r) = \lim_{n \to \infty} 2\pi i^{-n} \int_0^\infty \tilde{f}_n(q) J_n(qr) q dq \neq 0. \tag{3.62}$$

Again, any general band limit with $\tilde{f}_n(q > B_k) = 0$ may be assumed without rendering $\lim_{n \to \infty} f_n(r) = 0$, i.e., changing the azimuthal band limit. A visual representation illustrating the independence of the overall band limit and the azimuthal one is given in Fig. 3.14. Taking into account the independence of azimuthal and radial band widths, the tomographic studies shown in Chapter 6 employ radial and azimuthal sampling rates, which have been individually tested and verified against the corresponding band limits of the underlying functions, computed from the experimental tilt series.

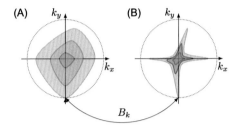

Figure 3.14 Independence of azimuthal and overall band limit. Both Fourier transformed functions $\tilde{f}$ possess the same band limit B_k. However, the azimuthal band limit B_φ in (B) is larger than in (A).

We close this subsection with one important remark. There is a contradiction between the fundamental support theorem stating that functions to be reconstructed tomographically must be zero outside of the reconstruction domain and the requirement of a finite band-limit permitting a faithful sampling of the projections in the experiment. Indeed, the Paley–Wiener theorem, which states that functions with compact support do not posses a finite band limit, prevents that both requirements are fulfilled simultaneously. Consequently, aliasing artifacts in the projected data, leading to certain reconstruction errors, cannot be completely avoided in tomography. Moreover, since the error amplification in a tomographic reconstruction is determined by the regularization, discussed in the next section, the impact of aliasing errors can significantly vary. Consequently, the regularization has to be chosen such to sufficiently suppress aliasing artifacts or the sampling has to be chosen sufficiently fine to suppress aliasing at a given regularization level in practice.

3.5. REGULARIZATION

The previously derived expressions for the tomographic reconstruction (either as discrete inversion formulas or as error minimization problem) are bound to fail under experimental conditions for the following reason. Experimental projection data inevitably contain deviations from the true projection. These inconsistencies are introduced by, for instance, a noisy detector, experimental and object instabilities occurring during the tilt series, or aliasing. Unfortunately, the inverse Radon transformation has the property to amplify such inconsistencies, in particular if they contain large spatial frequencies because of the following. Different functions with large spatial frequencies can lead to similar or practically indistinguishable projections because of the smoothing introduced by the projection integral. Reconstructing from such projection data can therefore easily lead to different results even though the projections look similar. Such a behavior is dangerous for a trustworthy tomographic reconstruction, since small errors in the projection data might lead to completely different reconstructions. Indeed, the Radon transformation shares this property with a large class of related problems (e.g., general Fredholm integral equations of the first kind or the numerical derivative), which are referred to as (mildly) ill-conditioned (Hansen, 1987; Natterer & Wübbeling, 2001; Louis, 2013), therefore.

Large and ongoing efforts, subsumed under the terminus regularization, seek a mitigation of this error amplification. Indeed, *every* tomographic reconstruction is regularized in practice and many algorithms basically differ in the type of regularization strategy applied. For instance, FBP reconstruction algorithms can be regularized by multiplying the $|q|$ filter with an additional low-pass reducing the amplification of artificial large frequency noise. One main result of this active area of research is that there is no such thing like the perfect regularization strategy and different regularization schemes are better adapted to different projection data.

In the following we shortly recapitulate, why the Radon transformation is classified (mildly) ill-conditioned. To this end the Singular Value Decomposition of the Radon transformation is introduced. Subsequently a small and incomplete choice of popular regularization schemes is introduced, which have turned out to work well in the field of Electron Tomography.

3.5.1 Singular Value Decomposition

The Singular Value Decomposition (SVD) permits to characterize linear systems such as the Radon transformation. In particular, it allows for a straightforward computation of the Moore–Penrose inverse (3.58) and a characterization of the error amplification upon inversion. The SVD of an arbitrary $M \times N$ matrix reads (Golub & Van Loan, 1996)

$$\mathbf{R} = \mathbf{U}\mathbf{\Sigma}\mathbf{V}^{T}, \tag{3.63}$$

where $\mathbf{U}$ and $\mathbf{V}$ are unitary matrices containing the eigenvectors of $\mathbf{R}^{T}\mathbf{R}$ and $\mathbf{R}\mathbf{R}^{T}$ in their columns, which can be easily checked by inserting (3.63) into the corresponding expressions. By the same token one sees that

$$
\boldsymbol{\Sigma} = \begin{pmatrix} \sigma_1 & 0 & 0 \\ 0 & \ddots & 0 \\ 0 & 0 & \sigma_N \\ 0 & 0 & 0 \end{pmatrix}
\tag{3.64}
$$

is a diagonal $M \times N$ matrix containing the eigenvalues of $\mathbf{R}^T\mathbf{R}$ (or $\mathbf{R}\mathbf{R}^T$) on the main diagonal. Because $\mathbf{R}^T\mathbf{R}$ is symmetric and positive semidefinite, these eigenvalues are real. They are typically sorted in decreasing order, i.e., $\sigma_1 \geq \sigma_2 \geq \ldots \geq \sigma_{\min(M,N)}$ on the diagonal of $\boldsymbol{\Sigma}$. Once the SVD is known, the Moore–Penrose inverse is simply obtained by

$$
\mathbf{R}^+ = \mathbf{U}^T \boldsymbol{\Sigma}^+ \mathbf{V},
\tag{3.65}
$$

where $\boldsymbol{\Sigma}^+$ is generated by transposing $\boldsymbol{\Sigma}$ and inverting the non-zero diagonal entries. This expression for the pseudoinverse permits a discussion of the error propagation in the tomographic reconstruction as follows.

A rough measure for the error amplification is given by the condition number κ of the Radon matrix

$$
\frac{\|\delta f^+\|}{\|f^+\|} \leq \kappa(\mathbf{R}) \left\| \frac{\delta F}{F} \right\|,
\tag{3.66}
$$

which provides an upper bound for the transfer of the relative error in the projection data to the reconstructed function. More accurate error bounds may be derived by taking into account additional information, such as the magnitude of the data error or the regularity of the function to be reconstructed (Louis, 2013). The following manipulations

$$
\begin{aligned}
\frac{\|\delta f^+\|}{\|f^+\|} &\leq \frac{\|\mathbf{R}^+\| \|\delta F\|}{\|\mathbf{R}^+ F\|} \\
&= \frac{\|\mathbf{R}\| \|\mathbf{R}^+\| \|\delta F\|}{\|\mathbf{R}\| \|\mathbf{R}^+ F\|} \\
&\leq \frac{\|\mathbf{R}\| \|\mathbf{R}^+\| \|\delta F\|}{\|\mathbf{R}\mathbf{R}^+ F\|} \\
&\leq \underbrace{\|\mathbf{R}\| \|\mathbf{R}^+\|}_{\kappa(\mathbf{R})} \left\| \frac{\delta F}{F} \right\|
\end{aligned}
\tag{3.67}
$$

show that the condition number of the Radon matrix is the product of the norms of the Radon matrix and its pseudoinverse. By noting the spectral norm of the Radon matrix $\|\mathbf{R}\| = \|\mathbf{U}\boldsymbol{\Sigma}\mathbf{V}^T\| = \|\boldsymbol{\Sigma}\| = \sigma_{\max}(\mathbf{R})$, this product can be expressed by the quotient of the largest and smallest singular value

$$\kappa(\mathbf{R}) = \frac{\sigma_{\max}(\mathbf{R})}{\sigma_{\min}(\mathbf{R})}. \tag{3.68}$$

Consequently, a large error amplification may occur, if the Radon transformation has very small singular values. It will be shown below that this is indeed the case.

As a consequence of the radial separability of the 2D Radon transformation its Singular Value Decomposition[2] depends on a radial "quantum" number m and an azimuthal quantum number l (Natterer & Wübbeling, 2001; Louis, 2013):

$$f = \sum_{m=0}^{\infty} \sigma_m \sum_{l=-m,\, l+m\epsilon 2\mathbb{Z}}^{l\leq m} \langle f, v_{ml}\rangle u_{ml}. \tag{3.69}$$

Here, the singular values

$$\sigma_m = \sqrt{\frac{4\pi}{m+1}} \tag{3.70}$$

depend only on the radial quantum number m. The singular functions in position space

$$v_{ml}(\mathbf{r}) = \sqrt{\frac{m+1}{\pi}}\, P_{\frac{m-|l|}{2}}^{(0,|l|)}\left(2|\mathbf{r}|^2 - 1\right)|\mathbf{r}|^{|l|}\, e^{il\varphi} \tag{3.71}$$

are composed of Jacobi polynomials ($P_n^{(\alpha,\beta)}$) of the order n with weights α and β, whereas the projection space singular functions

$$u_{ml}(p,\theta) = \frac{1}{\pi}\sqrt{1-p^2}\, U_m(p)\, e^{il\theta} \tag{3.72}$$

contain Chebyshev polynomials (U_m) in the radial coordinate. A visual representation of the first few terms of the SVD of the Radon transform is given in Fig. 3.15.

---

[2] The SVD is well-defined for compact operators on Hilbert spaces such as the Radon transformation.

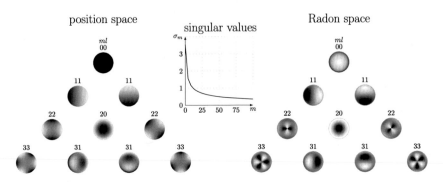

Figure 3.15 SVD of the Radon transform. The singular functions are plotted in a modified azimuthal basis, where the complex azimuthal harmonics are rearranged according to $e^{il\varphi} \pm e^{-il\varphi} \to \cos l\varphi, \sin l\varphi$. Note, furthermore, the non-trivial topology of the sinograms on the right-hand side (Fig. 3.5).

The SVD of the Radon transform readily reveals that the singular values (3.70) tend to zero in the limit of large m, which corresponds to an infinitely large condition number, $\kappa \to \infty$, of the continuous Radon transformation, indicating some sort of ill-conditioning. A further inspection of the singular values (3.70) reveals that the decay is of polynomial order, which is referred to as mildly ill-conditioned. Such problems are amenable to solutions under mild regularization measures, in contrast to strongly ill-conditioned problems featuring an exponential decay (Natterer & Wübbeling, 2001; Louis, 2013). The latter behavior may occur, if the projection data is incomplete, e.g., because of an incomplete tilt range ("missing wedge"; Natterer & Wübbeling, 2001). In the following, regularization schemes mainly designed for mildly ill-conditioned cases are discussed.

3.5.2 Regularization Methods

In order to cope with ill-conditioned problems one has to introduce auxiliary conditions, which are suited to reduce the error amplification. Formally, a regularization algorithm for an ill-conditioned problem ($F = \mathbf{R}f$ in our case) is a one-parameter family $\{\mathbf{R}_\lambda^+\}$ of operators such that

1. $\forall \lambda \, \mathbf{R}_\lambda^+$ is linear and continuous;
2. $\lim_{\lambda \to 0} \mathbf{R}_\lambda^+ = \mathbf{R}^+$.

The parameter λ is called regularization strength, because a large λ typically reduces the error amplification (condition number of $\mathbf{R}$). It is important to note, however, that the approximation $\mathbf{R}_\lambda$ introduces an additional error, the *regularization error*. Consequently, the regularization problem consists of finding the optimal λ_{opt} minimizing the total error consisting of the sum of the amplified input error and the regularization error in the reconstruction according to

$$\left\| \mathbf{R}^+_{\lambda_{opt}} \mathbf{F} - \mathbf{R}^+ \mathbf{R} \mathbf{f} \right\| \leq \left\| \mathbf{R}^+_{\lambda} \mathbf{F} - \mathbf{R}^+ \mathbf{R} \mathbf{f} \right\| . \tag{3.73}$$

Finding the optimum λ_{opt} is only possible, if the errors in the projection data $\mathbf{F}$ are sufficiently well characterized (Hansen, 1987; Neumaier, 1998). The latter requirement is typically violated in tomographic problems encountered in the TEM.[3] As a consequence, the particular choice of the regularization strength used in Electron Tomography studies is typically based on heuristics shrouded in a certain level of mystery. It is therefore particularly important that tomographic TEM studies, conducted under these circumstances, involve the inspection of a number of pseudosolutions with different regularization strengths λ to judge, which features in the regularized solution are real and which are reconstruction or regularization errors (cf. Section 6.2). In the following four prominent regularization strategies, namely (A) truncated SVD, (B) Tikhonov regularization (Tikhonov, 1963), (C) Tikhonov–Phillips regularization, and (D) compressive sensing, are introduced (Neumaier, 1998; Louis, 2013).

(A) The idea behind the truncated SVD is to limit the function space of the reconstruction to singular functions with singular values larger than some cut-off, i.e.,

$$\mathbf{R}^+_\lambda = \mathbf{U}^T \mathbf{\Sigma}^+_\lambda \mathbf{V}, \tag{3.74}$$

where

$$\mathbf{\Sigma}^+_\lambda = \begin{cases} 1/\sigma^n & \sigma_n > \lambda, \\ 0 & \text{otherwise.} \end{cases} \tag{3.75}$$

Consequently, the condition number

$$\kappa\left(\mathbf{R}_\lambda\right) = \frac{\sigma_1}{\lambda} < \frac{\sigma_1}{\sigma_{min}} = \kappa\left(\mathbf{R}\right) \tag{3.76}$$

is reduced. Computationally, applying the truncated SVD is cumbersome because it requires the computation of the SVD in a first step.

---

[3] To the best knowledge of the author there exists no electron tomographic study, where a quantitative analysis of the projection data's error, with the aim to derive an optimal regularization strength, has been conducted.

(B) The Tikhonov method solves this problem by damping the reconstruction of singular functions pertaining to small singular values with a Lorentzian according to

$$\mathbf{R}_\lambda^+ F = \sum_n \frac{\sigma_n}{\sigma_n^2 + \lambda} \langle F, v_n \rangle u_n. \tag{3.77}$$

Again the condition number

$$\kappa(\mathbf{R}_\lambda) = \frac{\left(\sigma_{min}^2 + \lambda\right)\sigma_1}{\left(\sigma_1^2 + \lambda\right)\sigma_{min}} < \frac{\sigma_1}{\sigma_{min}} = \kappa(\mathbf{R}) \tag{3.78}$$

is reduced. However, because

$$\mathbf{R}_\lambda^+ = \left(\mathbf{R}^T \mathbf{R} + \lambda \mathbf{I}\right)^{-1} \mathbf{R}^T \tag{3.79}$$

the regularized solution now minimizes

$$\mathbf{f}_\lambda^+ = \arg\min \left(\|\mathbf{R}f - F\|^2 + \underbrace{\lambda \|f\|^2}_{\text{penalty term}} \right). \tag{3.80}$$

No explicit knowledge of the SVD is required to apply this type of regularization in practice. One only has to enhance the objective function to be minimized by an additional penalty term.

(C) An extension of the Tikhonov method is obtained by generalizing the penalty term to

$$f_\lambda^+ = \arg\min \left(\|\mathbf{R}f - F\|^2 + \lambda \|\mathbf{D}f\|^2 \right) \tag{3.81}$$

with arbitrary matrices $\mathbf{D}$. A widely employed method, referred to as Tikhonov–Phillips, uses the discrete approximation to the derivative here, i.e., $\mathbf{D} \approx \nabla$ (Louis, 2013). Consequently, Tikhonov–Phillips regularized pseudosolutions tend to be more regular (smoother) than others, which is often a sensible condition in practice. Note, however, that the degree of smoothness determined by the value of λ, is arbitrary and non-unique in the absence of additional information about the underlying function. Thus, an evaluation of a set of solutions pertaining to different λ's is required in practice.

(D) A further generalization of the Tikhonov–Phillips method, somewhat loosely referred to as compressive sensing in the following, is obtained

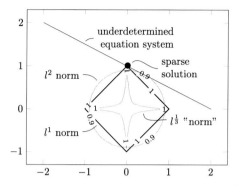

Figure 3.16 Compressive sensing and l_1-norm in $\mathbb{R}^2$. In this setting, one linear equation defines a set of infinite solutions represented by a line in $\mathbb{R}^2$. Three different norms, namely the Euclidean l_2, the sum norm l_1, and a pseudonorm $l_{1/3}$, serving as proxy for the difficult to visualize l_0 norm, are represented by their level sets at indicated values. The sketch illustrates that the sparsest solution (represented by a vector along x or y) is obtained by minimizing the l_1-norm for almost all underdetermined solutions (lines). Accordingly, the l_1-norm serves as proxy for the nonhomogeneous (nonconvex) $l_{p<1}$-pseudonorms (including l_0) facilitating the efficient location of the sparse solution with the help of convex optimization algorithms.

by replacing the Euclidean l_2-norm with the l_1-norm in the above penalty term, i.e.,

$$f_\lambda^+ = \arg\min\left(\left\|\mathbf{R}f - F\right\|^2 + \lambda\left\|\mathbf{D}f\right\|_1\right). \tag{3.82}$$

Phenomenologically, the l_1-norm penalizes vectors with many small non-zero components stronger than vectors containing a few large non-zeros, which promotes sparse pseudosolutions in the $\mathbf{D}f$ space. Consequently, this type of regularization is useful, if the function to be reconstructed is known to be sparse in some predefined basis. A short derivation of this non-obvious behavior is given below.

To illustrate the sparsity promotion by the l_1-norm, we consider the related problem of finding the sparsest solution to an under-determined system of linear equations. This is generally a formidable task, which mainly derives from the unfavorable properties of the so-called l_0-pseudonorm, counting the number of non-zero entries of the vector (in some basis). Here, the notion of a pseudonorm refers to the missing convexity of the space of functions bounded by some value of said pseudonorm in contrast to proper norms confining convex regions. The non-convexity renders the norm non-smooth with respect to the minimization procedure (note the cusps in Fig. 3.16). This complicates the efficient minimization of such pseudonorms with computationally effective convex optimization algorithms. To solve this problem, it has been proposed to use the l_1 norm as a proxy for the l_0-pseudonorm, which is possible under very general conditions (Candes, Romberg, & Tao, 2006; Donoho, 2006; Starck, Murtagh, & Fadili, 2010) (i.e., only a zero set of underdetermined equations may not be optimized). A visual representation this issue is shown in Fig. 3.16. The retrieval of unique solutions from under-determined systems by minimizing the l_1 norm in some basis, referred to as compressive sensing in the literature, is currently applied in such diverse fields as Magneto-Resonance Imaging, Radio Interferometry, or Electron Tomography (Lustig, Donoho, & Pauly, 2007; Velikina, Leng, & Chen, 2007; Chan et al., 2008; Wiaux, Jacques, Puy, Scaife, & Vanderghevost, 2009; Holland, Malioutov, Blake, Sederman, & Gladden, 2010; Holland, Bostock, Gladden, & Nietlispach, 2011; Saghi et al., 2011; Goris, Bals, et al., 2012; Goris, Van den Broek, Batenburg, Heidari Mezerji, & Bals, 2012).

In the context of Electron Tomography, a regularization method, referred to as Discrete Algebraic Reconstruction Technique, represents a special case of compressive sensing, where the domain of the reconstructed function is restricted to a set of discrete values (Herman & Kuba, 2007). Similar assumptions play a role in the emerging field of atomic resolution tomography, where sparsity constraints may promote atomicity of the scattering potential. Last but no least, the l_1 norm penalty applied to $\mathbf{D} \approx \nabla f$, referred to as total variation minimization, may be favorably used, if reconstructing a function with large constant regions and relatively few abrupt steps. It is emphasized, however, that the optimal magnitude of the regularization parameter λ remains essentially unknown, necessitating the inspection of differently regularized solutions akin to other generic regularization techniques.

3.5.3 Regularized Tomography

Both the Landweber as well as conjugate gradient methods discussed below posses a built-in regularization parameter, the inverse of the iteration number n. Consequently, the larger the iteration number the weaker the regularization. It is difficult to characterize the regularization in detail, however, a behavior similar to that of the Tikhonov regularization may be observed in both reconstruction methods. In order to regularize the FBP, one typically adds some low-pass to the original $|q|$-filter. The ratio behind such a filtering is that singular functions belonging to small singular

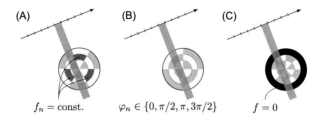

Figure 3.17 Three examples of spatial constraints applied to the polar sampling scheme: (A) 3-fold symmetric function, (B) azimuthal band-limited function, (C) predefined support of function.

values of the Radon transformation have a larger band-width (i.e., more oscillations in the higher order Jacobi polynomials (3.71)) than those belonging to small singular values. Consequently, a low pass is damping them stronger and the solution is regularized.

Last but not least, a large number of additional conditions can be used to regularize the reconstruction, apart from generic strategies such as the (generalized) Tikhonov method. They typically consist of additional restrictions on the function space allowed in the reconstruction, which typically are inferred from some previous knowledge. The most important are shape constraints (e.g., predefined support of f; Twitchett-Harrison et al., 2007), function value constraints (e.g., positive semidefinite or discrete-valued f; Batenburg et al., 2009), symmetry constraints (e.g., three-fold symmetric f; Lubk, Wolf, Prete, et al., 2014), and band-limit constraints (e.g., bounded harmonic expansion of f; Lubk, Wolf, Prete, et al., 2014; Lubk, Wolf, Simon, et al., 2014; Phatak et al., 2016). The latter two are particularly well adapted to the polar sampling grid discussed above. A visual representation of these constraints is shown in Fig. 3.17.

The corresponding Radon matrices generated for a polar grid incorporating no symmetry, three-fold, and full radial symmetry are depicted in Fig. 3.18. One readily observes the reduction of position space coordinates (i.e., number of rows) when using the polar grid with an increasing symmetry. Besides, the symmetry constraints lead to a reduction of sparsity since one representative of a symmetric position space pixel is projected more often into one detector coordinate. In the case of three-fold symmetry, one observes a three-fold superposition of projection lines. In the case of the full radial symmetry a whole fan of projections is superimposing. Accordingly, the Radon matrices subject to symmetry constraints are obtained by appropriate superpositions of the unsymmetrical one.

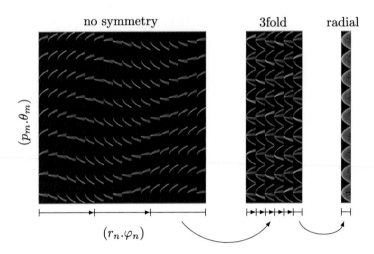

Figure 3.18 Radon matrices for polar grids with different symmetry constraints. The superposition of position space coordinates due to symmetry is indicated by converging arrows.

3.6. RECONSTRUCTION ALGORITHMS

For (generalized) matrix inversions a large number of numerical methods have been devised (Natterer, 2001). They circumvent the computationally expensive SVD discussed previously, e.g., by employing iterative procedures, which approximate the solution with increasing accuracy. In the course of this work three popular matrix inversion algorithms and extensions of which are considered, which have been proved particularly useful for Electron Holography and Tomography. The first one has been originally discovered by the Polish mathematician Kaczmarz (Kaczmarz, 1937) before being introduced as Algebraic Reconstruction Technique (ART) in the field of tomography (Gordon et al., 1970). The second one (cf. Section 3.6.2), referred to as Landweber iteration (Landweber, 1951), has met a similar fate and is referred to as Simultaneous Iterative Reconstruction Technique (SIRT) in the tomography community. Both schemes possess a particularly straightforward implementation and built-in regularization parametrized by the number of iterations (cf. Section 3.5). However, they can be suboptimal with respect to the convergence speed. Therefore, we also consider a conjugate gradient method (cf. Section 3.6.3), which significantly reduces the number of iterations and thus computing time.

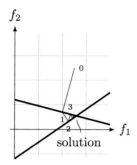

Figure 3.19 Kaczmarz method for a system of $m = 2$ equations and 2 unknowns.

3.6.1 Kaczmarz Method

The Kaczmarz method is applicable to any linear system of equations $\mathbf{A}x = b$. The basic idea of the algorithm is to iteratively project the solution onto hyperplanes defined by the above system of equations. If the system is consistent, i.e., admits a unique solution, the iteration converges to that solution (Fig. 3.19). We therefore solve the normal equation $\mathbf{R}^T\mathbf{R}f^+ = \mathbf{R}^T F$ instead of the typically (e.g., due to noise) inconsistent $\mathbf{R}f = F$. If denoting the number of rows of $\mathbf{A} = \mathbf{R}^T\mathbf{R}$ with m, the ith row of $\mathbf{R}^T\mathbf{R}$ with a_i, the ith element of $\mathbf{R}^T F$ with b_i, and the starting guess for f^+ with f^0, the kth iteration reads

$$f^{k+1} = f^k + \frac{b_i - \langle a_i, f^k \rangle}{|a_i|^2} a_i, \tag{3.83}$$

with the row index $i = k \bmod m + 1$. Inspecting Fig. 3.19 one observes that the convergence speed of the algorithm depends on the relative orientation of the hyperplanes (Lewis & Malick, 2008). Moreover, the convergence also depends on the starting guess. The algorithm is particularly fast if the system is sparse, which reduces the number of mathematical operations that have to be carried out. More recently, it has been noted that a randomized version of the Kaczmarz method (Strohmer & Vershynin, 2008), in which the ith equation is selected randomly with probability proportional to $|a_i|^2$, may converge faster compared to other orders, with the rate of convergence only depending on the condition number $\kappa(\mathbf{A})$ (Gower & Richtarik, 2015).

3.6.2 Landweber Method

The Landweber iteration may be considered as an extension of the Kaczmarz method, where all rows of the equation system are projected in

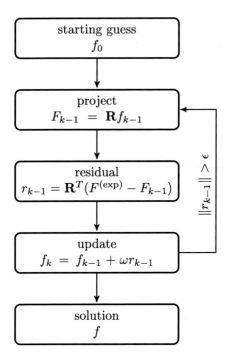

Figure 3.20 Simultaneous iterative reconstruction technique.

parallel to obtain the iterated solution. From a tomographer's perspective that amounts to an iterative improvement following a cycle of projections and back-projections as depicted in Fig. 3.20. Accordingly, the iteration begins with a re-projection of the current solution along the same directions as in the experimental tilt series. The reprojected solution is then subtracted from the experimental sinogram (denoted by $F^{(\mathrm{exp})}$) yielding the difference sinogram. This is back-projected and projected and finally added to the previous projection. In order to ensure convergence it is crucial for iterative reconstruction schemes (Natterer & Wübbeling, 2001) to incorporate a weighting factor ω as discussed below. The recursion formula describing this iteration cycle reads

$$f_k = f_{k-1} + \omega \mathbf{R}^T \left(F^{(\mathrm{exp})} - \mathbf{R} f_{k-1} \right), \tag{3.84}$$

which may be written in projection space by multiplying with $\mathbf{R}$ from the left

$$F_k = F_{k-1} + \omega \mathbf{R}\mathbf{R}^T \left(F^{(\mathrm{exp})} - F_{k-1} \right). \tag{3.85}$$

Introducing a measure for the deviation between the projection of some reconstructed function f and the sinogram

$$\epsilon\left(f\right) = \left\|\mathbf{R}f - F^{(\text{exp})}\right\|^2 \tag{3.86}$$

one may rewrite the iteration in terms of the reconstructed function as

$$f_k = f_{k-1} - \frac{1}{2}\omega\nabla\epsilon\left(f\right), \tag{3.87}$$

which shows that the above iteration is a special case of gradient descent (Fig. 3.21). Thus, the Landweber iteration automatically mitigates problems related to inconsistent data because it seeks a minimization of the residual in a steepest descent manner.

With the help of (3.85) the residual $r_k = F_k - F^{(\text{exp})}$ after k iteration cycles reads

$$r_k = \underbrace{\left(\mathbf{I} - \omega\mathbf{R}\mathbf{R}^T\right)}_{\mathbf{Q}}r_{k-1} = \mathbf{Q}^k r_0, \tag{3.88}$$

where $\mathbf{I}$ denotes the identity matrix. Thus, the (Euclidean) norm of the residual after k iterations, referred to as the projection error

$$\epsilon_k = \|r_k\| = \left\|\mathbf{Q}^k r_0\right\| \leq \|\mathbf{Q}\|^k \epsilon_0, \tag{3.89}$$

is bounded by the kth power of the matrix norm of $\mathbf{Q}$. If this norm is smaller than one, i.e., $\|\mathbf{Q}\| < 1$, the error can be minimized iteratively, at least as long as the residual is in the orthogonal complement of the kernel of the product of projection and back-projection matrix, i.e., $r \in \ker^T\left(\mathbf{R}\mathbf{R}^T\right)$. The latter expression ensures that the residual remains bigger than zero after a subsequent back-projection and projection.

It is shown below that with a suitable choice of the damping parameter ω, one can guarantee $\|\mathbf{Q}\| \leq 1$ and consequently the convergence of the Landweber iteration. In order to simplify the following calculations, the scalar factors C_k

$$C_k := \frac{\left\|\mathbf{R}\mathbf{R}^T r_k\right\|^2}{\|r_k\|^2}, \tag{3.90}$$

and c_k

$$c_k := \frac{\left\langle r_k, \mathbf{R}\mathbf{R}^T r_k\right\rangle}{\|r_k\|^2}, \tag{3.91}$$

are defined. With these abbreviations the following relation between subsequent projections errors holds

$$\epsilon_{k+1} = \|r_{k+1}\| = \left\|\left(I - \omega_k \mathbf{RR}^T\right) r_k\right\|$$ (3.92)

$$= \sqrt{\epsilon_k^2 - 2\omega_k \left(r_k, \mathbf{RR}^T r_k\right) + \omega_k^2 \left(\mathbf{RR}^T r_k, \mathbf{RR}^T r_k\right)}$$

$$\leq \sqrt{\left(1 - 2\omega_k c_k + \omega_k^2 C_k\right)} \epsilon_k .$$

Therefore, if ω_k is chosen such that

$$0 < \sqrt{\left(1 - 2\omega_k c_k + \omega_k^2 C_k\right)} < 1 ,$$ (3.93)

the error decreases. The left part of this inequality is always true because the parabola determined by the bracket under the square root never crosses zero from above. In other words, the zeros of the quadratic equation determined by

$$\omega_k^{(0)} = \frac{c_k}{C_k} \pm \sqrt{\frac{c_k^2 - C_k}{C_k^2}}$$ (3.94)

are complex, which can be proved with the help of the Cauchy–Schwarz inequality, (3.90), and (3.91)

$$\left(r_k, \mathbf{RR}^T r_k\right)^2 \leq \|r_k\|^2 \left\|\mathbf{RR}^T r_k\right\|^2$$
$$\Rightarrow c_k^2 \|r_k\|^4 \leq C_k \|r_k\|^4 .$$ (3.95)

Consequently, the projection errors decrease and the reconstruction converges, if the right hand side of the inequality (3.93)

$$2\omega_k c_k - \omega_k^2 C_k > 0$$ (3.96)

is fulfilled, yielding the following condition for the damping parameter

$$\omega_k < \omega_k^{max}, \quad \text{with } \omega_k^{max} = 2c_k/C_k,$$ (3.97)

in each iteration step. The projection error stops to decrease if $r_k \in \ker\left(\mathbf{RR}^T\right)$. Similar to the Kaczmarz method, the Landweber iteration is particularly fast if the matrix $\mathbf{R}$ is sparse, which is often the case for Radon matrices in practice.

To increase the suboptimal speed of the steepest descent convergence one may replace the back-projection $\mathbf{R}^T$ by a more suitable operator. Here,

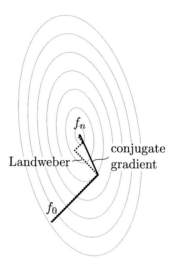

Figure 3.21 Convergence of Landweber and Conjugate Gradient algorithm in a two-dimensional optimization problem (level sets of the error metric indicated as thin lines). Each Landweber iteration corresponds to a gradient descent, with the step size determined by the weighting factor ω. Compared to the Landweber method, the Conjugate Gradient iteration converges faster and terminates after a finite number of iterations.

the optimal choice is given by the Moore–Penrose inverse $\mathbf{R}^+$ of the projection operator $\mathbf{R}$ (see (3.58)), facilitating convergence after one iteration. However, $\mathbf{R}^+$ has to be regularized in practice, which is not straightforward (cf. Section 3.5). It is therefore convenient to use an easily computable ill-conditioned approximation for $\mathbf{R}^+$, which is regularized subsequently by the iterative Landweber scheme. One straightforward approach consists of using the FBP instead of the simple back-projection in the Landweber iteration, which we coined Weighted SIRT (WSIRT) (Wolf, Lubk, & Lichte, 2014). We employ the WSIRT algorithm to reconstruct magnetic fields in Section 6.3.

3.6.3 Conjugate Gradient Method

Both the Kaczmarz and the Landweber method do not seek an optimal search direction in each iteration, because information from previous iterations is neglected. To remedy this drawback, Conjugate Gradient methods compute an optimal search direction (Fig. 3.21) by subtracting previous search directions from the residual $\mathbf{r}_n$, used within one Landweber iteration, according to

$$\mathbf{p}_n = \mathbf{r}_n - \sum_{m<n} \frac{\mathbf{p}_m^{\mathrm{T}} \mathbf{R}^T \mathbf{R} \mathbf{r}_n}{\mathbf{p}_m^{\mathrm{T}} \mathbf{R}^T \mathbf{R} \mathbf{p}_m} \mathbf{p}_m, \tag{3.98}$$

yielding the modified update

$$\mathbf{f}_{n+1} = \mathbf{f}_n + \omega_n \mathbf{p}_n. \tag{3.99}$$

The weighting parameter may then be obtained from

$$\omega_n = \frac{\mathbf{p}_n^{\mathrm{T}} \mathbf{R}^T \mathbf{F}_{(\exp)}}{\mathbf{p}_n^{\mathrm{T}} \mathbf{R}^T \mathbf{R} \mathbf{p}_n} = \frac{\mathbf{p}_n^{\mathrm{T}} (\mathbf{r}_{n-1} + \mathbf{R}^{\mathrm{T}} \mathbf{R} \mathbf{f}_{n-1})}{\mathbf{p}_n^{\mathrm{T}} \mathbf{R}^T \mathbf{R} \mathbf{p}_n} = \frac{\mathbf{p}_n^{\mathrm{T}} \mathbf{r}_{n-1}}{\mathbf{p}_n^{\mathrm{T}} \mathbf{R}^T \mathbf{R} \mathbf{p}_n}, \tag{3.100}$$

which takes into account that the set of $\mathbf{p}_n$s form a complete basis in the space of solutions. Consequently, after maximally $\dim \mathbf{R}^{\mathrm{T}} \mathbf{R}$ steps the algorithm has converged (Fig. 3.21), which is why Conjugate Gradient methods may be considered as direct instead of iterative. The above instance of the Conjugate Gradient algorithm may be further improved and adapted to the particular shape of Radon matrices, which represents an active line of research in particular if very large systems of projections are considered (Saad, 2003).

In one Conjugate Gradient iteration, the expansion into the $\mathbf{p}_n$s is ordered such to minimize the projection error as fast as possible, i.e., one computes first the most stable parts (large singular values) of $\mathbf{f}$ and the iteration roughly proceeds along the ordered Singular Value Decomposition[4] with the iteration number n serving as inverse regularization strength (Louis, 2013). For the (mildly)-ill conditioned Radon transformation, typically a small number of iterations ($\mathcal{O}(10)$) suffices to obtain a reasonably regularized solution. Similar to the previous methods, the iteration speed increases with the sparsity of the Radon matrix. We employ a conjugate gradient algorithm as implemented in the LSQR package (Paige & Saunders, 1982) of MATLAB for the tomographic reconstruction of electric potentials in Sections 6.1 and 6.2 as well as attenuation coefficients in Section 6.4.

---

[4] Indeed, the regularization is complicated and follows a non-linear behavior, which may not be described by a simple singular value filter (Louis, 2013).

REFERENCES

Batenburg, K. J., Bals, S., Sijbers, J., Kübel, C., Midgley, P., Hernandez, J., ... Kaiser, U. (2009). 3D imaging of nanomaterials by discrete tomography. *Ultramicroscopy, 109*(6), 730–740.

Bracewell, R. N., & Riddle, A. C. (1967). Inversion of fan beam scans in radio astronomy. *Astrophysical Journal, 150,* 427–434.

Candes, E. J., Romberg, J., & Tao, T. (2006). Robust uncertainty principles: Exact signal reconstruction from highly incomplete frequency information. *IEEE Transactions on Information Theory, 52*(2), 489–509.

Carazo, J.-M. (1992). The fidelity of 3D reconstructions from incomplete data and the use of restoration methods. In J. Frank (Ed.), *Electron tomography* (pp. 197–204). New York: Plenum.

Chan, W. L., Charan, K., Takhar, D., Kelly, K. F., Baraniuk, R. G., & Mittleman, D. M. (2008). A single-pixel terahertz imaging system based on compressed sensing. *Applied Physics Letters, 93,* 121105.

Chapman, C. H., & Cary, P. W. (1986). The circular harmonic Radon transform. *Inverse Problems, 2*(1), 23–49.

Cormack, A. M. (1963). Representation of a function by its line integrals, with some radiological applications. *Journal of Applied Physics, 34,* 2722–2727.

Cormack, A. M. (1964). Representation of a function by its line integrals, with some radiological applications. II. *Journal of Applied Physics, 35*(10), 2908–2913.

Crowther, R. A., Amos, L. A., Finch, J. T., De Rosier, D. J., & Klug, A. (1970). Three dimensional reconstructions of spherical viruses by Fourier synthesis from electron micrographs. *Nature, 226*(5244), 421–425.

Crowther, R. A., DeRosier, D. J., & Klug, A. (1970). The reconstruction of a three-dimensional structure from projections and its application to electron microscopy. *Proceedings of the Royal Society of London Series A, 317,* 319–340.

Deans, S. R. (1983). *The Radon transform and some of its applications.* New York: John Wiley & Sons.

Donoho, D. L. (2006). Compressed sensing. *IEEE Transactions on Information Theory, 52*(4), 1289–1306.

Fernandez, J.-J., Lawrence, A. F., Roca, J., Garcia, I., Ellisman, M. H., & Carazo, J.-M. (2002). High-performance electron tomography of complex biologic specimen. *Journal of Structural Biology, 138,* 6–20.

Funk, P. (1913). Über Flächen mit lauter geschlossenen geodätischen Linien. *Mathematische Annalen, 74.*

Gilbert, P. F. C. (1972). The reconstruction of a three-dimensional structure from projections and its application to electron microscopy. II: Direct methods. *Proceedings of the Royal Society of London Series B, 182,* 89–102.

Golub, G. H., & Van Loan, C. F. (1996). *Matrix computations. Johns Hopkins studies in the mathematical sciences.* Johns Hopkins University Press.

Gordon, R., Bender, R., & Herman, G. T. (1970). Algebraic reconstruction techniques (ART) for three-dimensional electron microscopy and X-ray photography. *Journal of Theoretical Biology, 29*(3), 471–481.

Goris, B., Bals, S., Van den Broek, W., Carbó-Argibay, E., Gómez-Graña, S., Liz-Marzán, L. M., & Van Tendeloo, G. (2012). Atomic-scale determination of surface facets in gold nanorods. *Nature Materials, 11*(11), 930–935.

Goris, B., Van den Broek, W., Batenburg, K., Heidari Mezerji, H., & Bals, S. (2012). Electron tomography based on a total variation minimization reconstruction technique. *Ultramicroscopy, 113*, 120–130.

Gower, R., & Richtarik, P. (2015). Randomized iterative methods for linear systems. arXiv:1506.03296.

Hansen, P. C. (1987). *Rank-deficient and discrete ill-posed problems: Numerical aspects of linear inversion*. SIAM.

Harauz, G., & van Heel, M. (1986). Exact filters for general geometry three dimensional reconstruction. *Optics, 73*, 146–156.

Helgason, S. (2011). *Integral geometry and Radon transforms*. New York: Springer.

Herman, G. T. (2009). *Fundamentals of computerized tomography*. Springer.

Herman, G. T., & Kuba, A. (2007). *Advances in discrete tomography and its applications*.

Holland, D. J., Bostock, M. J., Gladden, L. F., & Nietlispach, D. (2011). Fast multidimensional NMR spectroscopy using compressed sensing. *Angewandte Chemie, International Edition, 50*, 6548.

Holland, D. J., Malioutov, D. M., Blake, A., Sederman, A. J., & Gladden, L. F. (2010). Reducing data acquisition times in phase-encoded velocity imaging using compressed sensing. *Journal of Magnetic Resonance, 203*, 236.

Hounsfield, G. N. (1973). Computerized transverse axial scanning (tomography): Part 1. Description of system. *British Journal of Radiology, 46*(552), 1016–1022.

Hsieh, J. (2003). *Computed tomography: Principles, design, artifacts, and recent advances*. SPIE Press monograph. SPIE Press.

Kaczmarz, S. (1937). Angenäherte Auflösung von Systemen linearer Gleichungen. *Bulletin International de l'Académie Polonaise des Sciences et des Lettres. Classe des Sciences Mathématiques et Naturelles. Série A, Sciences Mathématiques, 35*, 355–357.

Kak, A. C., & Slaney, M. (1988). *Principles of computerized tomographic imaging*. IEEE Press.

Klug, A., Crick, F. H. C., & Wyckoff, H. W. (1958). Diffraction by helical structures. *Acta Crystallographica, 1*, 199–213.

Kuba, A., & Herman, G. T. (2008). Some mathematical concepts for tomographic reconstructions. In J. Banhart (Ed.), *Advanced tomographic methods in materials research and engineering* (p. 19). Oxford: Oxford University Press.

Landweber, L. (1951). An iteration formula for Fredholm integral equations of the first kind. *American Journal of Mathematics, 73*, 615–624.

Lewis, A. S., & Malick, J. (2008). Alternating projections on manifolds. *Mathematics of Operations Research, 33*(1), 216–234.

Louis, A. K. (2013). *Inverse und schlecht gestellte Probleme. Teubner Studienbücher Mathematik*. Vieweg+Teubner Verlag.

Lubk, A., Wolf, D., Prete, P., Lovergine, N., Niermann, T., Sturm, S., & Lichte, H. (2014). Nanometer-scale tomographic reconstruction of three-dimensional electrostatic potentials in GaAs/AlGaAs core–shell nanowires. *Physical Review B, 90*, 125404.

Lubk, A., Wolf, D., Simon, P., Wang, C., Sturm, S., & Felser, C. (2014). Nanoscale three-dimensional reconstruction of electric and magnetic stray fields around nanowires. *Applied Physics Letters, 105*(17), 173110.

Lustig, M., Donoho, D., & Pauly, J. M. (2007). Sparse MRI: The application of compressed sensing for rapid MR imaging. *Magnetic Resonance in Medicine, 58*(6), 1182.

Matteucci, G., Missiroli, G. F., Nichelatti, E., Migliori, A., Vanzi, M., & Pozzi, G. (1991). Electron holography of long-range electric and magnetic fields. *Journal of Applied Physics, 69*, 1835.

Miao, J., Förster, F., & Levi, O. (2005). Equally sloped tomography with oversampling reconstruction. *Physical Review B*, *72*(5), 052103.

Natterer, F. (2001). *The mathematics of computerized tomography. Classics in applied mathematics: Vol. 32.* Philadelphia: Society for Industrial and Applied Mathematics.

Natterer, F., & Wübbeling, F. (2001). *Mathematical methods in image reconstruction.* Society for Industrial and Applied Mathematics.

Neumaier, A. (1998). Solving ill-conditioned and singular linear systems: A tutorial on regularization. *SIAM Review*, *40*(3), 636–666.

Paige, C. C., & Saunders, M. A. (1982). LSQR: An algorithm for sparse linear equations and sparse least squares. *ACM Transactions on Mathematical Software*, *8*(1), 43–71.

Perry, R. M. (1975). Reconstructing a function by circular harmonic analysis of its line integrals. In *Image processing for 2-D and 3-D reconstruction from projections: Theory and practice in medicine and the physical sciences. Technical digest* (pp. 61–64). Washington, DC: Optical Society of America.

Phatak, C., de Knoop, L., Houdellier, F., Gatel, C., Hÿtch, M., & Masseboeuf, A. (2016). Quantitative 3D electromagnetic field determination of 1D nanostructures from single projection. *Ultramicroscopy*, *164*, 24–30.

Poularikas, A. D. (2010). *Transforms and applications handbook* (third edition). *Electrical engineering handbook.* CRC Press.

Radermacher, M. (2006). Weighted back-projection methods. In J. Frank (Ed.), *Electron tomography, methods for three-dimensional visualization of structures in the cell* (pp. 245–274). Berlin: Springer.

Radon, J. (1917). Über die Bestimmung von Funktionen durch ihre Integralwerte längs gewisser Mannigfaltigkeiten. *Sächsische Akadademie der Wissenschaften*, *69*, 262–277.

Saad, Y. (2003). *Iterative methods for sparse linear systems* (second edition). Society for Industrial and Applied Mathematics.

Saghi, Z., Holland, D. J., Leary, R., Falqui, A., Bertoni, G., Sederman, A. J., . . . Midgley, P. A. (2011). Three-dimensional morphology of iron oxide nanoparticles with reactive concave surfaces. A compressed sensing-electron tomography (CS-ET) approach. *Nano Letters*, *11*(11), 4666–4673.

Smith, P. R., Peters, T. M., & Bates, R. H. T. (1973). Image reconstruction from finite numbers of projections. *Journal of Physics A: Mathematical, Nuclear and General*, *6*(3), 361–382.

Starck, J. L., Murtagh, F., & Fadili, J. M. (2010). *Sparse image and signal processing: Wavelets, curvelets, morphological diversity.* Cambridge University Press.

Stiefel, E. (1952). Über einige Methoden der Relaxationsrechnung. *Zeitschrift für Angewandte Mathematik und Physik*, *3*(1), 1–33.

Strohmer, T., & Vershynin, R. (2008). A randomized Kaczmarz algorithm with exponential convergence. *The Journal of Fourier Analysis and Applications*, *15*(2), 262–278.

Tikhonov, A. N. (1963). Solution of incorrectly formulated problems and the regularization method. *Soviet Mathematics. Doklady*, *4*, 1035–1038.

Twitchett-Harrison, A. C., Yates, T. J. V., Newcomb, S. B., Dunin-Borkowski, R. E., & Midgley, P. A. (2007). High-resolution three-dimensional mapping of semiconductor dopant potentials. *Nano Letters*, *7*(7), 2020–2023.

Vainshtein, B. K. (1970). Finding the structure of objects from projections. *Soviet Physics. Crystallography*, *15*(5), 781–787.

Velikina, J., Leng, S., & Chen, G. (2007). Limited view angle tomographic image reconstruction via total variation minimization. *Proceedings of SPIE*, *6510*, 651020.

Wiaux, Y., Jacques, L., Puy, G., Scaife, A. M. M., & Vanderghevost, P. (2009). Compressed sensing imaging techniques for radio interferometry. *Monthly Notices of the Royal Astronomical Society, 395*, 1733.

Wolf, D., Lubk, A., & Lichte, H. (2014). Weighted simultaneous iterative reconstruction technique for single-axis tomography. *Ultramicroscopy, 136*, 15–25.

CHAPTER FOUR

Electron Optics in Phase Space

Axel Lubk

Institute for Structure Physics, Physics Department, Faculty of Mathematics and Natural Sciences, Technical University of Dresden, Dresden, Germany
e-mail address: axel.lubk@tu-dresden.de

Contents

"It seems as though we must use sometimes the one theory [pertaining to particles] and sometimes the other [pertaining to waves], while at times we may use either."

(Albert Einstein)

Transmission Electron Microscopy (TEM) looks back on a long history of developments in its theoretical, methodological and instrumental aspects; the sum of which facilitating a multitude of imaging modes in a modern TEM, which allow probing various specimen properties down to the sub-angstrom spatial resolution regime. There is a considerable number of excellent monographs on TEM with varying scope and focus permitting to retrace that development and comprehend the current status of the technique. In times of a vast and overflowing literature on the topic, these works provide an invaluable overview and condensation for a practitioner of TEM.

The books of Reimer and Kohl (2008) and Spence (2013) are classics, which cover more or less the whole range of TEM techniques. In particular they review the foundations of TEM, which are sometimes threatened to be forgotten in the modern digital era. De Graef's book (De Graef, 2003) contains a comprehensive introduction to the fundamentals of conventional (CTEM) with as strong emphasis on crystalline solids. A more recent publication with focus on high-resolution imaging and aberration

Advances in Imaging and Electron Physics, Volume 206
ISSN 1076-5670
https://doi.org/10.1016/bs.aiep.2018.05.004

correction has been provided by Erni (2010). A comprehensive introduction into spectroscopical techniques in the TEM is furnished by Egerton (1996). The monumental contribution of Hawkes and Kasper (Hawkes & Kasper, 1996, 1988; Kasper & Hawkes, 1995) still contains the most comprehensive and detailed treatise on electron optics. A more recent account of the field with a stronger focus on hardware aberration correction has been given by Rose (2009). Certainly, the above list is very incomplete and should be extended beyond the limited horizon of the author.

In the following an alternative approach to electron optics and holography that is particularly suited to the combination of which and only partially represented in the pertinent literature is presented. We seek an understanding of the transfer of the paraxial quantum state introduced in Sections 2.4 and 2.5 through the optical system of a TEM, because this quantum state thoroughly describes the electron beam including partial coherence. In what follows, the electron beam's quantum state will be mainly represented by its Wigner function, thereby exploiting a number of advantageous properties of this phase space representation. The Wigner function simultaneously describes the properties of the electron beam in position and diffraction (Fourier) space. In other words, inspecting the Wigner distribution gives immediate insight into the shape of the electron beam in both image and diffraction planes without necessitating Fourier transformations. Linear projections of the Wigner function correspond to experimentally measured intensity distributions in the image and diffraction plane or any other plane in between. Furthermore, the Wigner function can be considered to be the quantum mechanical analogue to the classical phase space density, which implies that a semiclassical limit is obtained in a straightforward manner. This close correspondence will become particularly obvious in the way aberrations affect the Wigner function, where one can distinguish between a classical deformation and an integral transformation of purely quantum mechanical origin. Accordingly, the Wigner distribution transfer in an ideal stigmatic system is exactly equivalent to that of the classical phase space density of the electron beam owing to the particular structure of the quantum Liouville equation.

We will derive the phase space formulation of TEM imaging in extension to wave optical principles, which have proved well-suited to describe the action of the optical elements in the TEM. Note, however, that an analytical wave optical transfer through most of the constituting elements of a modern TEM has not been established yet. Instead, one relies on very accurate (and hence sufficient) semiclassical approximations. Their

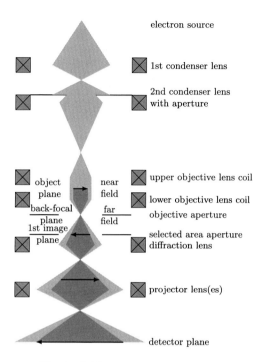

electron source

1st condenser lens

2nd condenser lens
with aperture

object plane / near field — upper objective lens coil

back-focal plane / far field — lower objective lens coil / objective aperture

1st image plane — selected area aperture / diffraction lens

projector lens(es)

detector plane

Figure 4.1 Conventional bright field TEM (CTEM) scheme with the electron source located in the far field (Fourier plane) of the object plane, which in turn is conjugated to the detector plane. Conjugated object planes are indicated by arrows. Note the typically larger extension of the illumination with respect to the observed field of view.

basis is the classical trajectories, around which quantum mechanical wave functions may be constructed by suitable formalisms. This procedure is sometimes described vividly as putting quantum mechanical flesh on classical bones. Thus, we begin with a short recapitulation of classical electron optics, which contains an introduction to geometric aberrations, playing a fundamental role in the imaging process. Subsequently, a well-known semiclassical procedure, Miller's semiclassical algebra, is used to "materialize" the wave function and its transfer properties. In a final step, the wave function transfer is generalized to that of the Wigner function, i.e., the partially coherent quantum state.

4.1. CLASSICAL OPTICS

A schematic of the ray path in a typical TEM is depicted in Fig. 4.1. Obviously, the lenses represent the most important building blocks and we will mainly focus on them in this subsection. The other principal elements, that is the electron source, apertures, and the detector are directly described quantum mechanically below.

It was the groundbreaking discovery of the focusing property of magnetic coils by Busch (1926), which laid the foundation for the invention of the TEM by Knoll and Ruska (1932). Soon afterward, Scherzer recognized that the spherical aberration cannot be avoided for rotationally symmetric, static lenses free of space charges (Scherzer, 1936), which is referred to as Scherzer theorem since then. These two achievements set the stage for the ensuing development of TEM. The original design of Ruska was improved and enhanced to illuminate more and more of the properties of matter at the nanoscale, while large and sustained efforts have been put into finding ways to correct for the aberrations. While Gabor's invention of holography (Gabor, 1948) permitted a-posteriori correction schemes, the efforts for an a-priori correction have been only fully rewarded after a journey of almost half a century (Scherzer, 1947; Seeliger, 1951; Möllenstedt, 1956; Beck, 1979; Crewe & Kopf, 1980; Rose, 1990; Haider et al., 1998; Krivanek, Dellbya, & Lupinic, 1999; Rose, 2008; Haider, Müller, & Uhlemann, 2008; Rose, 2009). State-of-the-art hardware correctors now correct a multitude of aberrations (Haider, Hartel, Müller, Uhlemann, & Zach, 2010) over increasingly large field of views (Müller et al., 2011) at decreasing acceleration voltages thereby pushing the resolution and the signal-to-noise ratios attainable with modern TEMs. Aberration correction triggered nothing short of a revolution in materials and life sciences as can be seen through the very incomplete compilation of results obtained by this technique (Nellist, Behan, Kirkland, & Hetherington, 2006; Dwyer, Kirkland, Hartel, Müller, & Haider, 2007; Jia et al., 2007; Urban, 2008; Girit et al., 2009; Alem et al., 2011; Gao et al., 2011; Jia, Urban, Alexe, Hesse, & Vrejoiu, 2011; Van Aert, Batenburg, Rossell, Erni, & Van Tendeloo, 2011; Bar-Sadan, Barthel, Shtrikman, & Houben, 2012; Yuk et al., 2012; Urban et al., 2013). Below, the perspective on hardware aberration correctors will be slightly shifted toward an optical device, which is capable of deliberately tuning aberrations with a very high precision (Clark et al., 2013; Guzzinati et al., 2015).

The remarkable success of aberration correction became only possible due to the development of concise computation schemes for classical electron paths and their deviations – the aberrations – from the ideal Gaussian trajectories (Hawkes, 1965; Rose, 1970, 1971, 1981). Aberrations research in optics has a very long tradition going back to the origins of light microscopy itself. A classification of aberrations of round lenses, for instance, has been developed by Seidel as early as in the middle of the 19th century. He showed that a perturbation expansion of the path deviations in polynomials of the beam distance and angle to the optical axis contains only a limited number of terms due to symmetry – the famous Seidel aberrations. Since then a considerable number of different aberration classification schemes has been developed, which mainly differ in the way boundary conditions on the trajectories are incorporated and how the perturbation is organized. In the course of this work, only the boundary conditions are important; the reader interested in the details of the perturbation expansion is referred to the abundant literature on the topic (Zeitler, 1990; Hawkes & Kasper, 1996; Müller, Uhlemann, Hartel, & Haider, 2008; Rose, 2009; Hawkes, 2013).

The magnetic lenses, employed in a TEM, either belong to the condenser, shaping the illumination, the objective, providing the principal imaging, or the projective, post-magnifying the first intermediate image. Their main feature is the capability to increasingly bend rays the larger their distance to the optical axis, eventually facilitating a common focus of a whole pencil of rays. The latter permits a stigmatic (i.e., one-to-one) mapping of the ray coordinates in some object plane to those pertaining to the image plane in the ideal case, as shown in the ray diagram in Fig. 4.2. Following standard conventions, rays emerging from the origin of the object plane ($\mathbf{r} = 0$) under a certain angle to the optical axis are referred to as axial and those having zero lateral momentum ($\mathbf{k} = 0$) in the object plane as field rays. Similar to the Gaussian rays, shown in Fig. 4.2, a classical paraxial electron trajectory may be uniquely defined by its position and momentum in a particular plane in the TEM (e.g., the object plane). This behavior is guaranteed by the Gaussian or paraxial path equations corresponding to a coupled system of two linear differential equations of the second order for the 2D ray coordinates $\mathbf{r}_e(z)$ perpendicular to the optical axis z. The two components may be decoupled by employing a rotating coordinate system absorbing the Larmor rotation introduced by the magnetic field of the lens. In the course of this work we shall always employ this Larmor frame of reference. Then, any solution of one of the decoupled paraxial path equations (being of second order) may be uniquely expanded into two fundamental (linear independent) solutions (rays), which are the axial and field ray defined above.

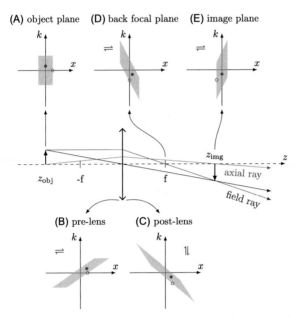

Figure 4.2 Stigmatic imaging at a thin lens in phase space. Accordingly, the initial rectangular phase space distribution in the object plane (A) passes through several (affine) shear transformations (B)–(D) yielding a 180° rotated and, along the momentum coordinate, sheared version in the image plane (E).

A comprehensive picture of the whole imaging process is obtained by analyzing the parameter space of the Gaussian rays, i.e., the evolution of their distance and angle to the optical axis. Within the paraxial regime (cf. Chapter 2) the angle is directly proportional to the lateral momentum $\mathbf{k} = \gamma k_0$; hence the parameter space corresponds to the phase space spanned by the lateral coordinate and momentum. Consequently, the imaging of an arbitrary set of beams corresponds to transforming their phase space distributions along the optical axis as schematically depicted in Fig. 4.2 in the case of imaging with a thin lens. The model of a thin lens represents a good approximation for the various weakly excited magnetic lenses in a TEM with the notable exception of the objective lens, which has to be treated as a complicated thick immersion lens, where the object resides within the field of the lens. According to Busch, the refraction power (reciprocal focal length) of a weak magnetic lens, given by

$$f^{-1} = C_f \, (NI)^2 \,, \tag{4.1}$$

is directly proportional to the squared ampere turns NI of the lens with the proportionality constant C_f determined by the geometry of the pole pieces.

In Fig. 4.2A, the initial phase space distribution of the beam in the object plane is modeled as a rectangular top hat function. This distribution is sheared along the position coordinate while freely propagating the beam to the lens plane, because the change of the distance to the optical axis is proportional to the initial lateral momentum (angle to the optical axis), which remains constant for each ray in free space (Fig. 4.2B). Subsequently, the ideal (thin) lens shears the phase space distribution along the momentum coordinate, because the amount of angular change depends on the distance to the optical axis (Fig. 4.2C). Then, the phase space distribution is sheared again along the position axis, which leads to distributions in the back focal plane and the image plane. In the former the originally rectangular shape is transformed to a parallelogram, where boundaries, which formerly were parallel to the position axis, are aligned parallel to the momentum axis (Fig. 4.2D). When adding up all beams with the same distance to the optical axis (i.e., integrating along k), one obtains the familiar result that the back focal plane contains the marginal lateral momentum distribution of the initial beam. The shape of the distribution in the image plane is also a parallelogram, where edges, which were parallel to the momentum axis in the object plane, are aligned to the momentum axis again (Fig. 4.2E). Consequently, all beams originating from the same point in the object plane intersect at the same point in the image plane, thereby forming a stigmatic image. Note that during the whole transfer process the phase space volume remains constant, which is referred to as Lagrange–Helmholtz relation in the optical literature.

Unfortunately, the famous Scherzer theorem states that spherical aberration cannot be avoided in rotationally symmetric and static electric or magnetic lenses containing no space charge (Scherzer, 1936). Consequently, the principal imaging lens of the TEM, the objective lens, does not permit a stigmatic (point-to-point) mapping from the object to image plane. To describe the deviations from the stigmatic paraxial paths, i.e., the aberrations, it is now crucial to select both, suitable paraxial reference paths and suitable planes, where to measure the deviation.

The most natural way is to define the aberrations as the image plane difference between an aberrated and the corresponding Gaussian trajectory starting from the same point under the same angle in the object plane (Fig. 4.3). This *position aberration vector* $\boldsymbol{\xi}$ can be computed efficiently from

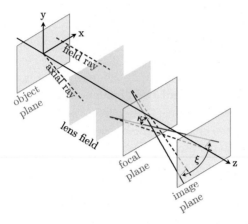

Figure 4.3 Aberrations in (thick) lens imaging defined as position and momentum deviations of the true ray in the image plane with respect to Gaussian reference trajectories originating from the same position with the same momentum in the object plane. The two fundamental Gaussian beams, i.e., the red axial ray and the blue field ray, are drawn dashed. The corresponding aberrated beams are indicated by continuous lines with lighter color. Accordingly, the position (momentum) aberration vector in the image plane ξ (κ) depends on the object plane phase space coordinates of the corresponding ray.

the projections of the path deviations on the field Gaussian ray as implemented in the perturbative scheme referred to as the path equation method (Haider et al., 2008). Computing the *momentum aberration vector* κ in the image plane on the other hand is more difficult as both Gaussian axial and field rays in the expansion of the aberrated ray contain momentum deviations in the image plane. Note, however, that the position aberration in the back focal plane is readily obtained by projecting the path deviations on the axial reference ray within this scheme (Fig. 4.3).

Alternatively, one can compute the difference between aberrated and Gaussian trajectories sharing the same object plane position and the same image plane momentum (Fig. 4.4). In that case the position deviation is taken in the image plane and the momentum deviation in the object plane. At first sight, this latter convention appears awkward as aberrations are defined in mixed planes and complicated Darboux transformations are required to relate aberration vectors from this convention to those pertaining to the first one. However, the latter approach is amenable to a reformulation of path deviations in terms of characteristic functions or eikonals, which allows for a considerable simplification of the perturbation expansion (Rose, 2009). Furthermore, the *momentum aberration vector* κ may be computed

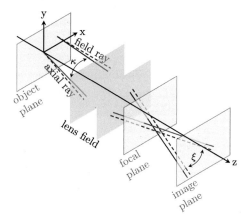

Figure 4.4 Aberrations in (thick) lens imaging defined as position (and momentum) deviations of the true ray in the image (object) plane with respect to Gaussian reference trajectories originating from the same position in the object plane and intersecting under the same angle the image plane. Accordingly, the position (momentum) aberration vector in the image (object) plane $\boldsymbol{\xi}$ ($\boldsymbol{\kappa}$) depends on the position (momentum) coordinate in the object (image) plane of the corresponding ray.

very efficiently by projecting the path deviations on the axial Gaussian ray in this scheme. The advantages and disadvantages of the different boundary conditions have been controversially discussed in the literature (Zeitler, 1990; Rose, 2009).

In the remainder of this work, the eikonal scheme is employed because of its close link to a particular semiclassical approximation, i.e., Miller's semiclassical algebra (Miller, 1974), which is subsequently utilized to describe wave optical aberrations. It has to be emphasized, however, that an equally well-supported semiclassical scheme, referred to as Frozen Gaussian Approximation (Heller, 1981; Herman & Kluk, 1984), can be used in combination with the first choice of boundary conditions (Lubk, Béché, & Verbeeck, 2015) (Fig. 4.3) and it remains an open question, which of the different semiclassical approximations is better suited for describing aberrations wave optically. In the case of the dominant isoplanatic aberrations both boundary conditions coincide as the aberration vector depends only on the momentum coordinate, i.e., $\boldsymbol{\xi} = \boldsymbol{\xi}(\mathbf{k})$, and $\boldsymbol{\kappa}$ is neglected. However, if imaging over large fields of view is considered, for instance in off-axis electron holography (Section 5.1) or large-field high-resolution studies, also off-axial dependencies, i.e., $\boldsymbol{\xi} = \boldsymbol{\xi}(\mathbf{r}, \mathbf{k})$, must be taken into account (Müller et al., 2011).

In the following, lateral position (canonical momentum) coordinates in the object plane and all conjugated planes are denoted by $\mathbf{r} = (x, y)^T$ ($\mathbf{k} = (k_x, k_y)^T$).[1] That is, we employ Seidel coordinates, where the magnification M has been absorbed. To distinguish between stigmatic and aberrated imaging, the corresponding quantities are indicated with s and a, respectively. Following this notation, an ensemble of classical rays, filling a certain 4D volume in phase space, gets distorted under the action of the lens including aberrations according to[2]

$$f_a(\mathbf{r}, \mathbf{k}) = f_s(\mathbf{r} - \boldsymbol{\xi}(\mathbf{r}, \mathbf{k}), \mathbf{k} + \boldsymbol{\kappa}(\mathbf{r}, \mathbf{k})). \tag{4.2}$$

We now proceed with noting the most important aberrations dropping out of the eikonal scheme. To this end, the aberration vectors are expanded in the following power series

$$\boldsymbol{\xi} = \sum_n \boldsymbol{\xi}^{(n)} \quad \text{and} \quad \boldsymbol{\kappa} = \sum_n \boldsymbol{\kappa}^{(n)}, \tag{4.3}$$

where the different terms are homogeneous polynomials of order n in phase space coordinates. Here, the position and momentum aberrations $\boldsymbol{\xi}$ and $\boldsymbol{\kappa}$ are not independent as will become obvious when introducing the eikonal further below.

The following list contains the most important isoplanatic position aberrations up to the third order including the notorious spherical aberration C_3, which hampered TEM for such a long time. For the sake of simplicity, the azimuthal orientation of all non-isotropic aberrations has been set to zero:

$$\boldsymbol{\xi}(\mathbf{k}) = \underbrace{\frac{C_1}{k_0} \begin{pmatrix} k_x \\ k_y \end{pmatrix}}_{\text{defocus}} + \underbrace{\frac{A_1}{k_0} \begin{pmatrix} k_x \\ -k_y \end{pmatrix}}_{\text{2-fold astigmatism}} \tag{4.4}$$

$$+ \underbrace{\frac{B_2}{k_0^2} \begin{pmatrix} 3k_x^2 + k_y^2 \\ 2k_x k_y \end{pmatrix}}_{\text{axial coma}} + \underbrace{\frac{A_2}{k_0^2} \begin{pmatrix} k_x^2 - k_y^2 \\ -2k_x k_y \end{pmatrix}}_{\text{3-fold astigmatism}}$$

---

[1] Note that $\hbar = 1$ in natural units.

[2] The different signs in front of the aberration vectors is consistent with our definition of aberration vectors applying in mixed planes (see also further below).

$$+ \frac{C_3 \left(k_x^2 + k_y^2 \right)}{k_0^3} \underbrace{\begin{pmatrix} k_x \\ k_y \end{pmatrix}}_{\text{spherical aberration}} + 4 \underbrace{\frac{S_3}{k_0^3} \begin{pmatrix} k_x^3 \\ -k_y^3 \end{pmatrix}}_{\text{star aberration}} + \underbrace{\frac{A_3}{k_0^3} \begin{pmatrix} k_x^3 - 3k_y^2 k_x \\ k_y^3 - 3k_x^2 k_y \end{pmatrix}}_{\text{4-fold astigmatism}}.$$

Arbitrarily oriented aberrations may be obtained by rotating the aberration vector. A similar expansion exists for the momentum aberration vector κ, where the coefficients are denoted by small letters, i.e., c_1, a_1, and so forth.

Non-isoplanatic aberrations, depending on both position and momentum coordinates, are typically small for the position space regions imaged in high-resolution TEM (small fields of view). However, the physical size of the field of view is successively increased by the projective lenses of the TEM. Moreover, the latest generation of TEMs, equipped with very large pixel detectors, aim at atomic resolution imaging over increasingly large fields of view. Similarly, within Off-Axis Holography as discussed in Section 5.1, beams are superimposed over large distances in the object plane. Therefore, the aberrations featuring a linear position dependency in the aberration vector $\boldsymbol{\xi}\left(\mathbf{r}, \mathbf{k}\right)$, which are referred to as generalized coma (Rose, 2003), become increasingly important for an accurate description of the imaging process in a TEM:

$$\boldsymbol{\xi}\left(\mathbf{r}, \mathbf{k}\right) = \underbrace{\frac{C_1^{(1)} \langle \mathbf{r}, \boldsymbol{\beta}_C \rangle}{k_0} \begin{pmatrix} k_x \\ k_y \end{pmatrix}}_{\text{off-axial defocus}} + \underbrace{\frac{A_1^{(1)} \langle \mathbf{r}, \boldsymbol{\beta}_A \rangle}{k_0} \begin{pmatrix} k_x \\ -k_y \end{pmatrix}}_{\text{off-axial astigmatism}} + \underbrace{\frac{B_2^{(1)} \langle \mathbf{r}, \boldsymbol{\beta}_B \rangle}{k_0^2} \begin{pmatrix} 3k_x^2 + k_y^2 \\ 2k_x k_y \end{pmatrix}}_{\text{off-axial coma}}.$$

$$(4.5)$$

Again, the azimuthal orientations have been deliberately aligned along the coordinate axis; and the βs denote unit vectors along which the spatial dependency of the aberrations is oriented. The familiar Seidel off-axial coma of round lenses corresponds to the third term. Aberrations of quadratic or higher order in the spatial coordinate $\mathbf{r}$ (e.g., field curvature, image distortion) are not considered since they play a minor role for the image formation in TEM (Haider et al., 2008; Rose, 2009).

Last but not least it is noted that all aberration coefficients carry an energy dependency. The most important chromatic aberration, introduced by the following electron energy dependence of the first-order aberration coefficient

$$C_1 \left(\delta E \right) = C_1 + C_c \frac{\delta E}{e U_A} \qquad (4.6)$$

focuses electrons of different energy in different planes, depending on the chromatic aberration coefficient C_c and the energy deviation δE from the mean energy E, normalized with the acceleration voltage.

We conclude our small tour through classical electron optics by introducing the notion of an eikonal or characteristic function providing the ratio behind the mixed boundary conditions used previously. It is well understood that classical mechanics can be described within various formalism, namely Lagrangian mechanics, Hamiltonian mechanics, and Hamilton–Jacobi theory. The latter constitutes a (nonlinear) partial differential equation for the classical action; the characteristics of which leading to the classical equations of motion. If the Hamiltonian does not explicitly depend on time, the time dependence may be separated from the action leading to the notion of the abbreviated action referred to as eikonal in optics. Furthermore, the classical action can be considered as a type-2 generating function for a canonical transformation of phase space coordinates, which in our case depends on the position in the object plane and the momentum in the image plane, denoted by $S\left(\mathbf{r}_{obj}, \mathbf{k}_{img}\right)$ or shortly $S\left(\mathbf{r}, \mathbf{k}\right)$ in the following. The partial derivatives of this mixed eikonal with respect to its arguments yield the momentum in the object and the position in the image plane, respectively. In the most simple case of stigmatic imaging the mixed eikonal reads $S\left(\mathbf{r}, \mathbf{k}\right) = \mathbf{r}\mathbf{k}$ ensuring that the image coordinate of a trajectory corresponds (up to magnification change) to the object coordinate

$$\mathbf{r}_{img} = \frac{\partial S\left(\mathbf{r}_{obj}, \mathbf{k}_{img}\right)}{\partial \mathbf{k}_{img}} = \mathbf{r}_{obj} . \tag{4.7}$$

Aberrations can be incorporated in that formalism by adding an aberration eikonal $\chi\left(\mathbf{r}, \mathbf{k}\right)$

$$S\left(\mathbf{r}, \mathbf{k}\right) = \mathbf{r}\mathbf{k} + \chi\left(\mathbf{r}, \mathbf{k}\right) . \tag{4.8}$$

Then, the position and momentum aberration vectors are defined by the mixed aberration eikonal in the following way

$$\boldsymbol{\xi}\left(\mathbf{r}, \mathbf{k}\right) = \nabla_k \chi\left(\mathbf{r}, \mathbf{k}\right) \tag{4.9}$$

and

$$\boldsymbol{\kappa}\left(\mathbf{r}, \mathbf{k}\right) = \nabla_r \chi\left(\mathbf{r}, \mathbf{k}\right) . \tag{4.10}$$

Thus, instead of two two-component aberration vectors $\boldsymbol{\xi}$ and $\boldsymbol{\kappa}$, only one mixed aberration eikonal is required to completely characterize non-stigmatic imaging. The aberration eikonal significantly simplifies computations and immediately exhibits all interdependencies between the position and momentum aberration vector.

Because the gradient relations (4.9) and (4.10) resemble those linking conservative forces and potentials in classical mechanics, the eikonals are also referred to as optical potentials, which can be computed from line integrals, similar to their mechanical counterparts. Up to third-order isoplanatic aberrations one obtains the following aberration eikonal (Fig. 4.5)

$$\chi\left(\mathbf{k}\right) = \underbrace{\frac{C_1}{2k_0}\left|\mathbf{k}\right|^2}_{\text{defocus}} + \underbrace{\frac{A_1}{2k_0}\left|\mathbf{k}\right|^2\cos 2\alpha}_{\text{2-fold astigmatism}} \tag{4.11}$$

$$+ \underbrace{\frac{B_2}{k_0^2}\left|\mathbf{k}\right|^3\cos\alpha}_{\text{axial coma}} + \underbrace{\frac{A_2}{3k_0^2}\left|\mathbf{k}\right|^3\cos 3\alpha}_{\text{3-fold astigmatism}}$$

$$+ \underbrace{\frac{C_3}{4k_0^3}\left|\mathbf{k}\right|^4}_{\text{spherical aberration}} + \underbrace{\frac{A_3}{4k_0^3}\left|\mathbf{k}\right|^4\cos 4\alpha}_{\text{4-fold astigmatism}} + \underbrace{\frac{S_3}{k_0^3}\left|\mathbf{k}\right|^4\cos 2\alpha}_{\text{star aberration}}.$$

Note that other representations of the aberrations may be generated by suitably rearranging the terms. Most importantly, in a Cartesian representation the various aberration orders are given by sums of homogeneous polynomials of the respective order. For instance, the first order terms in the Cartesian representation

$$\chi\left(\mathbf{k}\right) = \frac{C_1 + A_1}{2k_0}k_x^2 + \frac{C_1 - A_1}{2k_0}k_y^2 \tag{4.12}$$

correspond to two astigmatic line foci aligned along the x or y coordinate, if $C_1 = -A_1$ or $C_1 = A_1$, respectively. The aberration eikonals for the off-axial aberrations read

$$\chi\left(\mathbf{r},\mathbf{k}\right) = \underbrace{\frac{C_1^{(1)}\left\langle\mathbf{r},\boldsymbol{\beta}_C\right\rangle}{2k_0}\left|\mathbf{k}\right|^2}_{\text{off-axial defocus}} + \underbrace{\frac{A_1^{(1)}\left\langle\mathbf{r},\boldsymbol{\beta}_A\right\rangle}{2k_0}\left|\mathbf{k}\right|^2\cos 2\alpha}_{\text{off-axial astigmatism}} + \underbrace{\frac{B_2^{(1)}\left\langle\mathbf{r},\boldsymbol{\beta}_B\right\rangle}{k_0^2}\left|\mathbf{k}\right|^3\cos\alpha}_{\text{off-axial coma}}.$$

$$\tag{4.13}$$

These expressions are used in the next section to derive a semiclassical wave optical theory of aberrations.

4.2. SEMICLASSICAL WAVE OPTICS

We will now proceed with the second step of our program to derive a TEM imaging theory for arbitrary mixed quantum states. It consists of generating wave functions from classical trajectories with the help of the semiclassical algebra developed by Miller (1974). That formalism has its foundation in an approximation of unitary transformations, such as the paraxial propagation of a wave function from one plane to another, by means of stationary phase integrals. Within that framework a whole class of unitary transformations can be related to corresponding canonical transformations in classical mechanics inasmuch as the generating function of the transformation completely determines the stationary phase approximation. The scope of Miller's semiclassical algebra includes coordinate transformations, elastic propagation as well as inelastic scattering; in particular it encompasses WKB theory. It is heavily used in simulations of the dynamics and spectroscopy of complex molecular systems (Head-Gordon & Miller, 2012).

In case of electron optical imaging theory, elements of the semiclassical algebra, namely the WKB or eikonal approximation, have been developed prior to Miller's generalization. In particular, the dominant deviation from stigmatic imaging – the isoplanatic aberrations – has been incorporated in a wave optical imaging theory by multiplying the previously introduced aberration eikonal as phase factor in reciprocal space. Later, Rose showed how to treat nonisoplanatic aberrations within the eikonal approach (Rose, 2003), thereby providing a prescription identical to that obtained from the semiclassical algebra elaborated on in the following.

Following Abbe, the object plane, back–focal plane, and image plane are connected in an ideal (aberration-free) system wave optically by means of a Fourier and an inverse Fourier transformation, respectively. This result may be derived by propagating a wave given at some object distance z_{obj} through a thin lens with focal length f (Fig. 4.2)

$$\Psi\left(\mathbf{r}, z\right) = \frac{f}{i4\pi^2 k_0} \iint_{-\infty}^{\infty} d^2 k\, d^2 k'\, e^{i\mathbf{k}\mathbf{r}} \tilde{\Psi}\left(\mathbf{k}', z_{\mathrm{obj}}\right) \tag{4.14}$$

$$\times\, e^{-i\frac{z_{\mathrm{obj}}}{2k_0}\mathbf{k}'^2}\, e^{i\frac{f}{2k_0}(\mathbf{k}-\mathbf{k}')^2}\, e^{-i\frac{z}{2k_0}\mathbf{k}^2} \tag{4.15}$$

$$= \frac{f}{2\pi(f-z)} e^{-i\frac{k_0}{2(f-z)}\mathbf{r}^2}$$

$$\times \int_{-\infty}^{\infty} d^2 k'\, e^{\underbrace{\frac{i}{2k_0}\frac{z_{\mathrm{obj}}z - f(z_{\mathrm{obj}}+z)}{f-z}\mathbf{k}'^2}_{\delta z(z)}}\, \tilde{\Psi}\left(\mathbf{k}', z_{\mathrm{obj}}\right) e^{\overbrace{i\frac{f}{f-z}\mathbf{r}\mathbf{k}'}^{M^{-1}}},$$

where an effective free-space propagation length $\delta z\,(z)$ and magnification $M\,(z)$ with respect to a free-space propagation have been defined. It follows that the diffraction pattern $I(\mathbf{k}) = \left|\tilde{\Psi}(\mathbf{k})\right|^2$ forms in the back focal plane $(z=f)$ because (starting from the third line in (4.14))

$$
\begin{aligned}
\Psi\left(\mathbf{r}, f\right) &= \frac{f}{i4\pi^2 k_0} \int_{-\infty}^{\infty} \mathrm{d}^2 k'\, \tilde{\Psi}\left(\mathbf{k}', z_{\mathrm{obj}}\right) \\
&\quad \times \left(\int_{-\infty}^{\infty} \mathrm{d}^2 k\, e^{i\left(\mathbf{r} - \frac{f}{k_0}\mathbf{k}'\right)\mathbf{k}}\right) e^{i\frac{f-z_{\mathrm{obj}}}{2k_0}\mathbf{k}'^{\,2}} \\
&= \frac{f}{ik_0} \int_{-\infty}^{\infty} \mathrm{d}^2 k'\, \delta\left(\mathbf{r} - \frac{f}{k_0}\mathbf{k}'\right) e^{i\frac{f-z_{\mathrm{obj}}}{2k_0}\mathbf{k}'^{\,2}} \tilde{\Psi}\left(\mathbf{k}', z_{\mathrm{obj}}\right) \\
&= \frac{1}{i} e^{i\frac{k_0(f-z_{\mathrm{obj}})}{2f^2}\mathbf{r}^2} \tilde{\Psi}\left(\frac{k_0}{f}\mathbf{r}, z_{\mathrm{obj}}\right).
\end{aligned}
\tag{4.16}
$$

Accordingly, the propagation of the wave from the object plane to the back focal plane (4.16) corresponds to a unitary transformation.

Within Miller's semiclassical algebra (Miller, 1974) (and omitting the parabolic phase factor in (4.16) for the sake of simplicity) this representation change can be semiclassically approximated by a spatial integration of the wave function in position space weighted by a phase factor containing the generating function of the corresponding canonical transformation S and an amplitude factor determined by the determinant of S, i.e.,

$$
\tilde{\Psi}\left(\mathbf{r}, f\right) = \frac{1}{2\pi} \int_{-\infty}^{\infty} \det\left(\frac{\partial S\left(\mathbf{r}, \mathbf{r}'\right)}{\partial\mathbf{r}\partial\mathbf{r}'}\right)^{1/2} \Psi\left(\mathbf{r}, z_{\mathrm{obj}}\right) e^{-iS(\mathbf{r},\mathbf{r}')} \mathrm{d}^2 r \tag{4.17}
$$

with $\mathbf{r}' = k_0\mathbf{r}/f$. The type-1 (because the beams are parametrized by their positions in the object and back focal planes) generating function for the representation change reads $S\left(\mathbf{r}, \mathbf{r}'\right) = \mathbf{r}\mathbf{r}'$; hence the semiclassical approximation exactly reproduces the Fourier transformation noted previously in (4.16).

A magnified sharp image $I\left(M\mathbf{r}, z_{\mathrm{img}}\right) = I\left(\mathbf{r}, z_{\mathrm{obj}}\right)$ is obtained, if the effective defocus δz in (4.14) vanishes, which leads to the thin lens equation defining the image plane position z_{img}

$$
z_{\mathrm{obj}} z_{\mathrm{img}} - f(z_{\mathrm{obj}} + z_{\mathrm{img}}) = 0 \rightarrow \frac{1}{f} = \frac{1}{z_{\mathrm{obj}}} + \frac{1}{z_{\mathrm{img}}} \tag{4.18}
$$

and the image magnification

$$M = -\frac{z_{\text{img}}}{z_{\text{obj}}} = \frac{f - z_{\text{img}}}{f}. \tag{4.19}$$

Although the corresponding wave functions in the object and image plane are also connected by a unitary transformation (i.e., the identity transformation neglecting the magnification change and the parabolic phase factor), the semiclassical approximation of this transformation requires some additional care as the determinant of the corresponding type-1 generating function (point eikonal) diverges, where multiple rays emerging from the same point in the object plane intersect. This problem may be circumvented by considering the unitary transformation between the object wave function in position representation and the image wave function in momentum representation. In that case the generating function is of type 2 (mixed eikonal) and well-behaved, taking the ray's position coordinate in the object and momentum coordinate in the image plane as arguments. Again omitting the magnification change and the parabolic phase factor, the mixed eikonal reads $S(\mathbf{r}, \mathbf{k}) = \mathbf{rk}$ and, by comparison with (4.17), we note again that Miller's semiclassical algebra exactly reproduces the Fourier transformation.

Wave aberrations may now be introduced in a straightforward manner by adding the aberration eikonal (4.8) to the mixed eikonal, i.e., $S(\mathbf{r}, \mathbf{k}) = \mathbf{rk} + \chi(\mathbf{r}, \mathbf{k})$. If restricting ourselves for the moment to the isoplanatic case, the semiclassical wave propagation between object and image plane reads

$$\Psi_a(\mathbf{r}) = \frac{1}{4\pi^2} \iint_{-\infty}^{\infty} e^{-i\chi(\mathbf{k})} e^{-i\mathbf{kr}'} e^{i\mathbf{kr}} \Psi_s(\mathbf{r}') \, d^2r' d^2k \tag{4.20}$$

with the familiar *aberration function* $\chi(\mathbf{k})$ (Saxton, 1995) corresponding to the aberration eikonal depicted in Fig. 4.5.

In case of non–isoplanatic aberrations the mixed eikonal (4.13) also depends on the object plane coordinates. Consequently, the aberrated wave function has to be computed from the general expression

$$\Psi_a(\mathbf{r}) = \frac{1}{4\pi^2} \iint_{-\infty}^{\infty} \det\left(\frac{\partial S(\mathbf{r}', \mathbf{k})}{\partial \mathbf{r}' \partial \mathbf{k}}\right)^{1/2} \Psi_s(\mathbf{r}') \, e^{-iS(\mathbf{r}', \mathbf{k})} e^{i\mathbf{kr}} d^2r' d^2k \tag{4.21}$$

with the determinant taking into account the non-homogeneous distribution of classical trajectories in phase space. Points where the second

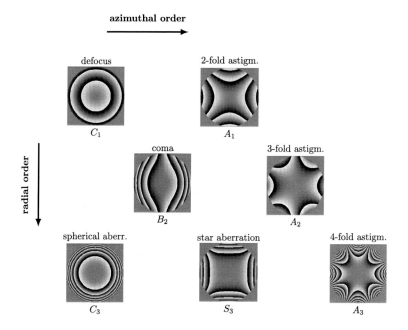

Figure 4.5 List of eikonals pertaining to isoplanatic aberrations.

derivative of the mixed eikonal tends to infinity physically correspond to caustics, where an infinite number of trajectories intersect. The eikonal approximation is difficult to apply in these cases.

Expression (4.21) has to be evaluated numerically for general non-isoplanatic aberrations. However, assuming small aberration coefficients, such as prevailing in modern aberration corrected TEMs, the Jacobi determinant can be approximated as is demonstrated in the following lines at the example of off-axial defocus

$$
\det\left(\frac{\partial S\left(\mathbf{r}',\mathbf{k}\right)}{\partial \mathbf{r}'\partial \mathbf{k}}\right)^{1/2} = \sqrt{\left| \begin{array}{cc} 1 + \frac{C_1^{(1)}}{k_0}k_x\cos\beta_C & \frac{C_1^{(1)}}{k_0}k_y\cos\beta_C \\ \frac{C_1^{(1)}}{k_0}k_x\sin\beta_C & 1 + \frac{C_1^{(1)}}{k_0}k_y\sin\beta_C \end{array} \right|} \tag{4.22}
$$

$$
\approx \sqrt{1 + \frac{C_1^{(1)}}{k_0}k_x\cos\beta_C + \frac{C_1^{(1)}}{k_0}k_y\sin\beta_C}
$$

$$
\approx 1 + \frac{\frac{C_1^{(1)}}{k_0}k_x\cos\beta_C + \frac{C_1^{(1)}}{k_0}k_y\sin\beta_C}{2}
$$

$$
= 1 + \frac{C_1^{(1)}}{2k_0}\left\langle \mathbf{k}, \beta_C \right\rangle.
$$

Similar computations may be carried out for the other non-isoplanatic aberrations.

A simplified eikonal approximation can be used to determine non-isoplanatic aberrations from Zemlin tableaus (Zemlin, 1979), which are routinely acquired for hardware aberration assessment and correction. Here, one assumes that the aberration function $\chi\left(\mathbf{r}', \mathbf{k}\right)$ only changes between subfields of the recorded images, i.e., $\chi\left(\mathbf{r}', \mathbf{k}\right) = \chi\left(\mathbf{k}\right)$ within each subfield. By evaluating Zemlin tableaus for these subfields one can assess their isoplanatic aberration coefficients and tune the corrector such to remove any differences between the patches (Müller et al., 2011). While this approach is suited to correct aberrations, it is problematic for determining non-isoplanatic aberrations quantitatively, because each subfield measurement remains disturbed by the non-isoplanaticity.

It has been mentioned previously that modern hardware aberration correctors permit not only the correction of the third order spherical aberration but also all other aberrations up to the third order including an increasing number of higher order aberrations. Moreover, they facilitate the adjustment of a defined aberrated state other than the fully corrected one. Therefore, hardware correctors can be used as additional optical degrees of freedom, opening pathways to an entire new class of electron optical experiments. For instance, combinations of 2, 3, and 4-fold astigmatisms can be used to create vortex beams (Clark et al., 2013); combinations of second order aberrations have been used to produce Airy beams (Guzzinati et al., 2015); and the symmetry breaking in a STEM probe introduced by a suitable 4-fold astigmatism proved advantageous in electron-energy-loss magnetic dichroism experiments on the atomic scale (Rusz, Idrobo, & Bhowmick, 2014).

In the course of this work we also capitalize on the minute control over the aberration coefficients, facilitating the imitation of certain differential equations in the TEM. The latter form the basis of novel holographic setups described in Chapter 5, where aspects of the beam electrons' quantum state are reconstructed from differential changes of 2-fold astigmatism and defocus. The link between differential changes of aberration coefficients and differential equations is most obvious in case of the isotropic defocus term. Here, a differential defocus change leads to a variation of the wave function in reciprocal space

$$\delta\Psi\left(\mathbf{k}\right) = \Psi\left(\mathbf{k}, \delta C_1\right) - \Psi\left(\mathbf{k}\right) = i\frac{\delta C_1}{2k_0}\left|\mathbf{k}\right|^2 \Psi\left(\mathbf{k}\right) \qquad (4.23)$$

corresponding the following differential in position space

$$\delta\Psi\left(\mathbf{r}\right) = \frac{\partial\Psi\left(\mathbf{r}, C_1\right)}{\partial C_1}\delta C_1 = -i\frac{\delta C_1}{2k}\triangle\Psi\left(\mathbf{r}\right) . \tag{4.24}$$

The latter is the well-known paraxial equation (2.21)

$$i\frac{\partial\Psi\left(\mathbf{r}, C_1\right)}{\partial C_1} = \frac{1}{2k_0}\triangle\Psi\left(\mathbf{r}\right) , \tag{4.25}$$

representing an inhomogeneous elliptic differential equation provided that the partial derivative of the wave function with respect to the defocus is given.

The same argument applied to the other aberrations yields a whole zoo of differential equations of different orders corresponding to the aberration order (Allen, Oxley, & Paganin, 2001). An instructive example is given by a differential change of the spherical aberration. Repeating the steps of the above derivation, the reciprocal space difference

$$\Psi\left(\mathbf{k}, C_3\right) - \Psi\left(\mathbf{k}\right) = -i\frac{C_3}{4k_0^3}\left|\mathbf{k}\right|^4 \tag{4.26}$$

translates to a position space derivative

$$i\frac{\partial\Psi\left(\mathbf{r}, C_3\right)}{\partial C_3} = \frac{1}{4k_0^3}\triangle^2\Psi\left(\mathbf{r}\right) , \tag{4.27}$$

where the right hand side contains the biharmonic Laplace operator.

4.3. WIGNER OPTICS

Based on the above principles of semiclassical wave optics the whole imaging process within a TEM can be described along the following lines. In a first step, a wave packet emanates from the source, its extension being determined by the emission process (e.g., cold field emission). In a second step it traverses the various optical elements, most notably lenses, apertures and multipole elements, serving the purpose to appropriately shape the beam in the object plane. After traversing the object, a second set of optical elements modulate the wave function, before finally reaching the detector. Partial coherence can be accounted for in this description by incoherently summing over a set of incoherent wave functions a posteriori. As a consequence, this approach is cumbersome, if the number of incoherent states is large.

Alternatively, the mixed partial coherent quantum state as emitted by the source can be transferred through the TEM as a whole. Such a quantum state description often permits to absorb at least a part of the above summation analytically, e.g., by exploiting symmetry, thereby abbreviating computations. Another advantage of such an approach is that the quantum state, once it is known in one plane of the microscope, can be easily computed for a different plane (Röder & Lubk, 2014). The cost of this simplification lies in the increased complexity of the quantum state, e.g., represented by a density matrix. In numerical computations, for instance, a 4D density matrix instead of a 2D wave function has to be kept in memory, which can easily exhaust the capabilities of modern computers and hence prevents a wide spread use of such numerical computations to date of this work. Therefore, quantum state transfer in the parlance of density matrices have been used predominantly in the past to analytically describe imaging after inelastic plasmon scattering (Verbeeck, Bertoni, & Schattschneider, 2008; Schattschneider & Verbeeck, 2008), core-loss scattering (Schattschneider, Nelhiebel, & Jouffrey, 1999), or frozen-lattice scattering (Rother, Gemming, & Lichte, 2009).

In the following the latter approach is taken to complete the imaging theory of quantum states. In contrast to the pertinent literature the quantum state is represented in phase space, i.e., as a Wigner function, and the optical transfer theory is reformulated accordingly. The reward is a closer correspondence to classical phase space permitting an alternative semiclassical recipe for quantizing classical results, yielding, e.g., a semiclassical quantum state of the source and a new perspective on isoplanatic and non-isoplanatic aberrations.

4.3.1 Source

Electron sources, employed in a TEM, are sharp tips immersed in an electric field, which emit a partial coherent electron beam, i.e., a mixed quantum state. After traversing the extraction field and the first beam shaping elements of the gun, an image of the source is formed in the focal plane of the condenser lens, which is subsequently considered as the effective source determining the imaging process. Consequently, the details of the associated quantum state's phase space distribution depend on the emission process as well as the probe forming elements. Both theoretical and experimental studies on electron sources employed in the TEM (Hawkes & Kasper, 1988) showed that it is possible to obtain a fairly accurate functional form by modeling the phase space distribution of the effective source as a so-called *Schell-model source* (Mandel & Wolf, 1995). The latter is characterized by a degree of coherence (2.71) $\mu_S(\mathbf{r}, \mathbf{r}') = \mu_S(\mathbf{r} - \mathbf{r}')$ in the source plane, which depends on the spatial difference between two coordinates only, leading to a density matrix of the following form

$$\rho_S(\mathbf{r}, \mathbf{r}') = \sqrt{I_S(\mathbf{r})\, I_S(\mathbf{r}')}\, \mu_S(\mathbf{r} - \mathbf{r}') . \tag{4.28}$$

Moreover, the width of typical electron sources is large in comparison to both the correlation length Δ (the effective width of $|\mu_S|$), allowing to approximate

$$I_S(\mathbf{r}) \approx I_S(\mathbf{r}') \approx I_S\left(\frac{\mathbf{r} + \mathbf{r}'}{2}\right) \tag{4.29}$$

within the correlation length. Such sources are referred to as quasi-homogeneous and can be used as an approximation for field-assisted thermal emitters (Schottky), the old fashioned thermionic guns, as well as LaB$_6$ sources and cold field emission guns, as long as their extension remains large with respect to their correlation length. A widely employed model for the source intensity is a Gaussian distribution of radial width σ_s, i.e.,

$$I_S(\mathbf{r}) = \frac{1}{\pi \sigma_s^2} e^{-\frac{r^2}{\sigma_s^2}} , \tag{4.30}$$

because it greatly simplifies analytic calculations.

Quasi-homogeneous sources are subject to the van Cittert–Zernike theorem (van Cittert, 1934; Zernike, 1938) as can be seen by noting the density matrix in the Fourier plane of the source corresponding to the object plane in the bright field TEM mode (cf. Fig. 4.1)

$$\tilde{\rho}_S(\mathbf{k}, \mathbf{k}') = \tilde{I}_S(\mathbf{k} - \mathbf{k}')\, \tilde{\mu}_S\left(-\frac{\mathbf{k}' + \mathbf{k}}{2}\right) . \tag{4.31}$$

Consequently, the coherence, i.e., the term depending on the difference $\mathbf{k} - \mathbf{k}'$, is the Fourier transform of the source's radiating intensity. The phase space distribution of the effective source is obtained by inserting (4.28) into the definition of the Wigner function (2.88)

$$W_S(\mathbf{r}, \mathbf{k}) = \frac{1}{4\pi^2} \int_{-\infty}^{\infty} I_S(\mathbf{r})\, \mu_S(\mathbf{r}')\, e^{-i\mathbf{k}\mathbf{r}'}\, d^2r' \tag{4.32}$$

$$= \frac{1}{2\pi} I_S(\mathbf{r})\, \tilde{\mu}_S(\mathbf{k}) .$$

Consequently, the phase space distribution of a quasi-homogeneous source separates into a product of a position and momentum distribution, with the latter being determined by the coherence in the source plane. Moreover, the shape of the intensity in the source plane is completely determined by $I_S(\mathbf{r})$, whereas that in the Fourier plane of the source, i.e., the object plane in the bright-field mode, solely depends on the Fourier transform of the degree of coherence in the source plane.

In order to discuss invariants of this beam, the Wigner function (4.32) is propagated to a plane at some distance $z > 0$ below the electron source. This is formally equivalent to a propagation of the density matrix as discussed previously in literature (Hawkes, 1978; Müller, Rose, & Schorsch, 1998). In the phase space representation the free space propagation is given by a shear of the Wigner function (2.111)

$$W_z(\mathbf{r}, \mathbf{k}) = \underbrace{W_S\left(\mathbf{r} - \frac{z}{k_0}\mathbf{k}, \mathbf{k}\right)}_{\text{sheared } W_S} = \frac{1}{2\pi} I_S\left(\mathbf{r} - \frac{z}{k_0}\mathbf{k}\right) \tilde{\mu}_S(\mathbf{k}) . \tag{4.33}$$

Consequently, a small coherence width Δ induces a strong illumination spread with increasing propagation distance, which can lead to conceptual and computational complications, if employing the model of a completely incoherent source (Hawkes, 1978; Rose, 1984; Pozzi, 1987; Müller et al., 1998).

The propagated Wigner function (4.33) exhibits several invariants, which may be used as figures of merit facilitating a straightforward comparison between the various types of emitters. First of all, the quantum mechanical purity ζ (2.104), also referred to as "coherence parameter" (Pozzi, 1987), is conserved upon propagation

$$\zeta_z = \iint_{-\infty}^{\infty} d^2 r d^2 k W_S^2\left(\mathbf{r} - \frac{z}{k_0}\mathbf{k}, \mathbf{k}\right) = \zeta_S . \tag{4.34}$$

Geometrically, this conservation law follows from the shear invariant phase space volume occupied by the quantum state. From a quantum perspective, this conservation reflects the unitary (reversible) evolution of the beam electrons during free space propagation. In the classical limit, this corresponds to the invariance of the geometrical "etendu" (Pozzi, 1987), which is inversely proportional to the squared purity. Note, however, that the classical "etendu" can become arbitrarily small, in contradiction to quantum mechanics, where the purity is bounded ($\zeta \leq 1$, cf. (2.104)).

In contrast to the free space propagation, other optical elements (e.g., apertures) limit the spatial or angular range of the electron beam, thereby changing the purity (i.e., the occupied phase space volume). It turns out that a second measure, the axial reduced brightness B_0, classically defined as the product between the phase space density and the overall current (Fink & Schumacher, 1974)

$$B_0 = I_0 W_S(0, 0) \, , \tag{4.35}$$

can be more robust in this case, because a large class of optical transformations does not change the origin of phase space. In the classical limit, the latter is strictly positive and gives the normalized number of electrons emitted from the area element of the effective source on the optical axis into a solid angle interval around the optical axis. In order to avoid unphysical negative values, the quantum mechanical analogue

$$B_0 = I_0 \iint_\Omega W_S(\mathbf{r}, \mathbf{k}) \, \mathrm{d}^2 r \mathrm{d}^2 k \tag{4.36}$$

deviates from the classical definition in that an integral of the Wigner function over the fundamental phase space cell Ω centered at the origin has to be incorporated. Typical electron sources currently employed in the TEM, however, attain phase space densities well below 0.001 (Lubk & Röder, 2015), hence may be described by a classical phase space distribution to a good approximation.

So far we have considered the phase space distribution pertaining to a mono-energetic source. In order to describe realistic electron sources emitting electrons within a finite energy interval of width σ_E, the Wigner function (or any other representation of the quantum state) has to carry an energy dependency. In the following, this dependency is modeled by a Gaussian distribution around the mean beam energy E according to

$$W_S(\mathbf{r}, \mathbf{k}, \delta E) = \underbrace{\frac{1}{\sqrt{\pi}\sigma_E} e^{-\frac{\delta E^2}{\sigma_E^2}}}_{f_E(\delta E)} W_S(\mathbf{r}, \mathbf{k}) \, . \tag{4.37}$$

Thereby, we implicitly assume that the energy distribution is completely incoherent and constant for each point of the emitter, which is a reasonable approximation for typical electron sources possessing a small coherent energy width only.

4.3.2 Aperture

Typically, apertures restrict the spatial or angular extension of the beam, corresponding to their location within the TEM. For instance, the objective aperture in the bright-field TEM mode (or the C2 aperture in the STEM mode) block the strongly aberrated beams with large transverse momentum, which can reduce the influence of aberrations. Classically, the action of such an aperture with radius R consists of multiplying the phase space distribution with a top-hat function

$$H_R(\mathbf{r}) = \begin{cases} 1 & |\mathbf{r}| \le R, \\ 0 & |\mathbf{r}| > R \end{cases} \tag{4.38}$$

in the plane of the aperture. Quantum mechanically that introduces an additional modulation of the phase distribution along the momentum coordinate originating from the scattering at the aperture rim (Feynman, Leighton, & Sands, 1965).

The effect of the aperture on the Wigner function may be derived in the following way. For an aperture in position space the ensuing density matrix is obtained by multiplying both arguments with a top hat function

$$\rho_a(\mathbf{r}, \mathbf{r}') = H_R(\mathbf{r}) \rho_s(\mathbf{r}, \mathbf{r}') H_R(\mathbf{r}') . \tag{4.39}$$

This expression is now inserted into the definition of the Wigner function (2.88), yielding

$$\begin{aligned} W_a(\mathbf{r}, \mathbf{k}) &= \frac{1}{4\pi^2} \int_{-\infty}^{\infty} d^2 r' e^{-i\mathbf{k}\mathbf{r}'} \rho_s\left(\mathbf{r} + \frac{1}{2}\mathbf{r}', \mathbf{r} - \frac{1}{2}\mathbf{r}'\right) \\ &\quad \times H_R\left(\mathbf{r} + \frac{1}{2}\mathbf{r}'\right) H_R\left(\mathbf{r} - \frac{1}{2}\mathbf{r}'\right) \\ &= W_s(\mathbf{r}, \mathbf{k}) \\ &\quad *_k \mathcal{F}_{r'}\left\{ H_R\left(\mathbf{r} + \frac{1}{2}\mathbf{r}'\right) H_R\left(\mathbf{r} - \frac{1}{2}\mathbf{r}'\right)\right\}(\mathbf{k}) \\ &= \underbrace{\left(\frac{2J_1(kR)}{k}\right)^2}_{\text{Airy disk}} *_k \underbrace{H_R(\mathbf{r})}_{\text{Aperture}} W_s(\mathbf{r}, \mathbf{k}) . \end{aligned} \tag{4.40}$$

Accordingly, the Wigner function is cut in position space with the top hat and convoluted in momentum space with the Airy disk, representing the well-known effect of a circular aperture on the intensity in Fourier space.

The corresponding expression for an aperture located in the back focal plane is obtained by interchanging the position and momentum coordinates.

4.3.3 Lens

The semiclassical action of lenses in phase space including aberrations can be derived from the eikonal approximation, noted previously, in the following way. In a first step the semiclassical expressions for the aberrated wave functions are inserted into the definition of the Wigner function (2.88), thereby mapping the stigmatic phase space distribution W_s into an aberrated one W_a. In a second step the ramifications of partial coherence can be incorporated by applying that transformation law to a mixed state. To illustrate the approach we first consider the first order isoplanatic aberrations (i.e., Gaussian optics) before proceeding to higher isoplanatic orders, non–isoplanatic aberrations and finally incoherent aberrations.

Isoplanatic First Order

A defocus corresponds to a free space propagation of the wave function. As already noted in (2.111), the induced transformation of the Wigner function is a shear, which is rederived in the following by inserting the corresponding eikonal transformation of the wave function (i.e., by inserting (4.11) into (4.20))

$$W_a(\mathbf{r}, \mathbf{k}) = \frac{1}{4\pi^2} \int_{-\infty}^{\infty} d^2k' \rho_s \left(\mathbf{k} + \frac{1}{2}\mathbf{k}', \mathbf{k} - \frac{1}{2}\mathbf{k}' \right) \tag{4.41}$$
$$\times e^{i\frac{C_1}{2k_0}\left(\mathbf{k} - \frac{1}{2}\mathbf{k}'\right)^2} e^{-i\frac{C_1}{2k_0}\left(\mathbf{k} + \frac{1}{2}\mathbf{k}'\right)^2} e^{i\mathbf{k}'\mathbf{r}}$$
$$= \frac{1}{4\pi^2} \int_{-\infty}^{\infty} d^2k' \rho_s \left(\mathbf{k} + \frac{1}{2}\mathbf{k}', \mathbf{k} - \frac{1}{2}\mathbf{k}' \right) e^{i\mathbf{k}'\left(\mathbf{r} - \frac{C_1}{k_0}\mathbf{k}\right)}$$
$$= W_s \left(\mathbf{r} - \frac{C_1}{k_0}\mathbf{k}, \mathbf{k} \right).$$

A similar calculation yields the corresponding Wigner function after 2-fold astigmatism ($\alpha_{A1} = 0$)

$$W_a(\mathbf{r}, \mathbf{k}) = W_s \left(\mathbf{r} - \frac{A_1}{k_0} \left(k_x \mathbf{e}_x - k_y \mathbf{e}_y \right), \mathbf{k} \right). \tag{4.42}$$

These expressions show that first order isoplanatic aberrations lead to classically distorted Wigner functions in the eikonal approximation. In other words, the quantum nature of the system is solely encoded in the underlying Wigner quasi-probability distribution. This behavior is analogous to the time evolution of quantum states in the harmonic oscillator, which is also governed by a classical Liouville equation (2.110) as spatial derivatives of the potential higher than the second order vanish. Thus, the analogy between the harmonic oscillator and Gaussian optics derives from their similar classical equations of motion; both are simple harmonic second order differential equations (2.24). For higher order aberrations, the exact analogy between classical and quantum mechanical phase space transformations is lost, as will be shown below.

Isoplanatic Second and Third Order

The second and third order isoplanatic position aberrations in phase space are obtained along the same lines as the first order expressions, i.e., by first aberrating the density matrix before computing the Wigner function. Below, the Wigner function from the aberrated density matrix, inserted in the first line, is noted in the second line and compared to the classical phase space distortion in the last line. The general second order expression comprising coma and threefold astigmatism reads (Lubk et al., 2015)

$$W_a(\mathbf{r}, \mathbf{k}) = \frac{1}{4\pi^2} \int_{-\infty}^{\infty} d^2 k' \rho_s \left(\mathbf{k} + \frac{1}{2}\mathbf{k}', \mathbf{k} - \frac{1}{2}\mathbf{k}'\right) e^{i\mathbf{k}'(\mathbf{r} - \boldsymbol{\xi}(\mathbf{k}))} e^{-\frac{i}{4}\chi(\mathbf{k}')} \quad (4.43)$$

$$= W_s(\mathbf{r} - \boldsymbol{\xi}(\mathbf{k}), \mathbf{k}) *_r \frac{1}{4\pi^2} \int_{-\infty}^{\infty} d^2 k' e^{-\frac{i}{4}\chi(\mathbf{k}')} e^{i\mathbf{k}'\mathbf{r}}$$

$$\overset{\chi<\pi}{\approx} W_s(\mathbf{r} - \boldsymbol{\xi}(\mathbf{k}), \mathbf{k}).$$

A similar expression is obtained in the third order comprising spherical aberration, 2-fold star aberration and 4-fold astigmatism

$$W_a(\mathbf{r}, \mathbf{k}) = \frac{1}{4\pi^2} \int_{-\infty}^{\infty} d^2 k' \rho_s \left(\mathbf{k} + \frac{1}{2}\mathbf{k}', \mathbf{k} - \frac{1}{2}\mathbf{k}'\right) e^{i\mathbf{k}'(\mathbf{r} - \boldsymbol{\xi}(\mathbf{k}))} e^{-\frac{i}{4}\boldsymbol{\xi}(\mathbf{k}')\mathbf{k}} \quad (4.44)$$

$$= W_s(\mathbf{r} - \boldsymbol{\xi}(\mathbf{k}), \mathbf{k}) *_r \frac{1}{4\pi^2} \int_{-\infty}^{\infty} d^2 k' e^{-\frac{i}{4}\boldsymbol{\xi}(\mathbf{k}')\mathbf{k}} e^{i\mathbf{k}'\mathbf{r}}$$

$$\overset{\boldsymbol{\xi}k<\pi}{\approx} W_s(\mathbf{r} - \boldsymbol{\xi}(\mathbf{k}), \mathbf{k}).$$

Upon inspecting the above expressions, one observes that the one-to-one correspondence between the eikonal semiclassics and a classical phase space distortions, present in Gaussian imaging, is lost (Rivera, Lozada-Cassou, Rodríguez, & Castaño, 2003). Remarkably, the semiclassical approximation still features a phase space distortion exactly resembling the classical one. However, there is an additional convolution with Airy-like functions in position space (denoted by $*_r$).[3] This is exemplary shown for the case of axial coma

$$W_a(\mathbf{r}, \mathbf{k}) = W_s\left(\mathbf{r} - \frac{B_2}{k_0^2}\begin{pmatrix} 3k_x^2 + k_y^2 \\ 2k_x k_y \end{pmatrix}, \mathbf{k}\right) \tag{4.45}$$

$$*_r \frac{1}{4\pi^2}\int_{-\infty}^{\infty} d^2 k' e^{-i\frac{B_2}{4k_0^3}\left(k_x'^2 + k_y'^2\right)k_x'} e^{i\mathbf{k'r}},$$

where the cubic dependency in $\mathbf{k'}$ is clearly visible.

Airy functions play a prominent role in the semiclassical WKB approximation, where they facilitate a uniform approximation of wave functions near a potential wall (Vallée & Soares, 2010). Thus, their appearance within a semiclassical theory of aberrations is not surprising. Indeed, the classical limit at short wavelengths ($k_0 \to \infty$) in the above expressions may be formally obtained by inserting the following representation of the δ-function $\delta(x) = \lim_{\epsilon \to 0} \epsilon^{-1} \text{Ai}(x/\epsilon)$. It is important to note that the classical limit is also approached, if the coherence length in reciprocal space, defined as the characteristic width of the momentum space density matrix along the anti-diagonal, is small (see first lines in (4.43) and (4.44)). This reflects an important aspect of quantum-to-classical transitions as noted in the theory of decoherence (Zurek, 2003; Schlosshauer, 2004). Classical observables emerge upon a decoherence process in originally delocalized quantum particles. In our case it is the loss of coherence in momentum space, which leads to a localization of the momentum observable and the emergence of classical behavior (i.e., classical aberrations). By deliberately tailoring the coherence of the illumination one may therefore suppress the quantum aberrations, which can lead to favorable imaging conditions (Rosenauer, Krause, Müller, Schowalter, & Mehrtens, 2014).

An interesting aspect of the phase space expression pertaining to second order aberrations is that the Airy transformation (i.e., the convolution with Airy functions) does not depend on the phase space coordinate $\mathbf{k}$. It follows that the Airy transformation commutes with the imaging process,

---

[3] The Airy function of the first kind is defined as $\text{Ai}(x) = \frac{1}{2\pi}\int_{-\infty}^{\infty} \exp\left(i\left(\frac{k^3}{3} + kx\right)\right) dk$. Convolutions with Airy functions are also referred to as Airy transformations (Vallée & Soares, 2010).

i.e., the projection of the Wigner function along the momentum coordinate $\mathbf{k}$. This allows to correct for the quantum effects (in contrast to the classical phase space distortions) of second order aberrations by means of an inverse Airy transformation of the recorded image intensity. Such a partial aberration correction does not work for the more important third order spherical aberration because of the momentum coordinate dependency of the corresponding Airy transformation.

Non-Isoplanatic Aberrations

A simple semiclassical approximation to non-isoplanatic aberrations may be obtained by classically distorting the Wigner function with a position aberration vector $\xi\,(\mathbf{r}, \mathbf{k})$ *and* a momentum aberration vector $\kappa\,(\mathbf{r}, \mathbf{k})$ depending on both the position and momentum according to (4.2)

$$W_{\mathrm{a}}(\mathbf{r}, \mathbf{k}) = W_{\mathrm{s}}\left(\mathbf{r} - \xi\,(\mathbf{r}, \mathbf{k})\,, \mathbf{k} + \kappa\,(\mathbf{r}, \mathbf{k})\right). \qquad (4.46)$$

Such an approach, however, would neglect the additional quantum corrections, i.e., the above Airy transformations, which prohibit an accurate semiclassical aberration theory by merely distorting quantum mechanical phase space. However, in case of first order isoplanatic aberrations a perfect agreement between semiclassical and classical aberrations has been established above, and one can hope that this result remains valid for non-isoplanatic aberrations being linear in both the position and momentum coordinate (i.e., the first two terms in (4.5) and (4.13)). The following lines show that it is indeed possible to derive this correspondence, if some additional approximations are applied.

As an example we discuss the position dependent defocus. First, the phase space weighting factor (Jacobian) in (4.21) has to be neglected, which corresponds to assuming that the non-isoplanatic aberrations are small. In this case the Wigner function reads

$$W_{\mathrm{a}}(\mathbf{r}, \mathbf{k}) = \frac{1}{16\pi^4} \iiint_{-\infty}^{\infty} \mathrm{d}^2 k' \mathrm{d}^2 r' \mathrm{d}^2 r'' e^{-i\frac{C_1^{(1)}(\mathbf{r}' - \mathbf{r}'', \beta_C)}{2k_0}\mathbf{k}^2} e^{-i\mathbf{k}(\mathbf{r}' - \mathbf{r}'')} \rho_{\mathrm{s}}\left(\mathbf{r}', \mathbf{r}''\right)$$

$$(4.47)$$

$$\times\, e^{i\mathbf{k}'\left(\mathbf{r} + \frac{1}{2}(\mathbf{r}' + \mathbf{r}'') + \frac{C_1^{(1)}(\mathbf{r}' + \mathbf{r}'', \beta_C)}{2k_0}\mathbf{k} + \frac{C_1^{(1)}(\mathbf{r}'' - \mathbf{r}', \beta_C)}{8k_0}\mathbf{k}'\right)}$$

$$= \frac{1}{4\pi^2} \iint_{-\infty}^{\infty} \mathrm{d}^2 k' \mathrm{d}^2 r' W_{\mathrm{s}}\left(\mathbf{r}', \mathbf{k} + \frac{C_1^{(1)}\beta_C}{2k_0}\left(\mathbf{k}^2 + \frac{1}{4}\mathbf{k}'^2\right)\right)$$

$$\times \, e^{i\mathbf{k}'\left(\mathbf{r}+\mathbf{r}'+\frac{C_1^{(1)}\langle\mathbf{r}',\boldsymbol{\beta}_C\rangle}{k_0}\mathbf{k}\right)}$$

$$\overset{C_1^{(1)}\ll 1}{\approx} \int_{-\infty}^{\infty} d^2r' \, W_s\left(\mathbf{r}', \mathbf{k}+\frac{C_1^{(1)}\boldsymbol{\beta}_C}{2k_0}\mathbf{k}^2\right)\delta\left(\mathbf{r}+\mathbf{r}'+\frac{C_1^{(1)}\langle\mathbf{r}',\boldsymbol{\beta}_C\rangle}{k_0}\mathbf{k}\right)$$

$$\overset{C_1^{(1)}\ll 1}{\approx} W_s\left(\mathbf{r}-\frac{C_1^{(1)}\langle\mathbf{r},\boldsymbol{\beta}_C\rangle}{k_0}\mathbf{k}, \mathbf{k}+\frac{C_1^{(1)}\boldsymbol{\beta}_C}{2k_0}\mathbf{k}^2\right)$$

$$= W_s\left(\mathbf{r}-\xi\left(\mathbf{r},\mathbf{k}\right), \mathbf{k}+\kappa\left(\mathbf{r},\mathbf{k}\right)\right).$$

In order to arrive at the simple classical phase space distortion, it had to be assumed repeatedly that the non-isoplanatic aberrations are small. Taking into account that modern hardware correctors permit a reduction of non-isoplanatic aberrations and typically fields of view employed in HRTEM are small, the above simplified semiclassical phase space prescription has a certain practical scope. A more detailed study of non-isoplanatic semiclassical aberrations in phase space including the deviations from the classical distortions is missing to date of this work.

Incoherent Aberrations

The above semiclassical theory of aberrations permits to incorporate so-called incoherent aberrations in the phase space formalism. One distinguishes between chromatic (longitudinal) and spatial (transversal) incoherence. The former is due to an energy dependence of the Wigner function (4.37) and the latter due to the extended size of the effective source (4.32). Their impact on the measurement is given by a summation over the incoherent ensemble of mono-energetic pure waves contained in the beam. Making the usual assumption that the field of view considered in HRTEM is small compared to the total extension of the beam (Fig. 4.1), one can neglect the spatial variation of the illumination in the object plane given by the Fourier transform of the degree of correlation in the source plane (4.32), reducing the incoherent summation to the energy distribution and the spatial extension of the source (corresponding to the angular spread of the illumination in the object plane in the bright-field TEM mode). Since the energy fluctuations as well as the angular spread in the object plane are typically too small to impart the electron scattering at the specimen itself, one can describe the impact on the bright-field TEM imaging process by summing Wigner functions resulting from polychromatic point emitters, which have been modulated by the object in the same way.

This approach is equivalent to the standard treatment of incoherent aberrations within the density matrix representation. After modifying the latter with the coherent aberrations and subsequently averaging over energy and source size one ends up with the notion of a *generalized* Transmission Cross Coefficient $\mathcal{T}$ as the principal transfer function of the density matrix (Röder & Lubk, 2014)

$$\rho_a\left(\mathbf{x}, \mathbf{x}'\right) = \frac{1}{4\pi^2} \iiint_{-\infty}^{\infty} d^2k\,d^2k'\,dE \rho_s\left(\mathbf{k}, \mathbf{k}', E\right) \mathcal{T}\left(\mathbf{k}, \mathbf{k}', \mathbf{x} - \mathbf{x}'\right) e^{i(\mathbf{k}\mathbf{x} - \mathbf{k}'\mathbf{x}')}$$

(4.48)

with

$$\mathcal{T}\left(\mathbf{k}, \mathbf{k}', \mathbf{d}\right) = \frac{1}{2\pi} \iint_{-\infty}^{\infty} d^2q\,dE\, e^{-i\chi(\mathbf{k}+\mathbf{q}, E)} e^{i\chi(\mathbf{k}'+\mathbf{q}, E)} I_S\left(\mathbf{q}\right) f_E\left(E\right) e^{i\mathbf{q}\mathbf{d}}.$$

(4.49)

In contrast to the conventional

$$\mathcal{T}_c\left(\mathbf{k}, \mathbf{k}'\right) = \mathcal{T}\left(\mathbf{k}, \mathbf{k}', \mathbf{d} = 0\right)$$

(4.50)

solely describing the intensity in the image plane (Frank, 1973; Rose, 1976; Wade & Frank, 1977; Wade & Jenkins, 1979; Ishizuka, 1980; Born & Wolf, 1999; Haider et al., 2010), this generalized Transmission Cross Coefficient incorporates the effect of spatial and temporal coherence on the diagonal ($\mathbf{d} = 0$) *and* off-diagonal ($\mathbf{d} \neq 0$) elements of the beam electron's density matrix in the image plane. In this general setting, the computations do not simplify by changing from the density matrix to the phase space representation. However, the spatial aberration-corrected imaging, most important for current applications in TEM, allows a simple analytic description of incoherent aberrations in phase space, which is shortly illustrated below.

In hardware-corrected TEM, defocus remains the main source of aberrations. In order to compute the impact of a finite source size on the Wigner function in the presence of defocus only, it suffices to incoherently sum up the underlying stigmatic Wigner function over the transverse momentum distribution of the source (the source size (4.32)). The resulting mixed quantum state reads

$$W_s(\mathbf{r}, \mathbf{k}) = \int_{-\infty}^{\infty} W_s^{(\mathrm{coh})}\left(\mathbf{r}, \mathbf{k} - \mathbf{k}'\right) I_S\left(\mathbf{k}'\right) d^2k',$$

(4.51)

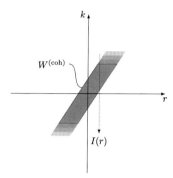

Figure 4.6 Impact of partial spatial coherence in combination with defocus on the average mixed quantum state. Spatial incoherence in combination with a defocus leads to a convolution in phase space, which may be deconvolved from the recorded image therefore.

which may be defocused in a second step

$$
W_a(\mathbf{r}, \mathbf{k}) = W_s\left(\mathbf{r} - \frac{C_1}{k_0}\mathbf{k}, \mathbf{k}\right) \tag{4.52}
$$

$$
= \int_{-\infty}^{\infty} W_s^{(\mathrm{coh})}\left(\mathbf{r} - \frac{C_1}{k_0}\mathbf{k}, \mathbf{k} - \mathbf{k}'\right) I_S\left(\mathbf{k}'\right) \mathrm{d}^2 k'.
$$

This convolution (Fig. 4.6) commutes with the projection of the Wigner function along the momentum coordinate (i.e., the measurement) according to

$$
I_a(\mathbf{r}) = \int_{-\infty}^{\infty} W_a(\mathbf{r}, \mathbf{k}) \mathrm{d}^2 k \tag{4.53}
$$

$$
= \iint_{-\infty}^{\infty} W_s^{(\mathrm{coh})}\left(\mathbf{r} - \frac{C_1}{k_0}\mathbf{k}, \mathbf{k} - \mathbf{k}'\right) I_S\left(\mathbf{k}'\right) \mathrm{d}^2 k \mathrm{d}^2 k'
$$

$$
= \int_{-\infty}^{\infty} I_a^{(\mathrm{coh})}\left(\mathbf{r} - \frac{C_1}{k_0}\mathbf{k}'\right) I_S\left(\mathbf{k}'\right) \mathrm{d}^2 k';
$$

and hence can be inverted (deconvolved) to a certain extent directly in the recorded image within the limits imposed by noise (Koch, 2008). From a phase space perspective, such a deconvolution corresponds to a purification of the quantum state, as the partial trace over a subspace of the total configuration space (i.e., the incoherent emitter surface) is reverted.

Carrying out the incoherent sum in the case of the chromatic aberration, one obtains the following aberrated Wigner function

$$W_{a}(\mathbf{r}, \mathbf{k}) = \int_{-\infty}^{\infty} f_{E}(\delta E) \, W_{s}\left(\mathbf{r} - C_{C}\frac{\delta E}{eU_{A}}\mathbf{k}, \mathbf{k}\right)d\delta E. \qquad (4.54)$$

This time, a convolution-like integral in position space parametrized by a momentum dependency is obtained. Therefore, the influence of chromatic aberration does not commute with the measurement process and can't be deconvoluted from a single recorded image. However, the influence of chromatic aberration may be corrected to some extent by an a-posteriori deconvolution of the chromatic aberration provided that the whole Wigner function of the quantum state is known. The next chapter elaborates on a class of quantum state reconstruction methods based on interferometric or tomographic techniques.

REFERENCES

Alem, N., Yazyev, O. V., Kisielowski, C., Denes, P., Dahmen, U., Hartel, P., ... Zettl, A. (2011). Probing the out-of-plane distortion of single point defects in atomically thin hexagonal boron nitride at the picometer scale. *Physical Review Letters, 106*(12), 126102.

Allen, L. J., Oxley, M. P., & Paganin, D. (2001). Computational aberration correction for an arbitrary linear imaging system. *Physical Review Letters, 87*(12), 123902.

Bar-Sadan, M., Barthel, J., Shtrikman, H., & Houben, L. (2012). Direct imaging of single Au atoms within GaAs nanowires. *Nano Letters, 12*(5), 2352–2356.

Beck, V. D. (1979). Hexapole-spherical-aberration corrector. *Optik, 53*, 241.

Born, M., & Wolf, E. (1999). *Principles of optics: Electromagnetic theory of propagation, interference and diffraction of light.* Cambridge: Cambridge University Press.

Busch, H. (1926). Calculation of the channel of the cathode rays in axial symmetric electromagnetic fields. *Annalen der Physik, 81*(25), 974–993.

Clark, L., Béché, A., Guzzinati, G., Lubk, A., Mazilu, M., Van Boxem, R., & Verbeeck, J. (2013). Exploiting lens aberrations to create electron-vortex beams. *Physical Review Letters, 111*(6), 064801.

Crewe, A. V., & Kopf, D. (1980). A sextupole system for the correction of spherical aberration. *Optik, 55*, 1–10.

De Graef, M. (2003). *Introduction to conventional transmission electron microscopy.* Cambridge: Cambridge University Press.

Dwyer, C., Kirkland, A. I., Hartel, P., Müller, H., & Haider, M. (2007). Electron nanodiffraction using sharply focused parallel probes. *Applied Physics Letters, 90*(15), 151104.

Egerton, R. F. (1996). *Electron energy-loss spectroscopy in the electron microscope.* Plenum Press.

Erni, R. (2010). *Aberration-corrected imaging in transmission electron microscopy.* London: Imperial College Press.

Feynman, R. P., Leighton, R. B., & Sands, M. (1965). *The Feynman lectures on physics.* Reading, MA: Addison-Wesley.

Fink, J. H., & Schumacher, B. W. (1974). The anatomy of a thermionic electron beam. *Optik, 39*, 543–557.

Frank, J. (1973). The envelope of electron microscopic transfer functions for partially coherent illumination. *Optik, 38*, 519–536.

Gabor, D. (1948). A new microscopic principle. *Nature, 161*, 777.

Gao, P., Nelson, C. T., Jokisaari, J. R., Baek, S.-H., Bark, C. W., Zhang, Y., … Pan, X. (2011). Revealing the role of defects in ferroelectric switching with atomic resolution. *Nature Communications, 2*, 591.

Girit, C. O., Meyer, J. C., Erni, R., Rossell, M. D., Kisielowski, C., Yang, L., … Zettl, A. (2009). Graphene at the edge: Stability and dynamics. *Science, 323*(5922), 1705–1708.

Guzzinati, G., Clark, L., Béché, A., Juchtmans, R., Van Boxem, R., Mazilu, M., & Verbeeck, J. (2015). Prospects for versatile phase manipulation in the TEM: Beyond aberration correction. *Ultramicroscopy, 151*, 85–93.

Haider, M., Hartel, P., Müller, H., Uhlemann, S., & Zach, J. (2010). Information transfer in a TEM corrected for spherical and chromatic aberration. *Microscopy and Microanalysis, 16*(04), 393–408.

Haider, M., Müller, H., & Uhlemann, S. (2008). Present and future hexapole aberration correctors for high-resolution electron microscopy. In P. W. Hawkes (Ed.), *Advances in imaging and electron physics: Vol. 153* (pp. 43–119).

Haider, M., Uhlemann, S., Schwan, E., Rose, H., Kabius, B., & Urban, K. (1998). Electron microscopy image enhanced. *Nature, 392*, 768.

Hawkes, P. W. (1965). The geometrical aberrations of general electron optical systems. *Philosophical Transactions of the Royal Society A, 257*, 479–552.

Hawkes, P. W. (1978). Coherence in electron optics. *Advances in Optical and Electron Microscopy, 7*, 101–184.

Hawkes, P. W. (Ed.). (2013). *Magnetic electron lenses. Topics in current physics*. Berlin, Heidelberg: Springer.

Hawkes, P. W., & Kasper, E. (1988). *Principles of electron optics: Vol. 2. Applied geometrical optics*. Elsevier Science.

Hawkes, P. W., & Kasper, E. (1996). *Principles of electron optics: Vol. 1. Basic geometrical optics*. Academic Press.

Head-Gordon, M., & Miller, W. H. (2012). *Quantum molecular dynamics: A festschrift in honour of William H. Miller. Molecular physics*. Taylor & Francis.

Heller, E. J. (1981). Frozen Gaussians: A very simple semiclassical approximation. *Journal of Chemical Physics, 75*(6), 2923–2931.

Herman, M. F., & Kluk, E. (1984). A semiclassical justification for the use of non-spreading wavepackets in dynamics calculations. *Chemical Physics, 91*(1), 27–34.

Ishizuka, K. (1980). Contrast transfer of crystal images in TEM. *Ultramicroscopy, 5*, 55–65.

Jia, C.-L., Nagarajan, V., He, J.-Q., Houben, L., Zhao, T., Ramesh, R., … Waser, R. (2007). Unit-cell scale mapping of ferroelectricity and tetragonality in epitaxial ultrathin ferroelectric films. *Nature Materials, 6*(1), 64–69.

Jia, C.-L., Urban, K. W., Alexe, M., Hesse, D., & Vrejoiu, I. (2011). Direct observation of continuous electric dipole rotation in flux-closure domains in ferroelectric Pb(Zr, Ti)O$_3$. *Science, 331*(6023), 1420–1423.

Kasper, E., & Hawkes, P. W. (1995). *Principles of electron optics: Vol. 3. Wave optics*. London: Academic Press.

Knoll, M., & Ruska, E. (1932). Das Elektronenmikroskop. *Zeitschrift für Physik, 78*, 318–339.

Koch, C. T. (2008). A flux-preserving non-linear inline holography reconstruction algorithm for partially coherent electrons. *Ultramicroscopy, 108*(2), 141–150.

Krivanek, O. L., Dellbya, N., & Lupinic, A. R. (1999). Towards sub-Å electron beams. *Ultramicroscopy*, *78*, 1–11.

Lubk, A., Béché, A., & Verbeeck, J. (2015). Electron microscopy of probability currents at atomic resolution. *Physical Review Letters*, *115*(17), 176101.

Lubk, A., & Röder, F. (2015). Semiclassical TEM image formation in phase space. *Ultramicroscopy*, *151*, 136–149.

Mandel, L., & Wolf, E. (1995). *Optical coherence and quantum optics*. Cambridge: Cambridge University Press.

Miller, W. H. (1974). Classical-limit quantum mechanics and the theory of molecular collisions. *Advances in Chemical Physics*, *25*, 69–177.

Möllenstedt, G. (1956). Elektronenmikroskopische Bilder mit einem nach O. Scherzer sphärisch korrigierten Objektiv. *Optik*, *13*, 209.

Müller, H., Maßmann, I., Uhlemann, S., Hartel, P., Zach, J., & Haider, M. (2011). Aplanatic imaging systems for the transmission electron microscope. *Nuclear Instruments & Methods in Physics Research. Section A, Accelerators, Spectrometers, Detectors and Associated Equipment*, *645*(1), 20–27.

Müller, H., Rose, H., & Schorsch, P. (1998). A coherence function approach to image simulation. *Journal of Microscopy*, *190*(1–2), 73–88.

Müller, H., Uhlemann, S., Hartel, P., & Haider, M. (2008). Aberration-corrected optics: From an idea to a device. *Physics Procedia*, *1*(1), 167–178.

Nellist, P. D., Behan, G., Kirkland, A. I., & Hetherington, C. J. D. (2006). Confocal operation of a transmission electron microscope with two aberration correctors. *Applied Physics Letters*, *89*(12), 124105.

Pozzi, G. (1987). Theoretical considerations on the spatial coherence in field emission electron microscopes. *Optik*, *77*, 69–73.

Reimer, L., & Kohl, H. (2008). *Transmission electron microscopy: Physics of image formation. Springer series in optical sciences*. Springer.

Rivera, A. L., Lozada-Cassou, M., Rodríguez, S., & Castaño, V. (2003). Wigner distribution function as an assessment of the paraxial character of an optical system. *Optics Communications*, *228*, 211–216.

Röder, F., & Lubk, A. (2014). Transfer and reconstruction of the density matrix in off-axis electron holography. *Ultramicroscopy*, *146*, 103–116.

Rose, H. (1970). Berechnung eines elektronenoptischen Apochromaten. *Optik*, *32*, 144–164.

Rose, H. (1971). Abbildungseigenschaften sphärisch korrigierter elektronenoptischer Achromate. *Optik*, *33*, 1–24.

Rose, H. (1976). Image formation by inelastically scattered electrons in electron microscopy. I. *Optik*, *45*, 139–158.

Rose, H. (1981). Correction of aperture aberrations in magnetic systems with threefold symmetry. *Nuclear Instruments and Methods*, *187*, 187–199.

Rose, H. (1984). Information transfer in transmission electron microscopy. *Ultramicroscopy*, *15*(3), 173–191.

Rose, H. (2003). High-resolution imaging and spectrometry of materials. In F. Ernst, & M. Rühle (Eds.), *Advances in electron optics* (pp. 189–269). Berlin: Springer.

Rose, H. (2008). History of direct aberration correction. In P. W. Hawkes (Ed.), *Advances in imaging and electron physics: Vol. 153. Aberration-corrected electron microscopy* (pp. 3–39). Elsevier.

Rose, H. H. (1990). Outline of a spherically corrected semiaplanatic medium-voltage transmission electron-microscope. *Optik, 85*, 19.

Rose, H. H. (2009). *Geometrical charged-particle optics. Springer series in optical sciences: Vol. 142.* Springer.

Rosenauer, A., Krause, F. F., Müller, K., Schowalter, M., & Mehrtens, T. (2014). Conventional transmission electron microscopy imaging beyond the diffraction and information limits. *Physical Review Letters, 113*(9), 096101.

Rother, A., Gemming, T., & Lichte, H. (2009). The statistics of the thermal motion of the atoms during imaging process in transmission electron microscopy and related techniques. *Ultramicroscopy, 109*(2), 139–146.

Rusz, J., Idrobo, J.-C., & Bhowmick, S. (2014). Achieving atomic resolution magnetic dichroism by controlling the phase symmetry of an electron probe. *Physical Review Letters, 113*(14), 145501.

Saxton, W. O. (1995). Observation of lens aberrations for very high-resolution electron microscopy. I. Theory. *Journal of Microscopy, 179*, 201–213.

Schattschneider, P., Nelhiebel, M., & Jouffrey, B. (1999). Density matrix of inelastically scattered fast electrons. *Physical Review B, 59*, 10959.

Schattschneider, P., & Verbeeck, J. (2008). Fringe contrast in inelastic LACBED holography. *Ultramicroscopy, 108*(5), 407–414.

Scherzer, O. (1936). Über einige Fehler von Elektronenlinsen. *Zeitschrift für Physik. A, Hadrons and Nuclei, 101*(9), 593–603.

Scherzer, O. (1947). Sphärische und chromatische Korrektur von Elektronen-Linsen. *Optik, 2*, 114–132.

Schlosshauer, M. (2004). Decoherence, the measurement problem, and interpretations of quantum mechanics. *Reviews of Modern Physics, 76*, 1267–1305.

Seeliger, P. (1951). Die Sphärische Korrektur von Elektronenlinsen mittels nicht-rotationssymmetrischer Abbildungselemente. *Optik, 8*, 311–317.

Spence, J. C. H. (2013). *High-resolution electron microscopy.* Oxford: OUP.

Urban, K. W. (2008). Studying atomic structures by aberration-corrected transmission electron microscopy. *Science, 321*, 506–510.

Urban, K. W., Mayer, J., Jinschek, J. R., Neish, M. J., Lugg, N. R., & Allen, L. J. (2013). Achromatic elemental mapping beyond the nanoscale in the transmission electron microscope. *Physical Review Letters, 110*(18), 185507.

Vallée, O., & Soares, M. (2010). *Airy functions and applications to physics.* Imperial College Press.

Van Aert, S., Batenburg, K. J., Rossell, M. D., Erni, R., & Van Tendeloo, G. (2011). Three-dimensional atomic imaging of crystalline nanoparticles. *Nature, 470*(7334).

van Cittert, P. H. (1934). Die wahrscheinliche Schwingungsverteilung in einer von einer Lichtquelle direkt oder mittels einer Linse beleuchteten Ebene. *Physica, 1*, 201–210.

Verbeeck, J., Bertoni, G., & Schattschneider, P. (2008). The Fresnel effect of a defocused biprism on the fringes in inelastic holography. *Ultramicroscopy, 108*(3), 263–269.

Wade, R. H., & Frank, J. (1977). Electron microscope transfer functions for partially coherent axial illumination and chromatic defocus spread. *Optik, 49*, 81–92.

Wade, R. H., & Jenkins, W. K. (1979). Tilted beam electron microscopy: The effective coherent aperture. *Optik, 50*, 1–17.

Yuk, J. M., Park, J., Ercius, P., Kim, K., Hellebusch, D. J., Crommie, M. F., . . . Alivisatos, A. P. (2012). High-resolution EM of colloidal nanocrystal growth using graphene liquid cells. *Science, 336*(6077), 61–64.

Zeitler, E. (1990). Analysis of an imaging magnetic energy filter. *Nuclear Instruments & Methods in Physics Research. Section A, Accelerators, Spectrometers, Detectors and Associated Equipment, 298,* 234–246.

Zemlin, F. (1979). A practical procedure for alignment of a high resolution electron microscope. *Ultramicroscopy, 4*(2), 241–245.

Zernike, F. (1938). The concept of degree of coherence and its application to optical problems. *Physica, 5,* 785–795.

Zurek, W. H. (2003). Decoherence, einselection, and the quantum origins of the classical. *Reviews of Modern Physics, 75*(3), 715–775.

CHAPTER FIVE

Electron Holography in Phase Space

Axel Lubk

Institute for Structure Physics, Physics Department, Faculty of Mathematics and Natural Sciences, Technical University of Dresden, Dresden, Germany
e-mail address: axel.lubk@tu-dresden.de

Contents

"... in quantum mechanics, one can know everything about a system and nothing about its individual parts ..."

(Leonard Susskind)

Holography concerns the reconstruction of a (quantum mechanical) wave function in amplitude and phase from a suitable interference pattern, the hologram. This is achieved by a two-step process consisting of recording the hologram from which the wave function is reconstructed subsequently. Gabor's original setup used an on-axis (inline) superposition of an undisturbed reference wave and a disturbed object wave to encode the phase information in the interference pattern. He recognized that the conventional intensity term arising from the two interfering waves superimposes the interference term, thereby hampering the reconstruction (Gabor & Goss, 1966). Moreover, the two conjugate terms ("twin images") forming

141

the interference term need to be separated in order to uniquely recon-struct a wave function. A large number of different setups emerged from the search for suitable holographic schemes facilitating the isolation of a unique wave function, which in case of electron holography have been partly summarized by Cowley (1992). In their seminal contribution, Leith and Upatnieks (1962, 1963) showed that an off-axial superposition of one partial wave, modulated by the object, with an undisturbed reference wave is particular well-suited to the task, because the spurious twin image and the conventional intensity can be removed completely under the weak pre-sumption that the wave function is band-limited. In the following, off-axis electron holography evolved to one of the most widespread techniques employed in the TEM (Tonomura, 1987; Tonomura, Allard, Pozzi, Joy, & Ono, 1995; Völkl, Allard, & Joy, 1999; McCartney et al., 2010; Kasama, Dunin-Borkowski, & Beleggia, 2011; Lichte et al., 2013; Pozzi, Beleggia, Kasama, & Dunin-Borkowski, 2014).

Frequently, however, the off-axis scheme is not well-adapted to a given problem, for instance because no suitable reference may be provided or be-cause the coherence within the electron beam is too small. To remove that obstacle, several advanced extensions to Gabor's original inline hologra-phy, most notably Focal Series Inline Holography (Haine & Mulvey, 1952; Coene, Janssen, Op de Beeck, & Van Dyck, 1992; Kirkland, 1984; Cowley, 1995; Coene, Thust, Op de Beeck, & Van Dyck, 1996; Thust, Coene, de Beeck, & Van Dyck, 1996; Koch, 2008) and the reconstruction of wave functions from two (or more) slightly defocused images referred to as Transport of Intensity Equation (TIE) phase retrieval (Teague, 1983), have been developed. Beyond the popular off-axis and inline schemes, a large number of other holographic setups has been proposed and realized. Table 5.1 gives a non-exhaustive overview including fields of application and some pertinent references.

From a broader perspective, holography aims at resolving a phase problem, which consist of reconstructing the argument of a complex wave field from one or multiple intensity measurements (Millane, 1990; Luke, Burke, & Lyon, 2002; Jaming, 2014). Such problems date back to the second half of the 19th century (Strutt, 1892) and became urgent with the advent of X-ray diffraction (Wolfke, 1920) and quantum me-chanics, when Pauli posed his famous question, whether it is possible to reconstruct a complex function from its modulus in position and Fourier space (i.e., object and diffraction plane in the parlance of optics) (Pauli, 1933). Phase problems arise in such diverse fields as X-ray crystallogra-

Table 5.1 Compilation of holographic schemes used in the TEM, including pertaining imaging modes and typical fields of application

Holographic setup	TEM modus	Fields of application
Off-axis (Möllenstedt & Wahl, 1968; Tonomura, 1987)	CTEM	EM^γ, strain fields, HR^δ imaging
TIE (Teague, 1983; Lubk & Zweck, 2015)	CTEM	EM fields
Focal series (Schiske, 1968, 1973; Kirkland, 1984; Coene et al., 1996; Thust et al., 1996; Koch, 2008)	CTEM	strain fields, HR imaging
DBH^α (Herring, Pozzi, Tanji, & Tonomura, 1993; Houdellier & Hÿtch, 2008)	$CBED^\beta$	scattering factors
STEM (Leuthner, Lichte, & Herrman, 1989; Takahashi, Yajima, Ichikawa, & Kuroda, 1994; Mankos, Cowley, & Scheinfein, 1996)	STEM	EM fields
DPC^ϵ (Rose, 1973; Dekkers & de Lang, 1974)	STEM	EM fields, HR imaging
Ptychography (Hoppe, 1969a, 1969b; Hoppe & Strube, 1969; Bates & Rodenburg, 1989; Rodenburg & Bates, 1992)	STEM	scattering factors

α Diffracted Beam Holography.
β Convergent Beam Electron Diffraction.
γ Electromagnetic.
δ High-Resolution.
ϵ Differential Phase Contrast.

phy, astronomy, quantum mechanics, seismology, and microscopy, which stirred a systematic investigation on the existence and uniqueness of solutions pertaining to the various phase reconstruction schemes including the previously mentioned holographic ones. For instance, the identification of unequal wave functions possessing the same modulus in position and Fourier space showed that Pauli's phase retrieval problem is not solvable in general (Jaming, 2014). However, a unique determination of the wave function from a limited number of intensity measurements in different planes may be possible, if imposing additional constraints, which narrow the dimensionality of the admitted function space. For instance, it has been shown that the ambiguity may be reduced, if the wave function to be reconstructed has a particular functional shape (Jaming, 2014), is strictly positive (Bruck & Sodin, 1979), has a finite support (Fienup, 1987;

Hurt, 2001; Bauschke, Combettes, & Luke, 2002; Latychevskaia & Fink, 2007), is analytic, i.e., globally representable by one convergent Taylor expansion (Huiser, Drenth, & Ferwerda, 1976; Huiser & Ferwerda, 1976; Huiser, van Toorn, & Ferwerda, 1977), or periodic and band-limited, i.e., representable by a finite Fourier series (Bruck & Sodin, 1979). The latter may be extended to generalized Fourier transforms, i.e., functions possessing a finite expansion in any basis.

These investigations also lead to the discovery that the propagation of a wave function from the near field to the far field could be considered as an increasing shear of the corresponding Wigner distribution (Fig. 5.1, (2.111)). It is shown in Section 5.3 that such a shear can be transformed into a rotation from $-90°$ (far field/diffraction plane at under-focus) via $0°$ (in focus) to $90°$ (far field/diffraction plane at over-focus) in phase space by suitably rescaling the phase space coordinates. The latter corresponds to the phase space evolution of a harmonic oscillator (Fig. 5.1). Therefore, a recording of a focal series ranging from the diffraction plane at under-focus to that at over-focus corresponds to a complete tomographic tilt series of the Wigner function in phase space. These data are in principle suited to recover the complete underlying phase space distribution of general (not necessarily pure) quantum states via phase space tomography (Smithey, Beck, Raymer, & Faridani, 1993; Breitenbach, Schiller, & Mlynek, 1997; Schleich, 2001). The field of phase space tomography and phase space reconstructions has evolved into a very active field of research, concerned with the reconstruction of both optical and matter (e.g., atomic) mixed quantum states (Raymer, Beck, & McAlister, 1994; Breitenbach et al., 1997; Leibfried, Pfau, & Monroe, 1998; Kienle, 2000; Cámara, 2015).

The phase space perspective also opened new perspectives on the respective requirements for a successful wave reconstruction by means of electron holography, because of the following two aspects (see Nugent, 2007 for a discussion of X-ray holography). First, conventional holography may be considered a phase space reconstruction of a pure state (wave) (Wolf & Rivera, 1997; Testorf & Lohmann, 2008), because the wave completely defines the phase space distribution via the definition of the Wigner function (2.88). In other words, conventional holography amounts to a phase space reconstruction under a purity constraint, given by one integral equation restricting the 4D phase space volume of the quantum state to $4\pi^2$ (2.104) (Nakajima, 1999; Candès, Eldar, Strohmer, & Voroninski, 2013). Such a constraint only marginally reduces the generally infinite dimensions

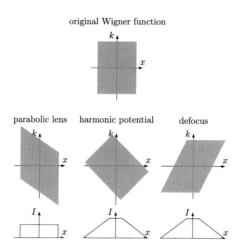

Figure 5.1 Electron optical phase space transformations. Suitably combining the action of lenses with free-space propagation facilitates a rotation of phase space.

of the functional space of all allowed quantum states by one.[1] Consequently, existence and uniqueness theorems for quantum state reconstructions may be transferred to wave reconstructions (Lvovsky & Raymer, 2009; Jaming, 2014). For instance, it is immediately obvious that only two projections (i.e., the image and diffraction plane intensity) are clearly insufficient for a unique tomographic reconstruction of a quantum state, which provides an alternative perspective on the incompleteness of Pauli's problem.

Second, technological and fundamental physical constraints (e.g., Boersch effect (Bergmann, Niedrig, & Eichler, 2004), quantum statistics of the detection process (Niermann, Lubk, & Röder, 2012)) limit the coherence of electron beams employed in TEM. Thus, electron holographic methods have to take into account partial coherence, typically by modeling its ramifications on the observed intensity distribution, in the hope to obtain a sufficiently unique wave reconstruction. From a phase space perspective that corresponds to inverting the pertaining modification (typically a blurring) of phase space (e.g., Fig. 4.6), which amounts to a purification of the quantum state. Such a purification typically involves some ill-conditioned operation, for instance the deconvolution of the spatial envelope (cf., e.g.,

[1] As an example, general mixed two-level quantum states (qubits) may be geometrically represented by the filled Bloch sphere, where pure states correspond to the surface only.

Fig. 4.6) employed in focal series reconstructions (Allen, McBride, O'Leary, & Oxley, 2004; Koch, 2008). Consequently, the scope of wave reconstructions from partially coherent (mixed) states is limited, even if the mixing process is well-known.

The above considerations illustrate that two holographic principles have to be distinguished. The first concerns the reconstruction of complex valued wave functions, whereas the second seeks a reconstruction of a general, possibly mixed, quantum state. Obviously, the latter is an object of much larger information content, since pure quantum states represent only a zero set within the set of all quantum states. Indeed, the quantum state represents the maximal information obtainable from a measurement in the reduced coordinates of the scattered electron (i.e., without determining the state of the object after interaction with the electron beam).

Reconstructing mixed instead of pure quantum states is experimentally more demanding, because a set of holograms, i.e., a "quorum" (Park & Band, 1971; Band & Park, 1971), needs to be recorded. The reward for the increased complexity is a linear reconstruction of the maximal information contained in the electron beam without imposing additional assumptions (e.g., on the purity), which might be violated in practice. Quantum state reconstructions have been extensively studied in light optics (Bastiaans, 1978, 1979; Atakishiyev, Chumakov, Rivera, & Wolf, 1996; Wolf, 1996; Wolf & Rivera, 1997; Mecklenbräuker & Hlawatsch, 1997; Lohmann, Testorf, & Ojeda-Castañeda, 2004; Testorf & Lohmann, 2008). The manifold analogies between light and electron optics allow transferring a considerable portion of these results. However, the particular properties of the electron, most notably its fundamentally different equation of motion, insert its own flavor, calling for an extension and adaption of methods used in other fields.

Quantum state reconstructions of electrons within the TEM have been rarely discussed so far. The notable exception pertains to inelastic off-axis electron holography (Harscher, Lichte, & Mayer, 1997; Lichte & Freitag, 2000; Schattschneider & Lichte, 2005; Sonnentag & Hasselbach, 2007; Röder & Lubk, 2014), which corresponds to the reconstruction of off-diagonals of the quantum state's density matrix (cf. Section 5.1). Furthermore, it has been also shown that Ptychography (Bates & Rodenburg, 1989; Rodenburg & Bates, 1992) corresponds to a deconvolution of the quantum state's Wigner function (Section 5.4). Hitherto, this result has been exploited in X-ray studies (Thibault & Menzel, 2013) only, whereas in TEM only pure states have been considered so far.

In the following, we will separately consider mixed state and pure state (wave) reconstructions, employing a variety of holographic setups in TEM, namely Off-Axis Holography, Transport of Intensity Reconstruction, Focal Series Inline Holography, Differential Phase Contrast, and Ptychography. Each holographic technique is in the first place considered as a general mixed quantum state reconstruction scheme, and the wave (pure state) reconstruction is treated as a special case of the former. This practice permits a comprehensive discussion of the ramifications of partial coherence on conventional wave reconstructions as well as a generalization toward partially coherent or incoherent signals such as resulting from inelastic scattering (cf. Section 2.3).

Consistent with this approach, the above choice of holographic schemes deliberately leaves out such phase retrieval methods, which do not permit a straightforward generalization to quantum state reconstructions. That particularly pertains to coherent diffractive imaging techniques (Zuo, Vartanyants, Gao, Zhang, & Nagahara, 2003; Morishita, Yamasaki, Nakamura, Kato, & Tanaka, 2008; De Caro, Carlino, Caputo, Cozzoli, & Giannini, 2010) seeking a reconstruction of complex wave functions from a single intensity measurement in the far field and some previous knowledge on the wave function or the scatterer.

5.1. OFF-AXIS HOLOGRAPHY

In 1954, Möllenstedt and Düker inserted a thin quartz filament, coated with gold, into an electron beam (Möllenstedt & Düker, 1955; Dücker & Möllenstedt, 1955). Upon applying a positive voltage to this filament they succeeded in tilting the two half waves to the left and right of the strand toward each other to interfere at some distance downstream the filament (Fig. 5.2). This working principle is analogous to that of a Fresnel biprism used for optical Off-Axis Holography (Leith & Upatnieks, 1962), hence the term electron biprism for the filament.

The step from interferometry to Off-Axis Holography (Bryngdahl & Lohmann, 1968), consisting of reversing the interferometric measurement by reconstructing the wave function from the hologram, was taken in 1968, again in the Möllenstedt group (Möllenstedt & Wahl, 1968). The most common off-axis setup employed within the TEM is depicted in Fig. 5.2. Here, the biprism is located slightly above the first intermediate image plane and the hologram appears in the intermediate image plane, which is magnified further by the following projective system.

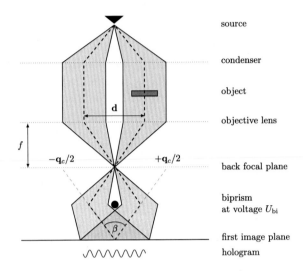

Figure 5.2 Simplified optical setup for off-axis electron holography. The holographic shear **d** is varied experimentally by changing the biprism voltage.

Consequently, the setup is particularly well suited to bring out modulations of the wave functions, which can be parametrized by the object coordinates. Most notably, that includes the important case of axial scattering into forward direction (cf. Section 2.3), containing the Phase Object Approximation as a special case. The latter links the holographically reconstructed phase with projected electric and magnetic potentials and holds under relative moderate imaging conditions for a large class of materials. Exploiting this simple relation, a large number of studies on electric and magnetic properties of materials has been reported (Tonomura et al., 1995; Völkl et al., 1999; McCartney & Smith, 2007; Thomas, Simpson, Kasama, & Dunin-Borkowski, 2008; Kasama et al., 2011; Lichte et al., 2013; Pozzi et al., 2014).

Furthermore, the built-up of an off-axis electron hologram from single electrons hitting the detector is analogous to that occurring in Young's double-slit experiment, which reveals the wave–particle duality of quantum particles in a particular straightforward way (Jönsson, 1974; Frabboni et al., 2012). Thus, a number of studies on quantum noise and the detection process has been conducted within the off-axial setup (Merli, Missiroli, & Pozzi, 1976). An important follow-up result was the accurate description of noise in the reconstructed quantities (i.e., amplitude and phase) resulting from the inevitable quantum noise (Lenz, 1988). A recent generalization of this formalism taking into account also the detector noise allows the ab-initio calculation of error bounds, pertaining to Off-Axis Holography, from detector transfer functions only (Niermann et al., 2012; Lubk et al., 2012; Röder, Lubk, Wolf, & Niermann, 2014).

In a different off-axial setup, referred to as Diffracted Beam Holography (DBH), the specimen is illuminated with a convergent beam, the biprism is placed in the back focal plane, and the interference pattern of the partial waves is recorded in the far field of the object (diffraction plane) (Herring et al., 1993). This setup has been used to measure isoplanatic aberrations and scattering factors (Houdellier & Hÿtch, 2008). Besides placing the biprism at different optical planes, considerable effort has been put into augmenting the capabilities of the interference geometry by inserting multiple biprisms in the electron beam (Bayh, 1962; Lichte & Möllenstedt, 1979; Harada, Togawa, Akashi, Matsuda, & Tonomura, 2004; Genz, Niermann, Buijsse, Freitag, & Lehmann, 2014). Using multiple biprisms has several advantages. First of all, the creation of interference patterns over varying fields of view with varying spatial resolution are greatly facilitated by independently controlling the hologram width and the fringe spacing. Here, it is possible to place the second biprism filament into the shadow of the first, magnified by an intermediate lens, which suppresses the spurious effect of Fresnel scattering at the defocused rim of the filament. Last but not least a larger magnification (Wang, Li, Domenicucci, & Bruley, 2013), e.g., facilitated by additional lenses, in between the biprism plane and the object plane allows for decreasing the biprism deflection angle and hence the biprism voltage including instabilities, which improves the hologram quality (Lichte, 1996).

In order to obtain an interference pattern permitting a separate reconstruction of one partial wave, the other one (reference wave) has to be known in advance. This can be achieved by inserting the object only into one of the half planes, leaving the wave in the other half plane untouched and therefore flat. This is realized in the standard off-axis setup employed for measuring electric (cf. Section 6.1) and magnetic (cf. Section 6.3) fields. A slight modification of this setup, referred to as dark-field electron holography (DFEH), permits the recording of holograms from beams diffracted at crystalline objects covering both the object and reference region (Hÿtch, Houdellier, Hue, & Snoeck, 2008; Hÿtch, Houdellier, Hüe, & Snoeck, 2011). If the crystal in the reference area is sufficiently uniform, DFEH may retrieve strain variations, which slightly modify the scattering direction in the object part, and hence the wave front of the diffracted beam. DFEH and its prospects for strain field tomography will be separately discussed in Section 6.5.

If an unknown object covers the whole object plane, thereby modulating both partial waves in an arbitrary way, it is not possible to reconstruct a unique object wave function anymore. However, a particular type of quantum state reconstruction uses a series of (energy-filtered) off-axis holograms to infer information on the inelastic interaction taking place in both half planes. This technique is referred to as inelastic electron holography (Lichte & Freitag, 2000); it has been used to characterize the coherence of electrons scattered inelastically on plasmons in the past (Potapov, Lichte, Verbeeck, Van Dyck, & Schattschneider, 2003; Schattschneider & Lichte, 2005; Verbeeck, Van Dyck, Lichte, Potapov, & Schattschneider, 2005; Röder & Lichte, 2011; Röder & Lubk, 2014). Following our program, we will first elaborate on the quantum state reconstruction scheme before passing to the pure state (wave) reconstructions as a special case.

5.1.1 Quantum State Reconstruction

To facilitate the description of Off-Axis Holography in terms of quantum states (i.e., density matrices or Wigner functions) it is necessary to reformulate the image formation process of partial coherent electron beams (mixed states), developed in Chapter 4, by incorporating the Möllenstedt biprism, which causes a mutual deflection of the two half beams in a plane slightly above the image plane (Schattschneider & Lichte, 2005; Verbeeck, Bertoni, & Schattschneider, 2008; Koch & Lubk, 2010; Röder et al., 2014) (Fig. 5.2). Neglecting the effect of Fresnel scattering at the biprism edge (e.g., by using the double biprism setup), one obtains the following density matrix in the hologram plane

$$\rho_h\left(\mathbf{x},\mathbf{x}'\right) = \rho_a\left(\mathbf{x}+\frac{\mathbf{d}}{2},\mathbf{x}'+\frac{\mathbf{d}}{2}\right)e^{-i\frac{\mathbf{q}_c}{2}(\mathbf{x}-\mathbf{x}')} \tag{5.1}$$

$$+ \rho_a\left(\mathbf{x}-\frac{\mathbf{d}}{2},\mathbf{x}'-\frac{\mathbf{d}}{2}\right)e^{i\frac{\mathbf{q}_c}{2}(\mathbf{x}-\mathbf{x}')}$$

$$+ \rho_a\left(\mathbf{x}+\frac{\mathbf{d}}{2},\mathbf{x}'-\frac{\mathbf{d}}{2}\right)e^{-i\frac{\mathbf{q}_c}{2}(\mathbf{x}+\mathbf{x}')}$$

$$+ \rho_a^*\left(\mathbf{x}-\frac{\mathbf{d}}{2},\mathbf{x}'+\frac{\mathbf{d}}{2}\right)e^{i\frac{\mathbf{q}_c}{2}(\mathbf{x}+\mathbf{x}')},$$

where the hologram carrier frequency $\mathbf{q}_c = k_0\boldsymbol{\beta}$ and the holographic shear $\mathbf{d}$ are indicated in Fig. 5.2. We have denoted the quantum state pertaining to the object (exit) plane with ρ_a to indicate a possible impact of aberrations (cf. Chapter 4). The first two terms are referred to as centerband (cb) and the remaining terms to as sidebands (sb), which derives from the location of the terms in the ambiguity function (reciprocal phase space) representation of the quantum state discussed below.

If describing the electron optical action of the biprism as a mutual shift and tilt in the hologram plane, one obtains the transfer of the density matrix incorporating the generalized Transmission Cross Coefficient (4.48) (Röder & Lubk, 2014). Accordingly, the hologram intensity separates into a centerband and sideband term

$$I_h\left(\mathbf{x}\right) = I_{cb}\left(\mathbf{x}\right) + I_{sb}\left(\mathbf{x}\right), \tag{5.2}$$

where the former is equivalent to a sum of conventional intensities pertaining to the left and right half of the object plane (4.48)

$$
\begin{aligned}
I_{cb}\left(\mathbf{x}\right) = {} & \rho_a\left(\mathbf{x} + \frac{\mathbf{d}}{2}, \mathbf{x} + \frac{\mathbf{d}}{2}\right) + \rho_a\left(\mathbf{x} - \frac{\mathbf{d}}{2}, \mathbf{x} - \frac{\mathbf{d}}{2}\right) \\
= {} & \frac{1}{4\pi^2}\iiint_{-\infty}^{\infty} d^2k d^2k' dE e^{i\left(\mathbf{x}+\frac{\mathbf{d}}{2}\right)(\mathbf{k}-\mathbf{k}')} \\
& \times \rho_s\left(\mathbf{k},\mathbf{k}',E\right)T_c\left(\mathbf{k},\mathbf{k}',E\right) \\
& + \frac{1}{4\pi^2}\iiint_{-\infty}^{\infty} d^2k d^2k' dE e^{i\left(\mathbf{x}-\frac{\mathbf{d}}{2}\right)(\mathbf{k}-\mathbf{k}')} \\
& \times \rho_s\left(\mathbf{k},\mathbf{k}',E\right)T_c\left(\mathbf{k},\mathbf{k}',E\right).
\end{aligned}
\tag{5.3}
$$

Note the appearance of the conventional Transmission Cross Coefficient T_c not depending on the holographic shear $\mathbf{d}$.

In contrast to that, the sideband intensity is modulated by the generalized Transmission Cross Coefficient (4.49)

$$
\begin{aligned}
I_{sb}\left(\mathbf{r}\right) = {} & \rho_a\left(\mathbf{x} + \frac{\mathbf{d}}{2}, \mathbf{x} - \frac{\mathbf{d}}{2}\right) \\
= {} & \iiint_{-\infty}^{\infty} d^2k d^2k' dE \rho_s\left(\mathbf{k},\mathbf{k}',E\right) \\
& \times T\left(\mathbf{k},\mathbf{k}',\mathbf{d},E\right) e^{2i\mathbf{k}\mathbf{d}}.
\end{aligned}
\tag{5.4}
$$

When discussing Off-Axis Holography of wave functions below, the interference term including the generalized transmission cross coefficient simplifies considerably, resulting in a linear transfer of the partial wave, passing the object half space through the objective lens. A more comprehensive discussion of off-axis holographic imaging including, e.g., the ramifications of Fresnel scattering has been given by Röder and Lubk (2014).

The first line in the above expression for the centerband intensity (5.3) reveals that the conventional object plane intensities are merely shifted in position, when changing the holographic shear (i.e., the biprism voltage). The sideband intensity, on the other hand, displays the conventionally inaccessible off-diagonals of the density matrix, which vary in their distance to the main diagonal upon changing the biprism voltage. Hence, the sideband intensity can be exploited for a reconstruction of the complete quantum state, which necessitates a separation of sideband and centerband intensities in the recorded hologram in a first step.

To facilitate this separation and to obtain a more comprehensive perspective on the quantum state reconstruction, it is useful to switch to the reciprocal phase space representation of the hologram's quantum state

$$
\begin{aligned}
\tilde{W}_{\mathrm{h}}\left(\mathbf{k}', \mathbf{r}'\right) = {} & \tilde{W}_{\mathrm{a}}\left(\mathbf{q}, \mathbf{r}'\right) e^{\frac{i}{2}\left(\mathbf{k}'\mathbf{d}+\mathbf{r}'\mathbf{q}_c\right)} \\
& + \tilde{W}_{\mathrm{a}}\left(\mathbf{k}', \mathbf{r}'\right) e^{-\frac{i}{2}\left(\mathbf{k}'\mathbf{d}+\mathbf{r}'\mathbf{q}_c\right)} \\
& + \tilde{W}_{\mathrm{a}}\left(\mathbf{k}'+\mathbf{q}_c, \mathbf{r}'+\mathbf{d}\right) \\
& + \tilde{W}_{\mathrm{a}}\left(\mathbf{k}'-\mathbf{q}_c, \mathbf{r}'-\mathbf{d}\right),
\end{aligned}
\tag{5.5}
$$

which contains the hologram's Fourier spectrum at the zero section $\mathbf{r}' = 0$ (2.98). In the ambiguity function the sideband terms (lines 3 and 4) are spatially shifted with respect to the centerband terms (lines 1 and 2). Consequently, the former may be separated from the latter by applying a numerical filter, if the phase space distribution obeys some band limit, i.e., $\tilde{W}_{s}(\mathbf{k}', \mathbf{r}') = 0$ for $|\mathbf{k}'| > q_B$ (Fig. 5.3). Moreover, by changing the holographic shear $\mathbf{d}$ (biprism voltage), different sections of the (sideband) ambiguity function are shifted to $\mathbf{r}' = 0$ eventually facilitating a tomographic reconstruction of the sideband's quantum state under the following restrictions.

In order to completely sample the 4D ambiguity function, it is necessary to vary the biprism voltage *and* its orientation. This represents a formidable experimental challenge and no complete 4D quantum state reconstruction by Off-Axis Holography has been reported to date of this work. The main problem of the technique pertains to the minimization of the biprism shadow and Fresnel scattering at its rim, hampering the reconstruction at small holographic shears. Additional complications arise from the modifications introduced by the generalized transmission cross coefficient (5.4)

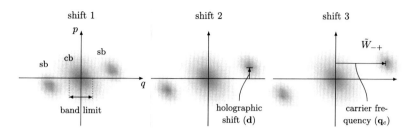

Figure 5.3 Quantum state (ambiguity function (5.5)) reconstruction by Off-Axis Holography measuring the cross-section $\tilde{W}(\mathbf{k}', \mathbf{r}' = 0)$ of the sideband (sb) for different holographic shears (**d**). A sufficiently large carrier frequency ($\mathbf{q}_c$) permits separation between center- (cb) and sidebands (sb).

and the influence of the energy filter, if seeking an energy-resolved determination of the quantum state (Röder et al., 2014).

Previous studies reduced the complexity of the method by exploiting additional symmetries. In case of homogeneous objects ($\rho(\mathbf{k}, \mathbf{k}') = I(\mathbf{k})\delta(\mathbf{k} - \mathbf{k}')$), for instance, the interference term (5.4) depends on relative distances only, and the influence of partial coherent illumination factorizes, because (cf. van Cittert–Zernike theorem (4.31))

$$\mathcal{T}\left(\mathbf{k}, \mathbf{k}, \mathbf{d}\right) = \mathcal{T}\left(\mathbf{0}, \mathbf{0}, \mathbf{d}\right) = \frac{1}{2\pi} \int_{-\infty}^{\infty} d^2 q I_S\left(\mathbf{q}\right) e^{i\mathbf{q}\mathbf{d}}, \tag{5.6}$$

yielding

$$\rho_h\left(\mathbf{x} - \mathbf{x}'\right) = \mathcal{T}\left(\mathbf{0}, \mathbf{0}, \mathbf{d}\right) \rho_a\left(\mathbf{x} - \mathbf{x}'\right). \tag{5.7}$$

That includes the special case of recording an empty hologram, where

$$\mu_c := \mathcal{T}\left(\mathbf{0}, \mathbf{0}, \mathbf{d}\right) \tag{5.8}$$

is referred to as the degree of coherence of the illuminating electron beam. Therefore, Off-Axis Holography can be employed to determine the precise shape of the effective source, by inverting the Fourier transform linking the degree of coherence and the incoherent source size (Völkl et al., 1999).

Similarly, 1D sections of the density matrix can be reconstructed without rotating the biprism for objects being homogeneous in normal direction to the biprism (e.g., planar surfaces). Such reconstruction have been used by various groups to study the inelastic excitation of plasmons by means of inelastic off-axis electron holography (Lichte &

Freitag, 2000; Verbeeck et al., 2005; Schattschneider & Lichte, 2005; Potapov, Lichte, Verbeeck, & van Dyck, 2006; Röder et al., 2014). To date, these experiments represent the only reported attempts of quantum state reconstructions in the TEM. They reveal the long-range delocalization of inelastic plasmon scattering and show that the influence of inelastic scattering on the formation of interference fringes reduces to a simple contrast damping for holographic shears $|\mathbf{d}| > 100$ nm, typically employed in medium resolution off-axis electron holography studies on electric and magnetic fields (Verbeeck, Bertoni, & Lichte, 2011). By virtue of this coherence filter property, (medium resolution) Off-Axis Holography can be regarded as efficient filter for elastically (i.e., coherently) scattered electrons, outperforming any hardware energy filter available. We will exploit that property in the next section considering the special case of pure quantum state (i.e., wave) reconstructions.

5.1.2 Wave Reconstruction

If two coherent plane waves Ψ with intensity $I_0 = |\Psi|^2 = \text{const.}$, mutually inclined under an angle β, are superimposed in the image plane, one obtains the following interference pattern with the carrier frequency $\mathbf{q}_c = k_0\boldsymbol{\beta}$ (Fig. 5.2 and (5.7))

$$I_h(\mathbf{r}) = 2I_0 + 2\mu_c(\mathbf{d}) I_0 \cos(\mathbf{q}_c\mathbf{r}) . \tag{5.9}$$

Here, the degree of coherence $\mu_c \leq 1$, defined in the previous section, dampens the fringe contrast of the sinusoidal interference term. Thus, Off-Axis Holography necessitates electron guns, which facilitate an acceptable degree of coherence over the desired holographic shears (Lichte, 2008). In spite of a further optical increase of the coherence by astigmatically demagnifying the source perpendicular to the biprism (i.e., by employing elliptic illumination), the coherency requirement imposes certain limits on the usable holographic field of view in Off-Axis Holography. Therefore, hologram widths beyond several tens of microns call for different holographic setups as discussed below.

Upon introducing an object (specimen) in the right half-plane, leaving the plane wave in the left half-plane undisturbed (Fig. 5.2), the wave in the object exit plane simplifies to

$$\Psi(\mathbf{r}) = \begin{cases} \Psi_a(\mathbf{r}) & x \geq 0, \\ 1 & x < 0. \end{cases} \tag{5.10}$$

Consequently, one obtains the modulated hologram

$$I_h\left(\mathbf{r}\right) = \underbrace{I_0 + I_a\left(\mathbf{r} - \frac{\mathbf{d}}{2}\right)}_{I_{cb}(\mathbf{r})} \qquad (5.11)$$

$$+ \int_{-\infty}^{\infty} \tilde{\Psi}_a\left(\mathbf{q}\right) \mathcal{T}\left(\mathbf{q}, 0, \mathbf{d}\right) e^{i\mathbf{q}(\mathbf{r}+\mathbf{d}/2)} d^2 q + c.c.$$

Using (4.49), the generalized Transmission Cross Coefficient simplifies to a product of a chromatic envelope, a spatial envelope and the wave transfer function, i.e.,

$$\mathcal{T}\left(\mathbf{q}, 0, \mathbf{d}\right) = \underbrace{e^{-\left(\frac{\pi}{2k_0} C_c \frac{\Delta E}{U_A}\right)^2 \mathbf{q}^4}}_{\text{chromatic envelope}} \qquad (5.12)$$

$$\times \underbrace{e^{-\left(\frac{1}{2}\frac{\partial \chi(\mathbf{q})}{\partial q_x} - d\right)^2 \sigma_S^2} e^{-\left(\frac{1}{2}\frac{\partial \chi(\mathbf{q})}{\partial q_y}\right)^2 \sigma_S^2}}_{\text{spatial envelope}}$$

$$\times e^{-i\chi(\mathbf{k})}.$$

The medium resolution setup allows two additional simplifications in the above expressions, which are crucial for the wave reconstructions shown below. First, the influence of the imaging system of the microscope (e.g., aberrations) can be ignored as they do not disturb the image formation at resolutions larger than 0.5 nm in state-of-the-art aberration corrected TEMs. Under these conditions, the hologram intensity reduces to

$$I_h\left(\mathbf{r}\right) = I_0 + I_s\left(\mathbf{r} - \frac{\mathbf{d}}{2}\right) + 2\mu_c\left(\mathbf{d}\right) I_{sb}\left(\mathbf{r} - \frac{\mathbf{d}}{2}\right) \cos\left(\mathbf{q}_c \mathbf{r} + \varphi\left(\mathbf{r} - \frac{\mathbf{d}}{2}\right)\right).$$

$$(5.13)$$

Here, it has been assumed that the wave function's band width is small compared to the carrier frequency, permitting the spatial frequency dependence of the fringe damping mechanism to be reduced to a single, carrier frequency dependent, prefactor $\mu_c(\mathbf{d}(\mathbf{q}_c))$. That approximation is mostly valid for the medium resolution setups used in combination with tomography in Chapter 6.

Under experimental conditions, the fringe contrast reduces further due to additional instabilities (e.g., ac-stray fields and drift of the biprism) and the modulation transfer of the camera (MTF) (Niermann et al., 2012; Lubk et al., 2012), yielding a reduced fringe contrast

$$\mu_c \rightarrow \text{MTF}\left(\mathbf{q}_c\right)\mu_{\text{inst}}\mu_c\left(\mathbf{d}\right). \tag{5.14}$$

The above off-axis hologram permits a separate reconstruction of the centerband intensity I_{cb}, the sideband intensity I_{sb}, and the phase difference
between image and reference wave φ by computing the diffractogram of I_h

$$\tilde{I}_h\left(\mathbf{k}\right) = I_0\delta\left(\mathbf{k}\right) + \tilde{I}\left(\mathbf{k}\right) \tag{5.15}$$
$$+ \mu_c\left(\tilde{I}_{sb}\left(\mathbf{k}\right)*\delta\left(\mathbf{k}-\mathbf{q}_c\right) + \tilde{I}_{sb}^*\left(-\mathbf{k}\right)*\delta\left(\mathbf{k}+\mathbf{q}_c\right)\right)$$

and cutting out (with an appropriate filter mask) the respective region (i.e.,
the centerband or one of the sidebands centered around the δ-functions)
(Lehmann & Lichte, 2002). In a last step, the masked regions are centered and inversely Fourier transformed to yield the above noted quantities.
Comprehensive accounts of the acquisition and reconstruction procedure
may be found in the abundant literature on the topic (Völkl et al., 1999).
   Second, the complicated scattering mechanisms of the electrons can be
simplified within the framework of the axial scattering approximation (cf.
Section 2.3). Accordingly, the mutual object transparency in the object
plane for the three relevant intensity terms forming the centerband and the
sidebands (5.11) can be written as

$$T\left(\mathbf{r}+\frac{\mathbf{d}}{2},\mathbf{r}+\frac{\mathbf{d}}{2}\right) = 1, \tag{5.16}$$
$$T\left(\mathbf{r}-\frac{\mathbf{d}}{2},\mathbf{r}-\frac{\mathbf{d}}{2}\right) = \exp\left(-\int_{-\infty}^{\infty}\mu_{\text{el}}^{(\alpha)}(\mathbf{r}-\frac{\mathbf{d}}{2},z)dz\right),$$

and

$$T\left(\mathbf{r}-\mathbf{d},\mathbf{r}+\mathbf{d}\right) = \exp\left(i\int_{-\infty}^{\infty}\frac{e}{v}\Phi(\mathbf{r}-\frac{\mathbf{d}}{2},z) - A_z(\mathbf{r}-\frac{\mathbf{d}}{2},z)dz\right) \tag{5.17}$$
$$\times \exp\left(-\frac{1}{2}\int_{-\infty}^{\infty}\mu_{\text{el}}^{(\alpha)}(\mathbf{r}-\frac{\mathbf{d}}{2},z)dz\right)$$
$$\times \exp\left(-\int_{-\infty}^{\infty}\mu_{\text{in}}(\mathbf{r}-\frac{\mathbf{d}}{2},\mathbf{r}+\frac{\mathbf{d}}{2},z)dz\right),$$

where vanishing potentials $\Phi = A_z = 0$ have been assumed (gauged) in the
field-free reference region. Note furthermore that α denotes the semi-angle
of the objective aperture intercepting electrons scattered into angles larger
than α.

It is now crucial that the coherence loss due to inelastic scattering reaches its maximum already at holographic shears $|\mathbf{d}| > 100$ nm employed in the medium resolution setup because the dominant low-loss inelastic interactions (e.g., plasmons) possess a coherence width small compared to such holographic shears (Verbeeck et al., 2011). This observation allows the following approximation (cf. Section 2.3)

$$\mu_{\text{in}}\left(\mathbf{r} - \frac{\mathbf{d}}{2}, \mathbf{r} + \frac{\mathbf{d}}{2}, z\right) = \mu_{\text{in}}\left(\mathbf{r} - \frac{\mathbf{d}}{2}, z\right). \tag{5.18}$$

Within the scope of the axial approximation taking into account elastic and inelastic scattering, the final hologram intensity then reads

$$I_{\text{h}}(\mathbf{r}) = I_0 + I_0 \exp\left(-\int_{-\infty}^{\infty} \mu_{\text{el}}^{(\alpha)}\left(\mathbf{r} - \frac{\mathbf{d}}{2}, z\right) + \mu_{\text{in}}^{(\alpha)}\left(\mathbf{r} - \frac{\mathbf{d}}{2}, z\right) dz\right) \tag{5.19}$$

$$+ 2\mu_c I_0 \exp\left(-\int_{-\infty}^{\infty} \left(\frac{1}{2}\mu_{\text{el}}^{(\alpha)}\left(\mathbf{r} - \frac{\mathbf{d}}{2}, z\right) + \mu_{\text{in}}\left(\mathbf{r} - \frac{\mathbf{d}}{2}, z\right)\right) dz\right)$$

$$\times \cos\left(\mathbf{q}_c \mathbf{r} + \frac{e}{v}\int_{-\infty}^{\infty} \Phi\left(\mathbf{r} - \frac{\mathbf{d}}{2}, z\right) dz - \int_{-\infty}^{\infty} A_z\left(\mathbf{r} - \frac{\mathbf{d}}{2}, z\right) dz\right).$$

Accordingly, the phase of the reconstructed wave is linearly proportional to the projected electrostatic and magnetostatic potentials in and around the target. Similarly the centerband and sideband attenuation follows a Lambert–Beer law containing projected attenuation coefficients due to elastic and inelastic scattering processes at the target as argument.

Both projection laws, the one pertaining to the potentials and the other to the attenuation coefficients, are line integrals of the respective quantity along the z-direction as discussed within the context of tomography in Chapter 3. Consequently, the 3D distribution of electrostatic and magnetostatic potentials as well as elastic and inelastic attenuation coefficients may be reconstructed from suitable hologram tilt series. This technique, referred to as Electron Holographic Tomography (EHT), is discussed at length in Chapter 3 of this volume. In the following, the information content of the intensity and phase reconstructed from a single hologram are discussed in detail. These considerations partly apply also to the other holographic techniques discussed below, where they will be used, if required.

Phase Information

Projected electrostatic and magnetostatic potentials are linked to the reconstructed phase through to the following simple relationship

$$\varphi\left(\mathbf{r}\right) = \mathrm{mod}_{2\pi} \left\{ \underbrace{\frac{e}{v} \int_{-\infty}^{\infty} \Phi(\mathbf{r} - \mathbf{d}, z)\mathrm{d}z}_{\varphi_{\mathrm{el}}(\mathbf{r})} - \underbrace{\int_{-\infty}^{\infty} A_z(\mathbf{r} - \mathbf{d}, z)\mathrm{d}z}_{\varphi_{\mathrm{mag}}(\mathbf{r})} \right\}, \quad (5.20)$$

where the modulus operation is referred to as phase wrapping. In the absence of phase singularities this wrapping may be uniquely inverted. If phase singularities are present in the wave itself or have been artificially introduced during holographic reconstruction (e.g., due to local under-sampling or insufficient fringe contrast), dedicated phase unwrapping algorithms, fixing the non-trivial topology of the reconstructed wave, have to be employed (Ghiglia & Pritt, 1998). A more comprehensive discussion of this crucial point is given in Section 6.1.

The accuracy and precision of off-axial potential measurements is typically superior to other electron holographic methods (Twitchett, Dunin-Borkowski, & Midgley, 2006; Koch & Lubk, 2010; Latychevskaia, Formánek, Koch, & Lubk, 2010; Niermann & Lehmann, 2016), mainly because of the unambiguous linear reconstruction procedure involving no additional assumptions other than the band limitation of the wave function. Moreover, the reconstruction procedure facilitates a straightforward computation of reconstruction errors due to quantum noise and the transfer properties of the detector (Röder et al., 2014). For other wave reconstruction schemes, in particular those discussed further below, the error analysis is significantly more involved, rendering the quantitative interpretation more complicated.

The interpretation of the reconstructed phase in terms of potentials is most straightforward, if electrostatic potentials are present only. A large number of holographic studies revealing the origin and properties of electrostatic potentials in solids have been conducted to date. Notable examples are those on drift-diffusion potentials in semiconductors (Frabboni, Matteucci, & Pozzi, 1985, 1987; McCartney & Smith, 1994; Rau, Schwander, Baumann, Höppner, & Ourmazd, 1999; McCartney, Ponce, & Cai, 2000; Twitchett, Dunin-Borkowski, & Midgley, 2002; Gribelyuk et al., 2002; Wang, Hirayama, Sasaki, Saka, & Kato, 2002; Cooper et al., 2011; He, Cho, Jung, Picraux, & Cumings, 2013; Pantzer et al., 2014), charge accumulations on nanoparticles (Komrska, 1971; Chen et al., 1989; Simon, Zahn,

Lichte, & Kniep, 2006; Latychevskaia et al., 2010; Polking et al., 2012; Gatel, Lubk, Pozzi, Snoeck, & Hÿtch, 2013) and nanowires (den Hertog et al., 2009; Beleggia, Kasama, Dunin-Borkowski, & Pozzi, 2010), stray field distributions around and charge distributions inside operating field emission tips (Cumings, Zettl, McCartney, & Spence, 2002; de Knoop et al., 2014), material modification (e.g., electron hole-pair generation or secondary electron generation) due to electron beam irradiation (McCartney, 2005; Cooper et al., 2006; Park et al., 2014), and Mean Inner Potential variations in organic materials (Harscher, 1999; Simon et al., 2002, 2003, 2008).

Here, the Mean Inner Potential (Φ_0) refers to a spatial average of the complete electrostatic potential over a region large compared to the interatomic distance of in the local atomic structure. Its value is positive due to the dominating contribution of the core potentials and it can be shown that it is correlated to the extension of the atomic electron shell, i.e., the atomic radius (Bethe, 1928). Therefore, solids composed of light elements have a lower averaged electrostatic potential than heavy ones, where the dependency on Z is heavily nonlinear, because the atomic orbitals are filled by virtue of quantum rules. Consequently, Φ_0 is tightly connected to the chemical composition and electronic structure of the material. Because of the complicated nature of correlated electron gases in solids, the computation of Mean Inner Potentials from first principles poses a formidable task, in particular if the required accuracy is in the range of 0.1 V (Kruse, Rosenauer, & Gerthsen, 2003; Schowalter, Titantah, Lamoen, & Kruse, 2005; Schowalter, Rosenauer, Lamoen, Kruse, & Gerthsen, 2006; Kruse, Schowalter, Lamoen, Rosenauer, & Gerthsen, 2006). Persisting challenges in these computations pertain to the correct choice of the exchange correlation functional, influencing the orbitals of the outer electrons and hence the atomic radius and Φ_0, and the effect of surface dipoles forming at specimen boundaries, which may influence the holographically and theoretically determined potentials in a different way (Fig. 5.4).

Therefore, accurate Mean Inner Potentials obtained from previous experimental or theoretical studies provide valuable reference points, when investigating new materials and an incomplete compilation of Mean Inner Potentials of different materials is shown in Appendix B. The compiled values may be also used to holographically reconstruct the thickness map of a known material by solving (5.20) for the thickness, i.e., $v\varphi\left(\mathbf{r}\right)/\left(e\Phi_0\right) = t(\mathbf{r} - \mathbf{d})$. In Section 6.1 we will exploit the link between MIP and chemical composition to study the structure of a GaAs–Al$_{0.33}$Ga$_{0.67}$As core–shell nanowire in 3D by means of Electron Holographic Tomography.

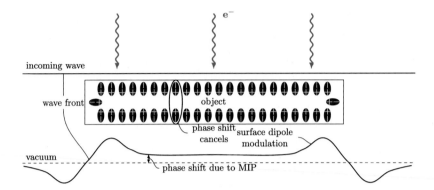

Figure 5.4 Mean inner potential and surface dipoles. Dipole layers on the side walls of the thin TEM specimen have a negligible influence on the projected potential in the middle because their potential quickly decays. Dipole layers at the top and the bottom do not exert any influence on the electric phase as their projected potentials integrate to zero. First principle calculations must be conducted in the same thin slab geometry to correctly model the holographic projection. Thus, results obtained from infinitely thick slabs (Kruse et al., 2003; Schowalter et al., 2005; Kruse et al., 2006; Schowalter et al., 2006; Pennington, Boothroyd, & Dunin-Borkowski, 2015) may overemphasize the influence of the dipole layer at the side walls (accumulating to an erroneous potential offset of the order of 0.1 V).

It is noteworthy that the projection law for the electric potential (5.20) also permits a straightforward determination of projected charges based on the following observation. When applying the 2D Laplace operator to the total electric phase image and integrating the result over a deliberately chosen region, one obtains the total charge Q within the chosen region according to

$$\int_{-\infty}^{\infty} \Delta \varphi_{\mathrm{el}}(\mathbf{r}) \, \mathrm{d}^2 r = \frac{e}{\nu} \iint_{-\infty}^{\infty} \Delta \left(\Phi \left(\mathbf{r} - \frac{\mathbf{d}}{2}, z \right) - \Phi \left(\mathbf{r} + \frac{\mathbf{d}}{2}, z \right) \right) \mathrm{d}^2 r \mathrm{d}z$$

$$\stackrel{(*)}{=} \frac{e}{\nu} \iint_{-\infty}^{\infty} \left(\Delta + \frac{\partial^2}{\partial z^2} \right) \Phi \left(\mathbf{r} - \frac{\mathbf{d}}{2}, z \right) \mathrm{d}^2 r \mathrm{d}z \qquad (5.21)$$

$$\stackrel{\substack{\text{Gauss} \\ \text{law}}}{=} -\frac{e}{\nu} \iint_{-\infty}^{\infty} \rho \left(\mathbf{r} - \mathbf{d}, z \right) \mathrm{d}^2 r \mathrm{d}z$$

$$= -\frac{e}{\nu} Q.$$

Here, we used $\lim_{|z| \to \infty} E_z = 0$ in the second step $(*)$.

Contrary to what is obtained when fitting a potential model to the reconstructed phase image (Latychevskaia et al., 2010), this charge measurement is not affected by a disturbed reference wave as long as the

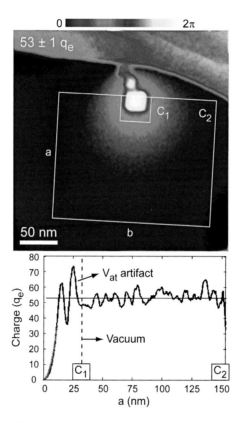

Figure 5.5 Holographic charge measurement. (Top) Reconstructed phase image of MgO nanocube, (bottom) charge enclosed by rectangular contour (side length $b \sim a$) as a function of the short side a. The constant fit outside of the particle, i.e., the total charge on the particle, is indicated by a black line. Within the particle a quadratic increase (indicated by a red line), i.e., $Q \sim a^2$, corresponding to a linear increase with the enclosed area, is observed.

reference region does not contain any charges. By the same token, neither any additional electric field nor any linear phase ramp introduced by the reconstruction process affects the measured charge, rendering this method very robust against artifacts afflicting off-axis electron holography. Evaluating holographically reconstructed phases in this way, elementary charges can be presently localized with a spatial resolution in the range of ten nanometers (Gatel et al., 2013). Fig. 5.5 shows an example of such a holographic contour charge evaluation on MgO nanocubes, positively charging up within the electron beam (for more details see Gatel et al., 2013).

A detailed analysis of the phase pertaining to a systematically diffracted beam scattered from a crystalline sample reveals that part of the phase shift may be traced back to small deviations from the perfect crystal structure, i.e., strain (cf. Section 6.5 for details). This phase shift is typically referred to as geometric phase[2] and can be measured by employing the off-axis setup in the dark-field imaging mode of the TEM (Hanszen, 1986; Hÿtch et al., 2008; Cooper, Barnes, Hartmann, Béché, & Rouvière, 2009; Hÿtch et al., 2011; Béché, Rouvière, Barnes, & Cooper, 2011; Lubk et al., 2014; Javon et al., 2014). In contrast to electrostatic and magnetic fields (5.20) at medium resolution imaging conditions, the projection of strain into the geometric phase follows a more complicated non-linear law, which has to be taken into account, when seeking the reconstruction of 3D strain field distributions (Lubk et al., 2014). We elaborate on this emerging technique in Section 6.5.

In the presence of both electric and magnetic fields the separation of electric and magnetic phase shifts requires the recording of a second holo-gram of the same specimen region with the orientation of the magnetic field and hence the sign of the corresponding magnetic phase shift flipped. Adding (subtracting) the reconstructed phase of the first hologram from that of the second then yields the projected electric (magnetic) potential (Tonomura, Matsuda, Endo, Arii, & Mihama, 1986). The orientation flip of the magnetic field can be achieved either by reversing the magnetization in the sample or by flipping the sample itself.

Slightly deviating from the electrostatic case, one typically seeks a recon-struction of the (projected) magnetic induction instead of the z-component of the magnetic vector potential in magnetic holography studies. To this end the magnetic phase is differentiated with respect to the in-plane coor-dinates x or y according to

$$\partial_{x,y}\varphi_{\mathrm{mag}}(\mathbf{r}) = \partial_{x,y}\oint \mathbf{A}\mathrm{d}s \qquad (5.22)$$

$$= \partial_{x,y}\oiint \mathbf{B}\mathrm{d}\mathbf{S}$$

$$= \int_{-\infty}^{\infty}\begin{pmatrix} B_y \\ -B_x \end{pmatrix}\mathrm{d}z\,,$$

---

[2] This geometric phase must not be confused with the geometric or Berry phases connected to non-trivial parameter spaces in quantum mechanics.

where Stoke's theorem was used in the second step to transform the original loop integral into an integral over the surface enclosed by the object and reference beam path. In practice, the derivative has to be performed numerically, which amplifies high-frequency noise in the recorded data. A low-pass is typically employed to suppress this effect.

Using the above reconstruction of magnetic phase shifts, a wealth of investigations has been conducted on magnetic materials, notably to study magnetic patterns forming in magnetic thin films (Tonomura, 1972; Lau & Pozzi, 1978; Dunin-Borkowski, McCartney, Smith, & Parkin, 1998; McCartney & Zhu, 1998; Murakami, Shindo, Oikawa, Kainuma, & Ishida, 2004; Kasama et al., 2005; Kasama, Dunin-Borkowski, & Eerenstein, 2006; Masseboeuf, Marty, Bayle-Guillemaud, Gatel, & Snoeck, 2009; Zhang, Ray, Petuskey, Smith, & McCartney, 2014; Marín et al., 2015; Röder et al., 2015), nanostructures (Dunin-Borkowski, McCartney, Kardynal, & Smith, 1998; Dunin-Borkowski et al., 2000, 2001; He, Smith, & McCartney, 2009a), or assemblies of nanoparticles (Seraphin et al., 1999; Dunin-Borkowski, Kasama, Wei, & Tripp, 2003; Dunin-Borkowski et al., 2004; Sugawara et al., 2007; Kasama et al., 2008; Yamamoto et al., 2008; Yamamoto, Hogg, Yamamuro, Hirayama, & Majetich, 2011; Varón et al., 2013), as well as magnetotactic bacteria (Dunin-Borkowski, McCartney, Frankel, et al., 1998; Kasama, Pósfai, et al., 2006), to reveal magnetic configuration in nanoparticles (Gao, Shindo, Bao, & Krishnan, 2007; Snoeck et al., 2008) and nanowires (Snoeck et al., 2003; Fujita, Hayashi, Tokunaga, & Yamamoto, 2006; Bromwich et al., 2006; Fujita et al., 2007; Junginger et al., 2007, 2008; He, Smith, & McCartney, 2009b; Akhtari-Zavareh et al., 2014), and to explore the polarization dynamics in magnetic read–write heads (Kim et al., 2008; Einsle et al., 2015), magnetic tunnel junctions (Xu et al., 2004; Javon, Gatel, Masseboeuf, & Snoeck, 2010; Park, Hirata, et al., 2012), fluxons in superconductors (Matsuda et al., 1989; Bonevich et al., 1993), or ferromagnetic shape memory alloys (Park, Murakami, Shindo, Chernenko, & Kanomata, 2003; Murakami et al., 2010; Park, Murakami, et al., 2012; Murakami et al., 2013).

Interestingly, the same integration law, which was used previously to compute charges from electric phase shifts, can also be applied to magnetic phases to measure the z-component of electric currents flowing in the sample according to

$$\iint \Delta \varphi_{\mathrm{mag}}(\mathbf{r}) \, d^2 r \quad = \int_{-\infty}^{\infty} \nabla \wedge \left(\begin{array}{c} -\partial_y \varphi(\mathbf{r}) \\ \partial_x \varphi(\mathbf{r}) \end{array} \right) d^2 r \qquad (5.23)$$

$$= \iint_{-\infty}^{\infty} \nabla \wedge \begin{pmatrix} B_x \left(\mathbf{r} - \frac{\mathbf{d}}{2}, z \right) \\ B_y \left(\mathbf{r} - \frac{\mathbf{d}}{2}, z \right) \end{pmatrix} d^2 r dz$$

$$\overset{\text{Ampere's}}{\underset{\text{law}}{=}} \mu_0 \iint_{-\infty}^{\infty} j_z \left(\mathbf{r} - \frac{\mathbf{d}}{2}, z \right) d^2 r dz$$

$$= \mu_0 \int_{-\infty}^{\infty} J_z(z) \, dz.$$

The sensitivity of such a measurement currently allows the determination of projected currents down to several $\mu A \times \mu m$ (Tavabi, Migunov, Savenko, & Dunin-Borkowski, 2014). Typical beam induced projected currents are several orders of magnitude smaller, hence do not affect a phase measurement.

If, however, the sample is magnetic, the pertaining current can easily become significant. Moreover, one may separate the current into a free and a bound component. The former may be neglected for non-metallic magnetic materials, yielding

$$J_z \overset{\text{rot}\mathbf{H}=0}{=} J_{z,b} = [\text{rot}\mathbf{M}]_z = \partial_x M_y - \partial_y M_x. \tag{5.24}$$

Consequently, the above evaluation allows the determination of the z-component of the bound magnetization current, independent of any magnetostatic model, usually required to link the holographically measured magnetic induction $\mathbf{B}$ to the magnetic properties of the sample. To date of this work bound current studies have not been reported in literature.

In a similar attempt, however, Beleggia, Kasama, and Dunin-Borkowski (2010) developed an ingenious method to directly determine the in-plane dipole moment of magnetic particles by evaluating the loop integral of the magnetic phase on *circles* around the nanoparticles according to

$$\mathbf{m} = \iint_{-\infty}^{\infty} \begin{pmatrix} M_x \\ M_y \end{pmatrix} d^2 r dz \tag{5.25}$$

$$\overset{(*)}{=} 2 \iint_{-\infty}^{\infty} \begin{pmatrix} B_x \\ B_y \end{pmatrix} d^2 r dz$$

$$= \frac{2}{\mu_0} \int_{-\infty}^{\infty} \left(-\partial_y \varphi(\mathbf{r}), \partial_x \varphi(\mathbf{r}) \right)^T d^2 r$$

$$= \frac{2}{\mu_0} \oint \varphi(\mathbf{r}) \, d\theta.$$

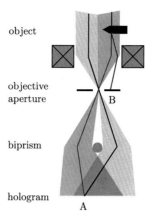

object

objective
aperture   B

biprism

hologram
            A

Figure 5.6 Off-axis holographic setup including "which-way" interpretation scheme. Both, the sideband intensity (contrast of interference fringe) and the centerband intensity (offset of interference fringe) of the black path interference are dampened in comparison to the gray path because of inelastic scattering (red) and scattering absorption (blue), respectively.

Contrary to the above integral methods for the charge and the current, the second equality (∗) holds exactly only if the *complete* magnetic material is enclosed by the *circle* and no spurious phase shifts due to residual phase wedges or stray fields leaking into the reference region are present. To overcome these restrictions Bellegia et al. suggested to evaluate contours of varying radius facilitating a removal of spurious contributions to $\mathbf{m}$ based on their characteristic polynomial dependency on the radius.

Intensity Information

In order to provide a physical explanation of the intensity attenuation terms occurring in (5.19), Off-Axis Holography is considered as a multiple parallel "which-way" experiment (Fig. 5.6). That is, each electron path from the image half plane superimposes with one particular path from the reference half plane to form the hologram intensity at the corresponding point on the detector. The laws of quantum mechanics imply that only those electrons interfere (i.e., create sideband intensity), which reach the detector without accumulating "which-way" information, permitting to decide, whether the electron had passed the biprism on the left or right hand side.

Accordingly, electrons, which have undergone inelastic scattering, do not contribute to the sidebands, because their paths have been fixed to the image half plane (red path in Fig. 5.6). The corresponding sideband intensity is attenuated at the inelastic scattering site with an associated attenuation coefficient μ_{in}. Additionally, the electrons may scatter elastically into angles sufficiently large to be intercepted by an aperture with opening semi-angle α. That process, referred to as scattering absorption (blue path in Fig. 5.6), also reduces the sideband intensity with an associated attenuation coefficient $\mu_{el}^{(\alpha)}$.

The conventional image intensity, on the other hand, is only dampened due to elastic and inelastic scattering into angles larger than the objective aperture, and consequently, inelastic scattering falling inside of the aperture still contributes to the image intensity.

Elastic and inelastic contributions can now be separated by combining the sideband intensity attenuation law (5.11)

$$I_{sb}(\mathbf{r}) = I_{sb,0} e^{-\int_{-\infty}^{\infty} \left(\frac{1}{2} \mu_{el}^{(\alpha)}(\mathbf{r},z) + \mu_{in}(\mathbf{r},z) \right) dz} \tag{5.26}$$

and the centerband intensity attenuation law (5.11)

$$I_{cb}(\mathbf{r}) = I_{cb,0} \left(\frac{1}{2} e^{-\int_{-\infty}^{\infty} \left(\mu_{el}^{(\alpha)}(\mathbf{r},z) + \mu_{in}^{(\alpha)}(\mathbf{r},z) \right) dz} + \frac{1}{2} \right) \tag{5.27}$$

along the following lines. Here, $I_{sb,0}$ and $I_{cb,0}$ denote the vacuum values, i.e., reference values in an empty hologram.

From the normalized logarithm of the sideband, one can extract the sum of the projected elastic and inelastic attenuation

$$\ln \frac{I_{sb}(\mathbf{r})}{I_{ref}} = -\frac{1}{2} \int_{-\infty}^{\infty} \left(\mu_{el}^{(\alpha)}(\mathbf{r}, z) + \mu_{in}(\mathbf{r}, z) \right) dz. \tag{5.28}$$

This Lambert–Beer type damping law for the sideband intensity has been experimentally verified in the pioneering work of McCartney and Smith (1994), when using Off-Axis Holography to determine the inelastic attenuation coefficient or its inverse, the inelastic mean free path length (MFPL) $\lambda = 1/\mu$, of bulk Si and MgO (cf. Table C.1).

The corresponding expression for the centerband intensity follows from subtracting the reference intensity, normalizing and taking the logarithm in (5.27)

$$\ln \left(\frac{I_{cb}(\mathbf{r})}{I_{ref}} - 1 \right) = -\int_{-\infty}^{\infty} \left(\mu_{el}^{(\alpha)}(\mathbf{r}, z) + \mu_{in}^{(\alpha)}(\mathbf{r}, z) \right) dz, \tag{5.29}$$

which is the well-known scattering absorption for conventional bright field imaging (Lenz, 1954; Zhang, Egerton, & Malac, 2010, 2012). From the above two expressions (5.28) and (5.29), the damping due to inelastic scattering into angles smaller than the objective aperture semi-angle can be evaluated to

$$
\ln\left(\frac{I_{cb}(\mathbf{r})}{I_{ref}} - 1\right) - 2\ln\frac{I_{sb}(\mathbf{r})}{I_{sb,0}} \tag{5.30}
$$
$$
= \int_{-\infty}^{\infty}\left(\mu_{in}(\mathbf{r}, z) - \mu_{in}^{(\alpha)}(\mathbf{r}, z)\right)dz.
$$

An inspection of (5.29) and (5.30) shows that elastic and inelastic attenuation coefficients can be separately determined from the holographically reconstructed center- and sideband intensities, only if the objective aperture semiangle α is sufficiently large to contain all inelastically scattered electrons, i.e.,

$$
\mu_{in}^{(\alpha)} \ll \mu_{el}^{(\alpha)} \quad \text{and} \quad \mu_{in}^{(\alpha)} \ll \mu_{in}. \tag{5.31}
$$

Thus, in order to be in the linear damping regime and to facilitate separability of elastic and inelastic damping, large acceleration voltages and large objective apertures should be employed in Off-Axis Holography.

The projection laws (5.29) and (5.30) together with the approximations (5.31) form the basis for the tomographic reconstruction of elastic and inelastic attenuation coefficients in a GaAs–Al$_{0.33}$Ga$_{0.67}$As core–shell nanowire detailed in Section 6.4. Moreover, the linear relationships (5.29), (5.30) in combination with simple theoretical expressions for the elastic (2.53) and inelastic (2.60) attenuation coefficients, facilitate an alternative TEM specimen thickness evaluation important for the quantitative determination of potentials or high-resolution (S)TEM contrasts, if other methods like the log-ratio method, the Kramers–Kronig sum rule method (Egerton, 1996), or Convergent Beam Electron Diffraction (Berriman, Bryan, Freeman, & Leonard, 1984) cannot be used due to instrumental or other limitations.

In the original work of McCartney and Smith (1994), the elastic contribution to the sideband intensity attenuation could be neglected, because a high-resolution TEM mode with a very large objective aperture has been employed. However, follow-up studies, employing Lorentz imaging TEM modes with small objective apertures and corresponding large elastic attenuation, also related the sideband damping to inelastic excitations only, thereby overestimating (underestimating) the linear inelastic attenuation co-

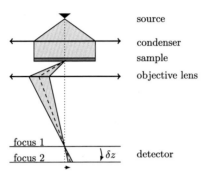

Figure 5.7 TIE setup in one lateral dimension. The defocus step is denoted as δz. The differential shift of the intensity with respect to the neighboring image point is indicated by an arrow.

efficients (mean free path lengths) in comparison to theory or values measured by Electron Energy Loss Spectroscopy (EELS). Appendix C contains a compilation of previously determined MFPL including a reevaluation taking into account the elastic damping. An improved agreement of the measured (in)elastic attenuation coefficients with theory or EEL measurements is clearly visible, demonstrating the quantitative character of off-axis electron holography.

5.2. TRANSPORT OF INTENSITY HOLOGRAPHY

In Off-Axis Holography the interference pattern is formed by two mutually inclined beams. An object wave function can be reconstructed from that hologram, if the object wave is band limited and an undisturbed reference wave is provided. In another class of holographic techniques, elaborated on in this section, the wave's phase is the solution to a differential equation, whose coefficients are determined experimentally (Allen, Oxley, & Paganin, 2001). One could say that the role of the reference wave is played by the Green's function involved in the solution to the differential equation.

The so-called Transport of Intensity Equation (TIE) method is the archetype member of this holographic principle (Teague, 1983). Its holographic data consists of at least two slightly defocused images, typically recorded in the near field of the object plane (Fig. 5.7), which are used to approximate the partial z-derivative of the intensity

$$\frac{\partial I(z)}{\partial z} = \frac{I(z + \delta z/2) - I(z - \delta z/2)}{\delta z} + \mathcal{O}\left(\delta z^2\right) \qquad (5.32)$$

and the intensity itself

$$I(z) = \frac{I(z + \delta z/2) + I(z - \delta z/2)}{2} + \mathcal{O}\left(\delta z^2\right). \qquad (5.33)$$

In order to minimize the $\mathcal{O}\left(\delta z^2\right)$ error, a focal step δz as small as possible is employed, where some lower limit is imposed by instrumental instabilities and the noise level (Ishizuka & Allman, 2005). A further improvement of the experimental input data may be achieved by employing higher-order discrete approximations of the z-derivative (Zheng, Xue, Xue, Bai, & Zhou, 2012).

The above experimental data may be used to transform the paraxial continuity equation (2.23a)

$$\frac{\partial I(\mathbf{r}, z)}{\partial z} = -\frac{1}{k_0} \nabla \cdot \mathbf{j}(\mathbf{r}, z) \qquad (5.34)$$

$$\overset{\text{pure}}{\underset{\text{state}}{=}} -\frac{1}{k_0} \nabla \cdot \left(I(\mathbf{r}, z) \nabla \varphi(\mathbf{r}, z)\right), \qquad (5.35)$$

into a first order partial differential equation for the probability current $\mathbf{j}$ or an elliptic partial differential equation for the wave's phase, depending on adopting a mixed state or pure state point of view.

Obviously, the two Cartesian components of the probability current cannot be solved from the continuity equation without further assumptions. The phase, however, may be retrieved, if appropriate boundary conditions for the latter can be provided (Lubk, Guzzinati, Börrnert, & Verbeeck, 2013). TIE phase retrievals have been reported for waves consisting of atoms (Fox et al., 2002), neutrons (Allman et al., 2000), X-rays (Nugent, 2010), electrons (Ishizuka & Allman, 2005), and visible light (Barty, Nugent, Paganin, & Roberts, 1998).

Within the TEM, a multitude of medium resolution studies on electric fields in, e.g., pn-junctions (Petersen, Keast, Johnson, & Duvall, 2007), nanoparticles (Petersen, Keast, & Paganin, 2008), grain boundaries (Schofield, Beleggia, Zhu, Guth, & Jooss, 2004; Dietrich et al., 2014), and phase plates (Pinhasi, Alimi, Perelmutter, & Eliezer, 2010) are reported. Magnetic field studies focused, e.g., on magnetic vortex states (Kohn,

Habibi, & Mayo, 2016) and magnetic nanoparticles (Petford-Long & De Graef, 2002). More recently, also strain field studies in the dark field setup have been conducted (Song, Shin, Kim, Oh, & Koch, 2013). To the best knowledge of the author atomic resolution reconstructions are still pending.

In contrast to Off-Axis Holography, TIE reconstructions neither require an additional optical element (the biprism) nor a vacuum reference region, which extends the scope of possible specimen geometries and lowers the coherency requirements for the pure state reconstruction. These advantages are opposed by the difficulties in accurately adjusting the defocus step and registering the defocused images, in particular in the high-resolution setting, and the problem of finding correct boundary conditions for the unknown phase. The latter is particularly problematic, if amplitude zeros, e.g., phase vortices (Lubk, Guzzinati, et al., 2013), are present in the wave field, because additional boundary conditions around these zeros need to be defined in this case (Lubk, Guzzinati, et al., 2013). Moreover, TIE reconstructions are mildly ill-conditioned for the reconstruction of low spatial frequency information, which complicates the study of long range phase variations, e.g., due to magnetic fields.

Similar to other holographic schemes, the presence of (partial) incoherence obscures the meaning of reconstructed phases (Gureyev, Roberts, & Nugent, 1995a; Zysk, Schoonover, Carney, & Anastasio, 2010; Schmalz, Gureyev, Paganin, & Pavlov, 2011), eventually rendering the TIE (5.35) and the corresponding phase retrieval invalid. It has been pointed out, however, that the TIE also permits to solve for the conservative probability current (Lubk, Guzzinati, et al., 2013), which corresponds to the first moment of the Wigner function along the momentum coordinate (2.94). Indeed, a considerable number of TIE phase reconstructions reported in literature pertain to the phase associated to the conservative probability current instead of the complete wave's phase as is shown below. In the following we discuss, how the reconstruction of the conservative probability current can be extended to the complete probability current and to higher-order moments of the Wigner function, which eventually amounts to a reconstruction of the complete quantum state. Again, we begin with a discussion of the mixed state case before proceeding to the wave reconstruction as a special case.

5.2.1 Quantum State Reconstruction

The Helmholtz decomposition of the probability current[3] into a conservative (curl-free) and solenoidal (divergence-free) part reads[4]

$$\mathbf{j}(\mathbf{r}, z) = \nabla \alpha(\mathbf{r}, z) + \begin{pmatrix} 1 \\ 1 \end{pmatrix} \wedge \nabla \beta(\mathbf{r}, z) . \tag{5.36}$$

When inserting the decomposed current into the TIE (5.35), the latter reduces to a simple Poisson equation (Teague, 1983; Paganin & Nugent, 1998; Allen, Faulkner, Oxley, & Paganin, 2001; Ishizuka & Allman, 2005)

$$\frac{\partial I(\mathbf{r}, z)}{\partial z} = -\frac{1}{k_0} \nabla \cdot \left(\nabla \alpha(\mathbf{r}, z) + \left(\begin{pmatrix} 1 \\ 1 \end{pmatrix} \wedge \nabla \beta(\mathbf{r}, z) \right) \right) \tag{5.37}$$

$$= -\frac{1}{k_0} \Delta \alpha(\mathbf{r}, z) .$$

Consequently, the Poisson formulation of the TIE reconstructs the conservative (curl-free) probability current $\mathbf{j}_\alpha = \nabla \alpha$ (Petruccelli, Tian, & Barbastathis, 2013; Lubk, Guzzinati, et al., 2013). Note, however, that a large class of quantum states, formed in the image plane of the TEM, does not possess conservative probability currents, which limits the significance of the Poisson formulation. For instance, phase singularities ($\nabla \wedge \nabla \varphi \neq 0$) present in the beam are always accompanied by solenoidal currents.

By employing two line defoci, $z_x := C_1 = -A_1$ and $z_y := C_1 = A_1$ (4.12), instead of one isotropic defocus ($z := C_1, A_1 = 0$), this obstacle may be removed, because the original 2D problem separates into two decoupled 1D continuity equations (Lubk & Röder, 2015b)

$$\frac{\partial I(\mathbf{r}, z)}{\partial z_{x,y}} = -\frac{1}{k_0} \partial_{x,y} j_{x,y}(\mathbf{r}, z) . \tag{5.38}$$

These two decoupled first order differential equations can be solved separately for the Cartesian components of the lateral probability current, if

---

[3] This is not to be confused with the Helmholtz decomposition of the normalized probability current $\mathbf{j}/I$ discussed in Paganin and Nugent (1998), Nugent and Paganin (2000), which is a fundamentally different quantity. For instance, $\mathbf{j}/I$ becomes singular at intensity zeros where $\mathbf{j} = 0$ remains well defined.

[4] Note that the divergence-free current may be obtained from a curl-free one by rotation of the local vectors around 90° in two dimensions.

giving suitable boundary conditions. Providing initial values on the boundary of the reconstruction domain yields

$$j_x(\mathbf{r}, z) = -k_0 \int_0^x \frac{\partial I(x', y, z)}{\partial z_x} dx' + j_x(x = 0, y, z) \qquad (5.39)$$

and

$$j_y(\mathbf{r}, z) = -k_0 \int_0^y \frac{\partial I(x, y', z)}{\partial z_y} dy' + j_y(x, y = 0, z). \qquad (5.40)$$

Alternatively, one may impose periodic boundary conditions, permitting the representation of the probability current as a Fourier series on the reconstruction domain (area Ω)

$$j_{x,y}(\mathbf{r}) = \sum_{\mathbf{k}} \tilde{j}_{x,y}(\mathbf{k}) e^{i\mathbf{kr}} \qquad (5.41)$$

with the Fourier coefficients

$$\tilde{j}_{x,y}(\mathbf{k}) = -i\frac{k_z}{k_{x,y}\Omega} \int_\Omega e^{-i\mathbf{kr}} \partial_{x,y} I(\mathbf{r}) \, d^2 r. \qquad (5.42)$$

In order to obtain a well-defined Fourier series one has to apply some regularization at $k_{x,y} \approx 0$ to suppress a large amplification of the inevitable experimental errors at small spatial frequencies (e.g., noise). Typically, a small region around $k_{x,y} \approx 0$ is regularized by multiplying with a suitable high-pass filter.

The probability current reconstructed with the help of the astigmatic TIE is directly proportional to projected electric and magnetic fields in and around the sample within the axial scattering regime (cf. Section 2.3), i.e.,

$$\mathbf{j}(\mathbf{r}, z) \sim -\frac{e}{v} \int_{-\infty}^{\infty} \begin{pmatrix} E_x(\mathbf{r}, z) \\ E_y(\mathbf{r}, z) \end{pmatrix} dz + \int_{-\infty}^{\infty} \begin{pmatrix} B_y(\mathbf{r}, z) \\ -B_x(\mathbf{r}, z) \end{pmatrix} dz. \qquad (5.43)$$

Moreover, a similar proportionality may be established for strain fields in strained crystal lattices, where the probability current pertaining to a systemically diffracted beam (reciprocal lattice vector $\mathbf{g}$) is proportional to the gradient of the displacement field $\mathbf{u}(\mathbf{r})$ according to $\mathbf{j_g} \sim -\nabla(\mathbf{g} \cdot \mathbf{u}(\mathbf{r}))$[5] (cf.

---

[5] If the displacement field also depends on z, a sinusoidal projection law has to be taken into account (cf. Section 6.5).

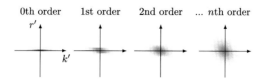

Figure 5.8 Transport of Intensity Quantum State Reconstruction. Moments with increasing order facilitate a Taylor expansion of the ambiguity function.

Section 6.5). Thus, astigmatic TIE reconstruction lends itself for mapping electric, magnetic and strain fields under partially coherent or incoherent illumination regimes. Moreover, the current pertaining to beam electrons, which have excited a dipole allowed transition of a core electron in the target, is linked to the difference probability of absorbing a virtual photon of left or right circular polarization. Therefore, the probability current may be also used to measure electron magnetic circular dichroic scattering in magnetic materials (Schattschneider, Stöger-Pollach, & Verbeeck, 2012).

A further generalization may be obtained by extending the reconstruction of both Cartesian components of the probability current to a complete quantum state reconstruction referred to as Transport of Intensity Quantum State Reconstruction in the following. This generalized TIE reconstruction principle is based on the observation that the density corresponds to the zeroth momentum $I(\mathbf{r}) = \int_{-\infty}^{\infty} W(\mathbf{r}, \mathbf{k}) d^2k$ and the probability current to the first momentum $\mathbf{j}(\mathbf{r}) = \int_{-\infty}^{\infty} \mathbf{k} W(\mathbf{r}, \mathbf{k}) d^2k$ of the phase space distribution along the momentum coordinate (cf. Section 2.5). By extension, all moments correspond to the derivatives of the ambiguity function (along $\mathbf{r}'$ at $\mathbf{r}' = 0$, cf. Section 2.5), and hence a given number of moments can be used to approximate the ambiguity function (Fig. 5.8) by its Taylor series.

Consequently, a generalization of the TIE reconstruction to higher-order moments eventually exposes the complete quantum state, which is experimentally realized by measuring higher-order derivatives of the intensity with respect to differently oriented line defoci. As will be shown below, higher-order phase space moments can be computed from that data by solving a hierarchy of continuity equations.

To illustrate the working principle, the corresponding expression for the second-order generalized current $j_{xx}^{(2)}$ is given as an example. First, we approximate the derivative of the probability current with respect to the line focus by the finite difference between two separate probability current reconstructions at two slightly different line foci, i.e.,

$$\frac{\partial j_x \left(\mathbf{r}, z \right)}{\partial z_x} \approx \frac{j_x \left(\mathbf{r}, z + \delta z_x/2 \right) - j_x \left(\mathbf{r}, z - \delta z_x/2 \right)}{\delta z_x}. \tag{5.44}$$

After inserting the definition of the probability current (2.94) and the free space 1D Liouville equation (2.110) into the above expression, one obtains

$$\frac{\partial j_x \left(\mathbf{r}, z \right)}{\partial z_x} = \frac{\partial \int_{-\infty}^{\infty} k_x W \left(\mathbf{r}, \mathbf{k}, z \right) \mathrm{d}^2 k}{\partial z_x} \tag{5.45}$$

$$= -\int_{-\infty}^{\infty} \frac{k_x}{k_0} k_x \partial_x W \left(\mathbf{r}, \mathbf{k}, z \right) \mathrm{d}^2 k$$

$$= -\frac{1}{k_0} \partial_x j_{xx}^{(2)} \left(\mathbf{r}, z \right),$$

a continuity equation for the $j_{xx}^{(2)}$ component of the generalized current. This equation is completely analogous to the TIE and can be integrated with the same methods, therefore. By induction one can show that there is a complete, so-called hydrodynamic, hierarchy of identical equations relating the nth order current to the $(n+1)$th order by virtue of a continuity equation (Trahan & Wyatt, 2006)

$$\frac{\partial j_{nxx}^{(n)} \left(\mathbf{r}, z \right)}{\partial z_x} = -\frac{1}{k_0} \partial_x j_{(n+1)\times x}^{(n+1)} \left(\mathbf{r}, z \right). \tag{5.46}$$

Other directional components, such as $j_{xy}^{(2)}$, may be derived along corresponding lines.

The experimental realization of the above reconstruction principle requires minute control of line defoci, which is challenging in particular in the high-resolution regime. It has been demonstrated, however, that hardware aberration correctors facilitate the recording of a line focal series with defocus steps down to 2 nm, which permits the reconstruction of probability currents down to atomic resolution (Lubk, Béché, & Verbeeck, 2015). These results indicate that the acquisition of a matrix of differential line foci with a step size of 2 nm, as required for a complete quantum state reconstruction at atomic resolution based on the hydrodynamic hierarchy of continuity equations, is indeed feasible. To date of this work an experimental proof of concept of this quantum state reconstruction principle is still pending. Its realization could overcome a number of principle problems of inelastic Off-Axis Holography, such as a limited holographic shear range or the recording and alignment of a biprism rotation series, thereby opening new avenues to the investigation of characteristic electron-energy-loss events within materials. For instance, the hitherto elusive determination of the, in general non-local, dielectric response tensor of inhomogeneous targets requires knowledge about the quantum state of the inelastically scattered electrons as provided by the Transport of Intensity Quantum State Reconstruction.

Before proceeding with pure state TIE reconstructions, the interested reader is invited to have a short glance back to the end of Section 4.2, where it was shown that the aberration correctors in modern TEMs can be used to mimic differential equations other than the paraxial equation discussed within the scope of the TIE method. For instance, a small variation of the spherical aberration corresponds to a biharmonic equation (4.27). Further insight into the significance of that equation may now be obtained by inserting the free-space paraxial Klein–Gordon–Fock equation (2.21) into the right hand side of (4.27) yielding

$$\frac{\partial \Psi (\mathbf{r}, C_3)}{\partial C_3} = \frac{i}{4k_0} \frac{\partial^2 \Psi (\mathbf{r})}{\partial^2 z}. \qquad (5.47)$$

Using this result for computing the partial derivative of the intensity with respect to the spherical aberration, i.e.,

$$\begin{aligned}
\frac{\partial I (\mathbf{r}, C_3)}{\partial C_3} &= \frac{i}{4k_0} \left(\Psi^* (\mathbf{r}) \partial_z^2 \Psi (\mathbf{r}) - \Psi (\mathbf{r}) \partial_z^2 \Psi^* (\mathbf{r}) \right) \qquad (5.48) \\
&= \frac{i}{4k_0} \partial_z \left(\Psi^* (\mathbf{r}) \partial_z \Psi (\mathbf{r}) - \Psi (\mathbf{r}) \partial_z \Psi^* (\mathbf{r}) \right) \\
&= -\frac{1}{4k_0} \partial_z j_z,
\end{aligned}$$

finally reveals a remarkable proportionality to the partial derivatives of the current density along z. This result remains valid also in the case of partial coherence, i.e., mixed quantum states, because the intensity I and the probability current j_z retain their meaning after performing the sum over all incoherent states on the left and right hand side of the above expression. As discussed in Lubk and Röder (2015a), similar arguments may be employed to facilitate a quantum state reconstruction by employing higher-order anisotropic aberrations instead of line foci.

5.2.2 Wave Reconstruction

As noted at the beginning of this section, the TIE (5.35) with given I and $\partial I/\partial z$ represents an inhomogeneous linear elliptic partial differential equation (PDE) for the wave's phase φ. The corresponding mathematical theory (Gilbarg & Trudinger, 1983; Evans, 1998) based on the Lax–Milgram theorem ensures existence and uniqueness of (weak) solutions, if the reconstruction domain D with strictly positive intensity $I > 0$ is *simply connected* and appropriate boundary conditions are provided. Possible choices are Dirichlet (Gureyev, 2003), Neumann (Volkov, Zhu, & De Graef, 2002), Robin (mixed Dirichlet and Neumann), or periodic boundary conditions (Van Dyck & Coene, 1987; Gureyev & Wilkins, 1998; Paganin & Nugent, 1998; Ishizuka & Allman, 2005).

These conditions put some restrictions on the applicability of TIE phase retrieval. First, the boundary conditions must be deducible from previous knowledge on the scattered wave. For instance, Dirichlet boundary conditions, $\varphi\,(\partial D) = 0$, setting the phase at the boundary ∂D to zero, are applicable, if the object is surrounded by vacuum. Similarly, beams that have been deliberately truncated by some aperture call for the application of Neumann boundary conditions, $\nabla_{\mathbf{n}}\varphi\,(\partial D) = 0$ (Gureyev, Roberts, & Nugent, 1995b; Zuo, Chen, & Asundi, 2014), setting the directional gradient of the phase (and hence the probability current) at the aperture to zero. Second, wave fields containing amplitude zeros $I = 0$, such as phase vortices, violate the simple connectivity of the reconstruction domain and have to be treated carefully. Indeed, the presence of amplitude zeros has been frequently completely excluded (Van Dyck & Coene, 1987; Gureyev & Wilkins, 1998; Ishizuka & Allman, 2005; Schmalz et al., 2011) or stated to yield non-unique or failing reconstructions (Gureyev et al., 1995a; Allen & Oxley, 2001; Peele et al., 2004) in previous TIE studies. Third, TIE reconstructions invoking Neumann or periodic boundary conditions are mildly ill-conditioned with respect to low spatial frequencies (Ishizuka & Allman, 2005). This property of TIE reconstructions originates from the small susceptibility of small spatial frequency wave components with respect to a defocus and has been noted already in the context of TIE quantum state reconstructions above.

In the following, TIE phase retrieval of arbitrary wave fields including amplitude zeros $I = 0$, e.g., phase vortices, is considered. Such zeros are ubiquitous features in wave fields (Nye & Berry, 1974; Allen, Faulkner, Oxley, et al., 2001), e.g., a complicated pattern of phase vortices already occurs upon interference of three plane waves (O'Holleran, Padgett, & Dennis,

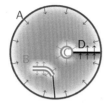

BC type	φ_l	$\nabla_\mathbf{n}\varphi_l$
A		$f(\partial D_A)$
B	$m\pi + \varphi_r$	$\nabla_\mathbf{n}\varphi_r$
C		$f(\partial D_C)$
D	$w2\pi + \varphi_r$	$\nabla_\mathbf{n}\varphi_r$

Figure 5.9 TIE boundary condition (BC) types with arrows indicating directional derivatives denoted by ∇_n, and $\varphi_{l,r}$ denoting the phases on the left/right hand side of the boundary (the latter is only defined at interior boundaries B, D): (A) outer Neumann boundary conditions, (B) normal zero with constant derivative and possible $m \cdot \pi$ jumps of the phase, (C) Neumann boundary conditions around vortex, and (D) phase sheet change with constant derivative and $w \cdot 2\pi$ jump of the phase.

2006). In order to solve the TIE in the presence of amplitude zeros, the latter have to be excluded from the reconstruction domain and appropriate boundary conditions have to be defined at the edges of the removed zeros. Since the remaining holey domain is not simply connected anymore, thereby violating the previously mentioned existence and uniqueness theorem, additional cuts from the zeros to the boundary, rendering the domain simply connected again, have to be introduced.

Therefore, three different inner boundary conditions have to be considered in addition to the outer boundary conditions, denoted by A (see Fig. 5.9). They occur (B) at lines of normal zeros, where the phase jumps integer multiples of π, depending on the order of the zero; (C) at zero dimensional phase singularities (with winding number $w \neq 0$ around singularity); and (D) at lines connecting the singular zeros (C) to the boundary, where phase jumps comprising integer multiples of 2π can occur (Fig. 5.9). The latter boundary conditions determine the topology, i.e., the winding number, of the respective phase singularity at (C). While any boundary conditions at (A) and (C), consistent with the topology fixed by the boundary conditions at (D), are allowed in principle, an often well-adapted choice to practical problems are Neumann boundary conditions because they do not impose any topology (unlike Dirichlet or periodic boundary conditions). It has to be noted, furthermore, that simple zeros and phase singularities might superimpose in arbitrary ways. Having identified the boundaries around the zeros, the central obstacle toward a TIE reconstruction consists of finding the correct boundary conditions to the unknown phase, which is equivalent to determining all possible m, w as well as the derivatives of the phase on (A) and (C). Three observations help to significantly reduce the (infinitely) large number of possibilities:

(i) Due to the linear nature of the TIE, solutions to almost exact boundary conditions are close (in a weak sense) to the exact solution. Therefore, Neumann boundary conditions $\nabla_{\mathbf{n}}\varphi\,(\partial D) = 0$ for type (A) and (C) boundaries are acceptable, if the wave function close to (A) and (C) is flat. That is often the case in practice, e.g., because the object is well localized within the limits of the reconstruction domain. The remaining set of possible boundary conditions is now reduced to all possible combinations of m_k (for a given set of K normal zero manifolds) and w_n (for a given set of N singularities and cuts connecting them to the outer boundary).

(ii) The different solutions of the TIE (5.35) pertaining to different boundary conditions show a different behavior upon large defoci. This observation provides the basis for a consistency check with a reference intensity $I\,(z)$ at a sufficiently large defocus z, singling out the correct solution (Allen, Faulkner, Oxley, et al., 2001). This procedure can be very efficient, when starting to test the small m_k and w_n first, because higher-order vortices are usually unstable and dissolve into first order ones (Freund, 1999; Lubk, Clark, Guzzinati, & Verbeeck, 2013).

(iii) Since TIE reconstructions may be performed at deliberately chosen defoci, one can pick a particular plane z with a minimal set of normal zeros, which in turn minimizes the set of test solutions in (ii). Such a plane exists, because normal zeros, in contrast to phase singularities, are not protected topologically under defocus. By the same token a large set of differently focused images can be used to identify the topologically protected phase singularities. The latter search may be further improved by noting that vortices show up as stable minima of both the intensity and the reconstructed curl-free current (5.37) in the focal series.

To give a vivid example of the above principles, a TIE phase reconstruction of an electron vortex beam is shown in Fig. 5.10. Such beams recently attracted considerable attention for their interesting physical properties and potential applications (Verbeeck, Tian, & Schattschneider, 2010; Uchida & Tonomura, 2010; McMorran et al., 2011; Lubk, Clark, et al., 2013; Juchtmans, Béché, Abakumov, Batuk, & Verbeeck, 2015; Edström, Lubk, & Rusz, 2016). Some details of that reconstruction are summarized in the figure caption; the reader interested in a comprehensive description of this experiment is referred to Lubk, Guzzinati, et al. (2013).

We finally elaborate on a more detailed discussion of reconstruction errors, due to experimentally recorded data being inconsistent with the

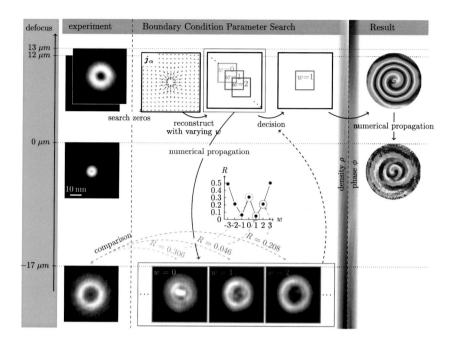

Figure 5.10 TIE reconstruction scheme. The left column shows selected recorded intensities from a focal series of a vortex beam. The boundary condition parameter search comprises a set of possible solutions with winding number $w = \ldots, 0, 1, 2, \ldots$. All solutions are propagated to a different plane and compared to the experimental image recorded at this particular defocus value. According to a R-factor consistency check, the $w = 1$ solution fits best. The result column on the right shows the reconstructed phase with a $w = 1$ singularity and the phase of the numerically focused reconstructed wave.

pure state continuity equation (5.35), because of partial coherence, noise and the $\mathcal{O}\left(\delta z^2\right)$ correction terms in (5.32), (5.33). To that end we consider the impact of small (incoherent) fluctuations δI around a (pure) state $I_p = \left|\Psi_p\right|^2$ with corresponding phase φ_p in the TIE reconstruction

$$\frac{\partial \left(I_p + \delta I\right)}{\partial z} = -\frac{1}{k_0} \nabla \cdot \left(I_p + \delta I\right) \nabla \left(\varphi_p + \delta\varphi\right) . \qquad (5.49)$$

After subtracting (5.35) with $I = I_p$ and $\varphi = \varphi_p$, one obtains an elliptic PDE for the phase error $\delta\varphi$

$$\frac{\partial \left(\delta I\right)}{\partial z} + \frac{1}{k_0} \nabla \cdot \delta I \nabla \varphi_p = -\frac{1}{k_0} \nabla \cdot I_p \nabla \delta\varphi . \qquad (5.50)$$

Accordingly, the right hand side (homogeneous part) of the above equation is equivalent to the right hand side of the pure state TIE (5.35), whereas the left hand side (the inhomogeneous term) is determined by both the fluctuations δI and the pure state phase φ_p.

Finding the solution of the above PDE, i.e., computing the reconstruction error in experimental TIE studies, is not as straightforward as in case of Off-Axis Holography, where simple error propagation laws apply. For instance, one has to know the exact (pure state) solution in order to compute the phase error, but only the defective result of the reconstruction is available. In connection with the problematic determination of the phase boundary conditions, the resulting unknown reconstruction error may occasionally hamper the applicability of TIE phase retrieval, in particular, when very accurate reconstructions of small spatial frequencies are required.

Nevertheless, the linear structure of the equation permits to draw a number of general conclusion independent of the exact structure of the underlying wave. Because of the affine structure of the solutions of an inhomogeneous linear PDE, $\delta\varphi$ can grow large, if the inhomogeneous perturbation term surmounts the pure state one, i.e., at sufficiently stationary regions, where

$$\frac{\partial I_p}{\partial z} \leq \frac{\partial (\delta I)}{\partial z} + \frac{1}{k_0} \nabla \cdot \delta I \nabla \varphi_p . \tag{5.51}$$

Noting that stationary regions correspond to wave components of small spatial frequencies the above analysis provides another perspective on the problematic reconstruction of small spatial frequencies with the TIE method. Thus, any source of inconsistency, such as the above noted $\mathcal{O}\left(\delta z^2\right)$ propagation errors, might contribute to large errors in stationary regions. Moreover, the I_p factor in the homogeneous part of (5.50) implies that large erroneous phase gradients may show up in regions of small intensities such as phase vortices. Consequently, TIE reconstructions from regions with small intensity ($I \ll 1$) suffer from increased phase noise. This behavior is similar to the phase noise in Off-Axis Holography (Lenz, 1988) and follows from more fundamental uncertainty relations between the intensity and the phase (Fick, 1988).

5.3. FOCAL SERIES INLINE HOLOGRAPHY

Transport of Intensity Quantum State Reconstruction reconstructs generalized probability currents in the electron beam and hence its quan-

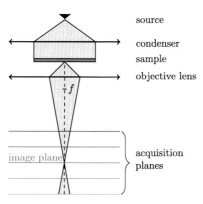

source

condenser

sample

objective lens

acquisition
planes

image plane

Figure 5.11 Principle optical setup of Focal Series Inline Holography. Accordingly, the defocus interval is ideally bound by the far field located below (i.e., focal plane f) and above the image plane.

tum state from a hierarchy of differential equations, whose coefficients are determined from a set of slightly defocused images recorded in the near field of the object. Focal Series Inline Holography, also referred to as quantum state tomography (Lubk & Röder, 2015a) in the typical setup employed in TEM (Fig. 5.11), intends to recover the quantum state from a set of defocused images taken over a larger focal range compared to TIE phase retrieval (Fig. 4.6). It is successfully employed to study mixed (i.e., incoherent) quantum states of matter (e.g., atoms) and light (Smithey et al., 1993; Breitenbach et al., 1997; Schleich, 2001). Quantum state reconstructions of electron beams in the TEM, however, have not been reported to date of this work.

Focal series reconstructions of wave functions (pure states) (Schiske, 1968, 1973; Kirkland, 1984; Coene et al., 1992, 1996; Thust et al., 1996; Koch, 2008), on the other hand, are widely-used to reconstruct wave functions with atomic resolution (Coene et al., 1992; Kawasaki, Takai, Ikuta, & Shimizu, 2001; Allen et al., 2004; Kirkland & Meyer, 2004; Hsieh, Chen, Kai, & Kirkland, 2004; Tillmann, Houben, Thust, & Urban, 2006). In the medium resolution regime focal series studies aimed at mapping electric potentials in grain boundaries (Dietrich et al., 2014) or strain distributions in transistor channels (Koch, Özdöl, & van Aken, 2010; Song et al., 2013). Additionally, the structure of phase vortices at caustics has been explored (Petersen et al., 2013). The widespread use of that technique is owed partially to the availability of focal series reconstruction packages (TrueImage – Kübel & Thust, 2006; IWFR – Allen et al., 2004), which permit a relatively comfortable and semi-automated reconstruction of wave functions from focal series acquired at a TEM. The reconstruction algorithms are founded in the principles of free wave propagation as discussed below. However, the reconstruction of unique wave functions from a finite set of defocused images remains somewhat elusive. In particular questions concerning the required focal range, the precise meaning of the reconstructed wave in the presence of incoherence and noise, or the prerequisites toward coherence, noise or spurious aberrations, are not conclusively answered as of today. Indeed, precise reconstruction error estimates are typically not available and non-unique reconstructions, e.g., depending on the starting guess or other parameters in typical iterative reconstruction algorithms are reported in literature (Fienup & Wackerman, 1986; Seldin & Fienup, 1990; Allen, Faulkner, Nugent, Oxley, & Paganin, 2001; Martin & Allen, 2007). We shall elaborate on these issues below.

The acquisition of an artifact-free focal series in a TEM that is free from variations other than defocus, represents a formidable experimental challenge and a multitude of approaches to manipulate the defocus, subject to different advantages and disadvantages have been devised. The thin lens model[6] introduced in Section 4.1 (4.14) indicates three ways to change the effective defocus δz (facilitating a focal series) by varying z_{obj}, f, or z_{img}, respectively. In a TEM, typically, the object plane (i.e., the specimen or some conjugated plane) and the image plane (i.e., the detector plane or some conjugated plane) remain fixed, however. Thus, using a single lens, focal series are typically recorded by varying the lens excitation, hence f.

It is important to note that by varying the focal length f not only in the effective defocus δz but also in the magnification M changes with respect to a free-space propagation. This rescaling has to be taken into account, when reconstructing wave functions from an experimental focal series. Moreover, changing the electron optical system induces additional image rotations, distortions, shifts as well as parasitic aberrations (Allen et al., 2004; Koch, 2008, 2014). The ramifications of the latter on the quantum state detected in the image plane are accounted for by the generalized Transmission Cross Coefficient (4.50) in case of the most important isoplanatic aberrations. The modulations of the quantum state induced by non-isoplanatic or higher-order chromatic aberrations require more elaborate computations (cf. Section 4.3). Similarly, the modulation transfer function of the camera

---

[6] Indeed, strongly excited electron lenses cannot be described by a thin lens anymore, and require more elaborate optical models instead.

(Niermann et al., 2012; Lubk et al., 2012) has to be taken into account in the restoration process (Meyer, 2002).

These additional modifications of the quantum state represent a considerable challenge to focal series reconstructions as compared to the previously discussed reconstruction schemes. Advances in the fields of hardware aberration correction and image processing helped in significantly reducing the impairing effect on the focal series. For instance, the varying influence of spatial coherence upon changing the defocus in a hardware-corrected TEM may be deconvolved from the recorded intensity to some extent (Koch, 2008) (Fig. 4.6). Moreover, several methods have been devised to correct for the effect of spurious aberrations and distortions developing throughout the focal series. For the following considerations, we shall assume that the defocus can be adjusted perfectly in the experiments by employing these methods. We proceed with the focal series reconstruction of general mixed quantum states (i.e., quantum state tomography) and consider the reconstruction of wave functions (pure states) subsequently.

5.3.1 Quantum State Reconstruction

In contrast to generalized Off-Axis Holography and Transport of Intensity holography, which allow a mixed quantum state reconstruction only after acquiring multiple holograms, a focal series already comprises the required information for a mixed quantum state reconstruction and no additional experimental effort is required. The quorum is given by the focal series itself. In the following, we elaborate on the ensuing reconstruction algorithm. The free-space dynamics of the Wigner function in the paraxial regime is governed by the free-space Liouville equation (2.110), and the Wigner function of an isotropically defocused quantum state (i.e., the solution of the Liouville equation) reads

$$W(\mathbf{r}, \mathbf{k}, z) = W\left(\mathbf{r} - \frac{z}{k_0}\mathbf{k}, \mathbf{k}, 0\right). \tag{5.52}$$

As stated previously, this is a shear in 4D phase space. Therefore, an isotropic defocus series corresponds to a shear series in 4D phase space (see Fig. 5.12)

$$I(\mathbf{r}, z) = \int_{-\infty}^{\infty} W\left(\mathbf{r} - \frac{z}{k_0}\mathbf{k}, \mathbf{k}\right) d^2k. \tag{5.53}$$

One can transform this shear series into a tomographic tilt series by introducing the following dimensionless phase space coordinates (scaling factor k_σ)

$$\mathbf{k}^s = \frac{k_\sigma^{-1}\mathbf{k}}{\cos\alpha} - k_\sigma \mathbf{r}\sin\alpha \quad \text{and} \quad \mathbf{r}^s = k_\sigma \mathbf{r}\cos\alpha. \tag{5.54}$$

Experimentally, the rescaling of the position coordinate may be achieved by numerically magnifying the recorded intensities. The shear along the momentum axis does not affect the recorded intensity and may be assumed implicitly therefore. In rescaled coordinates the focal series reads

$$I^s(\mathbf{r}^s, \alpha) = \int_{-\infty}^{\infty} W\big(k_\sigma^{-1}(\mathbf{r}^s\cos\alpha - \mathbf{k}^s\sin\alpha), k_\sigma(\mathbf{k}^s\cos\alpha + \mathbf{r}^s\sin\alpha)\big)\, d^2k^s$$

$$\tag{5.55}$$

with the rescaled recorded intensity

$$I^s(\mathbf{r}^s, \alpha) = k_\sigma^{-2}\frac{I\big(k_\sigma^{-1}\mathbf{r}^s\cos^{-1}\alpha, z\big)}{\cos^2\alpha}. \tag{5.56}$$

The above integral transformation corresponds to a generalized Radon transformation (cf. Section 3.2), referred to as a restricted K-plane transform K_{2r}^4 (see below for details)

$$I^s(\mathbf{r}^s, \alpha) = K_{2r}^4\big\{W(\mathbf{r}^s, \mathbf{k}^s)\big\}. \tag{5.57}$$

The dimensionless phase space tilt angle

$$\alpha = \arctan\underbrace{\frac{zk_\sigma^2}{k_0}}_{F^{-1}} \tag{5.58}$$

is equal to the arctan of the inverse Fresnel number F, which serves as a measure to distinguish between the near ($|F| \gg 1$) and the far field ($|F| \ll 1$) regime of propagation of pure states with phase space widths $k_\sigma \sim r_\sigma^{-1}$ related by the uncertainty principle. Accordingly, the modulus of the rotation angle α is smaller (larger) than $\pi/4$ in the near (far) field.

As a consequence of the Fourier slice theorem (cf. Section 3.2.3), the projections of the tilted Wigner function (5.61) correspond to zero sections of the tilted ambiguity function. Consequently, the latter may be tomographically reconstructed by appropriately assembling a sufficient number of sections, which can be done by tomographic reconstruction algorithms such as a filtered back projection or an algebraic reconstruction algorithm

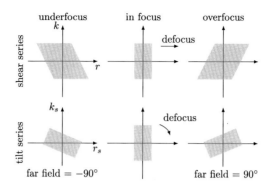

Figure 5.12 Inline holography as quantum state reconstruction. The shear series of the defocused Wigner function is shown in the first row. The corresponding tilt series obtained by suitably transforming the phase space (5.54) is depicted in the second row.

(cf. Section 3.6). The corresponding phase space diagram of the inline phase space reconstruction principle is depicted in Fig. 5.12.

From the above formulation of a focal series as tilt series in phase space we may directly derive a number of important principles, applying to the focal series reconstructions of both mixed and pure states. First, a tilt interval of $[-\pi/2, \pi/2[$ is required for a complete tomographic reconstruction (cf. Section 3.2), corresponding to a defocus interval ranging from the far field at $z \ll 0$ to the far field at $z \gg 0$. Second, the tomographic reconstruction is optimally performed with equally spaced tilt intervals (tilt increment $\delta\alpha = \text{const.}$), which translated to unequally space defocus intervals $\delta z(\alpha)$ scaled according to

$$\delta z(\alpha) = \frac{k_0}{k_\sigma^2}\Big(\tan(\alpha + \delta\alpha) - \tan\alpha\Big) \qquad (5.59)$$

$$\overset{\delta\alpha \ll 1}{\approx} \frac{k_0}{k_\sigma^2}\frac{2}{\cos(2\alpha) + 1}\delta\alpha$$

in the case of the free-space propagation discussed here. Therefore, a linear defocus sampling is optimal in the limit of small α (near field) only, whereas larger defocus steps should be used, when approaching the far field. Third, the tilt series lives in 4D phase space, that is, each individual defocused image is obtained by projecting along a 2D plane in phase space tilted by the defocus. Such a projection transformation represents a generalization of the 2D Radon transformation, referred to as K-plane transform K_2^4 (e.g., Keinert, 1989), if the projections are performed over all 2D planes in 4D phase space. An isotropic defocus, however, only facilitates a subset of projections. Similar to the incomplete tilt interval problem ("missing wedge") encountered in the 2D Radon transformation (cf. Section 3.2.3), the tomographic reconstruction of a mixed quantum state from an isotropic tilt series is ill-posed and suffers from artifacts due to the missing projections. This restriction may be lifted, if line foci of varying orientation (where one direction remains in focus) have to be employed (cf. Section 5.2), which permit arbitrary shears in 4D phase space. The complexity of the projection set is the same as for the related off-axis quantum state reconstruction synthesizing phase space from slices in the 4D density matrix representation (cf. Section 5.1). It poses a formidable challenge to record astigmatic focal series ranging from the near to the far field at presently available TEM instruments and the experimental realization of a 4D mixed quantum state reconstruction is still pending to date of this work. In the following, we focus on more simple cases, where the phase space distribution contains additional symmetries allowing a reduction of complexity.

We will now consider the case, where the in-focus intensity only depends on one spatial coordinate (x without loss of generality), i.e.,

$$W(\mathbf{r}, \mathbf{k}) = W(x, \mathbf{k}). \tag{5.60}$$

This permits a reduction of the original 4D problem to 2D in the following way. Performing the same coordinate transformation as above, one obtains the familiar 2D Radon transformation

$$k_\sigma^{-1} \frac{I\left(k_\sigma^{-1} \cos^{-1} \alpha x, z\right)}{\cos \alpha} = \mathcal{R}\left\{\overline{W}\left(x^s, k_x^s\right)\right\} \tag{5.61}$$

relating the reduced Wigner function

$$\overline{W}\left(x, k_x\right) = \int_{-\infty}^{\infty} W(x, \mathbf{k}) dk_y \tag{5.62}$$

with its defocused projections. As a consequence of the missing y-dependency the reduced Wigner function can be evaluated in terms of a 2D phase space spanned by x and k_x. Note that the average of an arbitrary Wigner function in y-direction can be reconstructed along the same lines, yielding a reduced Wigner function corresponding to the projection of the original one along y and k_y. Other symmetries permitting a dimensional reduction are radially symmetric quantum states or quantum states separable in Cartesian coordinates (Cámara, Alieva, Rodrigo, & Calvo, 2009).

5.3.2 Wave Reconstruction

In the following, we consider focal series wave reconstruction as a quantum state tomography restricted to pure states. Note, however, that a certain violation of the pure state assumption, e.g., due to inelastic scattering at the sample, the partial coherence of the electron emitter, or the point spread of the detector (Niermann et al., 2012), can never be completely avoided, and must be taken into account therefore (Martin et al., 2006). In a hardware-corrected TEM or at medium (nanometer) resolution the defocus remains as the main source of aberrations. We furthermore assume that the effects of temporal incoherence may be sufficiently suppressed due to chromatic aberration correction and energy filtering. In (4.53), we showed that the impact of a finite incoherent source size on the recorded intensity reduces to a mere convolution of the intensity in the presence of defocus only, i.e.,

$$I(\mathbf{r}, z) = \int_{-\infty}^{\infty} I^{(coh)}\left(\mathbf{r} - \frac{z}{k_0}\mathbf{k}', 0\right) I_S(\mathbf{k}') \, \mathrm{d}^2 k'. \qquad (5.63)$$

This smearing may be removed by deconvolving the recorded image with the source distribution $I_S(\mathbf{k}')$ within the limits determined by noise (Koch, 2008; Lubk & Röder, 2015a). From a phase space perspective, such a deconvolution corresponds to a purification of the quantum state, as the partial trace over a subspace of the total configuration space (i.e., the incoherent emitter surface) is reverted.

In contrast to the two previously discussed wave reconstruction schemes, off-axis EH and TIE, which permit a deterministic linear wave reconstruction in one step under the prerequisite of a known reference wave or known phase boundary conditions, respectively, a wave reconstruction from a focal series is generally a non-linear procedure in the sense that non-linear operators acting on the recorded intensity data are involved in the various reconstruction algorithms. These non-linearities are responsible for a considerable increase of complexity pertaining to the mathematical foundations and algorithmic implementations of focal series reconstruction of wave functions.

Moreover, *every* experimental focal series is inconsistent with respect to a single underlying wave function due to the presence of partial coherence (e.g., from the electron gun, inelastic interaction and thermal diffuse scattering; Rother, Gemming, & Lichte, 2009) or shot and detector noise (Niermann et al., 2012) as well as fluctuating geometric and chromatic aberrations depending on the defocus. Similarly, focal series are incomplete in practice because of a limited number and range of defocus values, typically restricted to the near field regime, and the use of isotropic foci, whereas astigmatic ones are required for an unambiguous reconstruction of an underlying wave function (cf. previous section). For instance, the problematic reconstruction of low spatial frequencies (Thust et al., 1996; Ophus & Ewalds, 2012; Niermann & Lehmann, 2016) can be traced back to missing focal series data in the far field. The design of non-linear focal series reconstruction algorithms behaving sufficiently well in the presence of inconsistent and incomplete data is highly non-trivial and subject to large and ongoing research efforts (Ophus & Ewalds, 2012).

The history of focal series phase retrieval (wave reconstruction) goes back to the first half of the 20th century. Pauli's famous question noted earlier can be considered as one particular instance of a reconstruction from a focal series consisting only of one in-focus and one far field image. In the wake of finding whole classes of functions (up to constant phase factors) sharing the same amplitude both in position space and Fourier space (Luke et al., 2002; Jaming, 2014), the prospects of such reconstructions have been considered rather pessimistic. For instance, if the Fourier transforms are symmetric, i.e., $\mathcal{F}\{\Psi\}(k) = \mathcal{F}\{\Psi\}(-k)$, the corresponding amplitudes are invariant under complex conjugation, i.e.,

$$|\Psi(x)| = |\Psi^*(x)| \tag{5.64}$$
$$|\mathcal{F}\{\Psi\}(k)| = |\mathcal{F}\{\Psi^*\}(k)| \, .$$

In a seminal paper Gerchberg and Saxton, however, proposed a simple projection algorithm, which produces surprisingly stable and well-behaved wave reconstructions from 2D image plane and diffraction plane intensities (Gerchberg & Saxton, 1972). Subsequently, modifications and extensions of the original Gerchberg–Saxton algorithm (Misell, 1973; Gonsalves, 1976; Boucher, 1980; Quatieri & Oppenheim, 1981; Fienup, 1982; Levi & Stark, 1984; Cederquist, Fienup, Wackerman, Robinson, & Kryskowski, 1989; Combettes & Trussell, 1990; Stark & Sezan, 1994; Yang, Dong, Gu, Zhuang, & Ersoy, 1994; Dong, Zhang, Gu, & Yang, 1997; Zou & Unbehauen, 1997; Ohneda, Baba, Miura, & Sakurai, 2001; Takajo, Takahashi, Itoh, & Fujisaki, 2002) have been successfully applied in a large number of different fields ranging from measuring the aberrations of telescopes (Fienup, Marron, Schulz, & Seldin, 1993; Lyon et al., 1997; Baba & Mutoh, 2001) to phase retrieval in electron microscopy (Allen et al., 2004; Koch, 2008, 2014; Allen & Oxley, 2001). It is interesting to note that in particular the unfortunate initial manufacturing error of the Hubble space telescope's optics (Burrows et al., 1991) and its subsequent characterization (Lyon, Miller, & Grusczak, 1991; Burrows, 1991; Fienup et al., 1993; Lyon, Dorband, & Hollis, 1997) lead to a boost in research on wave front reconstruction algorithms and their properties, which exhibits a remarkable parallel to Gabor's invention of holography that had been stimulated by the search for aberration-corrected electron optics in the TEM (Gabor, 1948).

The Gerchberg–Saxton algorithm consists of propagating a trial wave function into various focal planes, where its modulus is iteratively updated to the experimental value. If we start the iteration with a deliberately chosen guess for the phase $\Psi_0\,(z=0) = \sqrt{I(z)}\exp\,(i\varphi_0(\mathbf{r},z))$ at the initial plane $z = 0,$[7] the propagation over an infinitesimally small defocus step δz at some point $z = n\delta z$ along the optical axis reads

$$
\begin{aligned}
\Psi_n&(\mathbf{r}, z + \delta z) \\
&= \left(1 + \frac{i\delta z}{2k_0}\Delta\right)\Psi_n(\mathbf{r}, z) \\
&= \left(1 - \frac{\delta z}{2k_0}\left(\frac{\nabla A(\mathbf{r}, z)\nabla\varphi_n(\mathbf{r}, z)}{A(\mathbf{r}, z)} - i\frac{\Delta A(\mathbf{r}, z)}{A(\mathbf{r}, z)} + i(\nabla\varphi_n(\mathbf{r}, z))^2\right)\right) \\
&\quad\times A(\mathbf{r}, z)e^{i\varphi_n(\mathbf{r}, z)} \\
&= e^{-\frac{\delta z}{2k_0}\left(\frac{\nabla A(\mathbf{r}, z)\nabla\varphi_n(\mathbf{r}, z)}{A(\mathbf{r}, z)} - i\frac{\Delta A(\mathbf{r}, z)}{A(\mathbf{r}, z)} + i(\nabla\varphi_n(\mathbf{r}, z))^2\right)}A(\mathbf{r}, z)e^{i\varphi_n(\mathbf{r}, z)}.
\end{aligned}
\tag{5.65}
$$

In the last line, we used the smallness of the focal step δz to write the expression appearing in the bracket of the second line as an exponential. Note that this transformation is only possible if the amplitude $A \neq 0$. This reflects the fact that a large relative change of the wave function may occur upon small propagation steps at points, where the amplitude is zero.

The Gerchberg–Saxton algorithm proceeds by replacing the amplitude in the plane $z + \delta z$ with the experimental amplitude $A(\mathbf{r}, z + \delta z) = \sqrt{I(\mathbf{r}, z + \delta z)}$ in that plane

$$
\begin{aligned}
\Psi_{n+1}(\mathbf{r}, z + \delta z) &= e^{-i\frac{\delta z}{2k_0}\left(\frac{\Delta A(\mathbf{r}, z)}{A(\mathbf{r}, z)} - (\nabla\varphi_n(\mathbf{r}, z))^2\right)} \\
&\quad\times A(\mathbf{r}, z + \delta z)e^{i\varphi_n(\mathbf{r}, z)}.
\end{aligned}
\tag{5.66}
$$

Consequently, the iterated phase reads

---

[7] The amplitude is fixed by the experimental intensity I in that plane.

$$\varphi_{n+1}(\mathbf{r}, z + \delta z) = \varphi_n(\mathbf{r}, z) \tag{5.67}$$

$$- \frac{\delta z}{2k_0} \left(\frac{\Delta A(\mathbf{r}, z)}{A(\mathbf{r}, z)} - \left(\nabla \varphi_n(\mathbf{r}, z) \right)^2 \right)$$

$$= \varphi_n(\mathbf{r}, z) + \frac{\partial \varphi_n(\mathbf{r}, z)}{\partial z} \delta z,$$

where we inserted the quantum Hamilton–Jacobi equation (2.23b) in the last line.

The last equation reveals that the ordered Gerchberg–Saxton iteration over small focal step sizes corresponds to the numerical integration of the quantum Hamilton–Jacobi equation starting from some deliberately chosen initial wave function. Thus, the Gerchberg–Saxton algorithm may be considered as conjugate to the phase retrieval based on solving the Transport of Intensity Equation, since the coupled system of TIE and QHJE is equivalent to the paraxial Klein–Gordon–Fock equation (2.21). Noting that any starting guess φ_0 yields a solution to the above numerical integration, reconstructing the correct wave function underlying the observed intensities in the various focal planes corresponds to singling out the correct starting guess. The latter may be achieved by starting the integration at the far field at under-focus and integrating until reaching the far field at overfocus, because only the correct solution(s) may be identified in the far field at over-focus with the corresponding starting guess at under-focus after inversion of spatial coordinates. In other words, the far field data is necessary to identify the self-consistent solution(s) (and the pertaining periodic Bohm trajectories forming the characteristics of the quantum Hamilton–Jacobi equation).

Relating the Gerchberg–Saxton algorithm with the numerical integration of the (quantum) Hamilton–Jacobi equation allows to use the comprehensive Hamilton–Jacobi theory for characterizing the algorithm. Within the scope of this work, however, we are content with finding an alternative justification for the long focal range and leave questions, such as how many self-consistent solutions, i.e., how many wave function possess the same intensities in a long-range isotropic focal series to future work. Nevertheless, we will encounter an example below, where at least two different wave functions (i.e., not differing by a mere phase offset only) may not be distinguished from their intensities in isotropic focal planes.

In practice, the focal step employed in a focal series is too large to permit a linear approximation of the propagation as in (5.65). Therefore, the full (Fresnel) propagator has to be used to propagate the wave function from focal plane to focal plane. Under these circumstances, the direct interpretation of the algorithm as numerical integrator of the quantum Hamilton–Jacobi equation is violated to a certain extent and one may not prove in a rigorous way anymore that the true solution can be obtained by exploring the space of starting conditions. In fact, the larger defocus steps cause the Gerchberg–Saxton iteration to deviate from the QHJ solution pertaining to one particular starting wave by meandering in a complicated way through solutions corresponding to different starting waves, while iterating. To get more insight into this convergence behavior, we first note the following bound for the distance between iterated wave functions

$$\left\| \Psi_n(\mathbf{r}, z) - \Psi_{n-1}(\mathbf{r}, z) \right\|^2 = \left\| A_n(\mathbf{r}, z)e^{i\varphi_n(\mathbf{r}, z)} - A(\mathbf{r}, z)e^{i\varphi_n(\mathbf{r}, z)} \right\|^2 \qquad (5.68)$$

$$= \int_{-\infty}^{\infty} A_n^2(\mathbf{r}, z) + A^2(\mathbf{r}, z) - 2A_2(\mathbf{r}, z) A(\mathbf{r}, z) \, \mathrm{d}^2 r$$

$$\leq \int_{-\infty}^{\infty} A_n^2(\mathbf{r}, z) + A^2(\mathbf{r}, z) - 2A_2(\mathbf{r}, z) A(\mathbf{r}, z)$$

$$\times \cos\left(\varphi_n(\mathbf{r}, z) - \varphi(\mathbf{r}, z)\right) \mathrm{d}^2 r$$

$$= \left\| A_n(\mathbf{r}, z)e^{i\varphi_n(\mathbf{r}, z)} - A(\mathbf{r}, z)e^{i\varphi(\mathbf{r}, z)} \right\|^2 ,$$

which shows that one iteration step corresponds to an orthogonal projection on the closest representative from a set of wave functions (with arbitrary φ) sharing the same modulus A with the experimental value in that plane (Fienup, 1982). Consequently, the Gerchberg–Saxton algorithm can be considered as an iterative projection algorithm as used in convex optimization, with the crucial difference between both settings being the non-convexity of the set of wave functions sharing the same modulus in some optical plane (Fig. 5.13). This feature prevents the transfer of existence and convergence theorems from the convex setting to the typical long-focal step Gerchberg–Saxton iteration (see Fig. 5.14 for a visual representation of the problem). Accordingly, the algorithm may fail to converge, even if a consistent solution to the focal series exists (which is typically not the case due to noise, partial coherence, etc.). Note furthermore that the solution depends on the initial guess in the under-determined case (corresponding to incomplete focal series typically recorded in the TEM) even in the convex setting. These two characteristic properties underline the susceptibility of the Gerchberg–Saxton algorithm with respect to the starting guess, which has to be taken into account in the focal series reconstruction.

In spite of the above noted complicated convergence behavior, the large number of sensible solutions, which have been obtained previously, shows

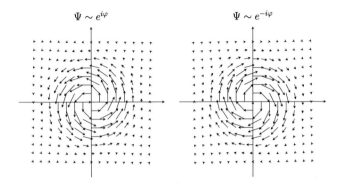

Figure 5.13 Polya plots of two vortex wave functions with the same amplitude $A = |\Psi|$, where each vector corresponds to a complex number in the complex plane. The complete set of wave functions with the same amplitude is obtained by arbitrarily changing the orientation of the vectors. Accordingly, linear combinations of wave functions from this set, $\Psi = \sum_i c_i \Psi_i$ with $\sum c_i = 1$, generally do not belong to this set. Consequently, the set of wave functions with the same modulus is not convex.

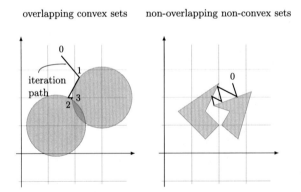

Figure 5.14 Projection on overlapping convex versus non-overlapping (i.e., inconsistent) non-convex sets. As long as the convex sets are overlapping the iterative projection converges to the intersection of the sets closest to the initial guess of the iteration in the here depicted under-determined case. In the non-convex setting the iteration may get trapped at points of close distance between the sets, not reaching other (and possibly better) solutions.

that increased focal step sizes do not necessarily invalidate the reconstruction algorithm as such. Indeed, various strategies and constraints, mainly derived from the phase space perspective, are suited to stabilize the algorithm to a certain extent as will be discussed below.

(A) First of all, we note that the phase space perspective is intimately connected to matrix completion strategies (Candès et al., 2013), which have been successfully employed to transfer projections on non-convex sets to underdetermined projections on convex sets, recently. The crucial point here is that the set of phase space distributions sharing the same projection in some direction is convex, which allows to lift the original Gerchberg–Saxton projection on non-convex sets of complex wave functions to convex sets of phase space distributions sharing the same modulus in some plane. However, the mixed state formulation of focal series reconstructions (i.e., quantum state tomography), discussed above, revealed that a set of line foci ranging from the far field at under-focus via the in-focus plane to the far field at over-focus and comprising all possible orientations of the line focus is necessary to obtain a unique quantum state reconstruction. Presently used stigmators do not permit the acquisition of a line focal series from the near to the far field, however. Thus, focal series typically comprise only isotropically defocused images, although it has been noted that the use of line foci in the near field reduces the ambiguity of the reconstruction (Petersen & Keast, 2007; Henderson, Williams, Peele, Quiney, & Nugent, 2009). Furthermore, in spite of significant progress being achieved in the fields of image registration (Saxton, 1994; Meyer, 2002), aberration correction and aberration assessment (Meyer, Kirkland, & Saxton, 2002, 2004), it remains a formidable challenge to record and register a long range isotropic defocus series free from spurious aberrations, distortions, rotations and magnification changes. To mitigate these issues, focal series are typically recorded in or close to the near field of the object taking into account that the limited focal range limits the focal series reconstruction of low spatial frequencies (Haigh, Jiang, Alloyeau, Kisielowski, & Kirkland, 2013; Niermann & Lehmann, 2016). More recently, also non–linear defocus variations extending further into the far field have been employed successfully in various inline holography studies (Song et al., 2013; Haigh et al., 2013; Koch, 2014) to increase the reconstructed spatial frequency band. Because equal tilt intervals in phase space correspond to non-equally spaced focal steps, a non–linear sampling of the defocus decreasing toward the far field according to (5.59) is optimal in these studies, if seeking a minimal number of images in the focal series.

(B) To fulfill the support theorem of tomography (Helgason, 2011), the complete electron wave to be reconstructed should be contained within the field of view throughout the entire tilt series. In practice, however, focal series reconstructions, in particular in the atomic resolution regime, are frequently carried out with electron beams being larger than the recorded field of view, thereby violating the tomographic support theorem. It has been noted that the corresponding artifacts in the reconstruction may be mitigated to some extent by numerically padding the intensity images with zeros, i.e., artificially introducing a boundary to the beam within the reconstructed domain (Lin, Chen, Chen, Tang, & Peng, 2006; Ophus & Ewalds, 2012).

(C) The alternating projection on convex sets in the phase space setting corresponds to the Kaczmarz (or ART) algorithm in the parlance of tomography (Wei, 2015). This analogy also suggests that a non-convergence of the Gerchberg–Saxton algorithm, e.g., due to ubiquitous inconsistencies in the recorded data, may be mitigated by combining all Kaczmarz iterations within one cycle, i.e., by projecting on all focal planes simultaneously. This strategy is an adaption of the Landweber or SIRT reconstruction well-known from tomography and forms the basis behind Allen's improved Gerchberg–Saxton algorithm (Allen et al., 2004). Fig. 5.15 shows a modified version of the latter, including the previously mentioned iteration over the starting guess and some additional processing step, which is used in the course of this work. Various TEM studies employed similar variants of the Gerchberg–Saxton algorithm and a commercial version of Allen's IFWR algorithm is available (Allen et al., 2004; Koch, 2008, 2014; Allen & Oxley, 2001).

The above lines show that typically recorded focal series are not sufficient to guarantee a unique and error-free reconstruction, which requires additional care in the interpretation of the results, in particular because error estimates are less straightforward compared to Off-Axis Holography. Therefore, the application of additional assumptions restricting the set of possible quantum states can be very useful to further improve the quality of the focal series reconstructions. Which assumptions can be used and how they may be incorporated into efficient focal series reconstruction algorithms represents an active area of research, bearing close analogies to the regularization strategies discussed within the context of tomography. In the following a short overview of possible strategies is given.

We first note that each step of the Gerchberg–Saxton algorithm does not modify the topology of a wave function, which is given by the total winding number (2.28) computed on the outer boundary of the electron beam because only the amplitude is updated in the differently defocused

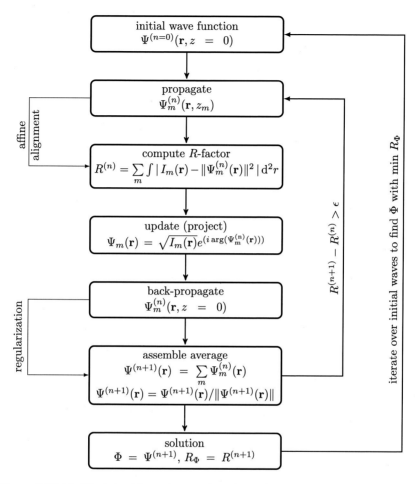

Figure 5.15 Modified Gerchberg–Saxton algorithm, where the wave functions Ψ_m in the various focal planes (index m) are computed simultaneously by propagating the actual iterated wave function (iteration index n) from some predefined (focal) plane (e.g., $z = 0$). The affine alignment of the recorded intensities with respect to the simulated wave functions may optionally be inserted to correct for spurious image rotations, magnification changes, distortions, etc. The regularization before computing the updated wave function may be applied to impose additional constraints, e.g., on the smoothness (regularity) of the wave. The iteration stops if some convergence criterion (e.g., R-factor limit ϵ or maximal iteration number) is reached.

planes.[8] Thus, the topology of the starting guess for the wave function is

[8] Here it is assumed that the wave function is well-contained within the reconstruction domain.

preserved throughout the iteration (Martin & Allen, 2007) and it is important to endow the starting guess with the correct topology to ensure convergence to the true solution. Unless employing electron vortex beams (Verbeeck et al., 2010; McMorran et al., 2011; Lubk, Clark, et al., 2013), the wave functions making up a conventional TEM beam possess a trivial topology (winding number equals zero), because elastic scattering at the object does not change the topology of the initial wave function. Consequently, a wave of trivial topology has to be used as starting guess under these typical circumstances. This implicit restriction of the solution space, present in almost all focal series reconstruction from TEM images reported to date of this publication, is one of the main reasons for reasonable reconstructions from restricted focal ranges. In case of an unknown wave topology, it may also be possible to single out the correct one by comparing the R-factors of the reconstructions pertaining to different topologies. Note, however, that such a test requires well-converged solutions for the different topologies, in order to ensure that the inconsistencies due to an erroneous topology are not overshadowed by partial coherence, alignment issues, noise, etc.

To illustrate the above principles, an exemplary long-range focal series reconstruction of a higher-order vortex beam ($w = 3$) based on the algorithm sketched in Fig. 5.15 is shown in Fig. 5.16. The electron vortex beam has been created by inserting a fork aperture (Verbeeck et al., 2010; McMorran et al., 2011; Grillo et al., 2014) into the condenser aperture of a FEI Titan[3] TEM operated at 300 kV and cutting out one of the $|w| = 3$ sidebands with a square aperture in the selected area plane. The focal series was recorded by varying the excitation of the diffraction lens (see Clark, Guzzinati, Béché, Lubk, & Verbeeck, 2016 for details) from 32% to 60%, yielding a series encompassing 20 images in total. The calibration of the effective propagation length and magnification was done by a dedicated calibration procedure (Lubk et al., 2016) and a comparison with a simulated reference series (see Clark et al., 2016 for details). In spite of using the whole range of diffraction lens excitations, the defocus values ranged from 0 (in-focus) to a large overfocus only. Thus, approximately 90° of the associated tilt series in phase space are missing, which translates into considerable missing information and needs to be taken into account, when evaluating the results.

The results of the reconstruction depicted in Fig. 5.16 show that the initial topology of the starting guess is indeed preserved throughout the Gerchberg–Saxton iteration, i.e., the winding number of the solution is

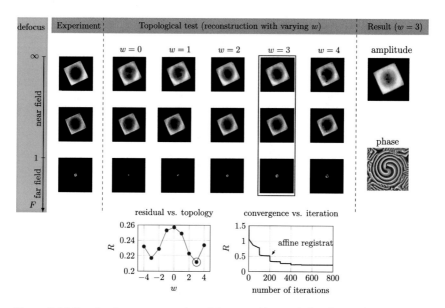

Figure 5.16 Focal series reconstruction of the vortex beam. Left column: Three recorded images from the defocus series. Central column: Solutions to $w = 0, 1, 2, 3, 4$ vortices with $w = 3$ fitting best to the experimental data according to the R-factor plotted for the different topologies (highlighted by small red circle). The convergence of the $w = 3$ solution exhibits large improvements after the first few affine image registration steps correcting for spurious misalignments of the magnification and rotation. Right column: Amplitude and phase in the aperture plane (selected area plane) pertaining to the $w = 3$ solution. The three phase vortices appear separated due to small manufacturing faults of the fork aperture, a small defocus of the impinging wave in the selected area plane, and the 3-fold symmetry breaking by the square aperture.

that of the starting guess. Based on this property, the correct topology of the wave could be determined by comparing reconstructions pertaining to different topologies with the help of the corresponding R-factors (central column in Fig. 5.16). This topology test fails, however, if the far field images are excluded from the series (see Lubk et al., 2016 for details), thereby removing the crucial low spatial frequency information. To clarify whether the topology constraint is sufficient for a unique reconstruction from a complete isotropic-focus series ranging from the far field at underfocus to the near field and again to the far field at overfocus is, however, beyond the scope of this work.

Additional restrictions other than topology may pertain to a possible sparsity of the wave function in some basis, possible smoothness (i.e., regularity) restrictions (Allen et al., 2004; Parvizi, Van den Broek, & Koch,

2016), support constraints of the wave function (Latychevskaia et al., 2010), or positivity of the phase shift. Similar to the previously discussed regularization of tomographic algorithms, these additional constraints may introduce additional regularization errors (e.g., at large gradients of the wave), which need to be balanced against the dampened reconstruction error. A similar situation, that is, the identification and exploration of suitable shape constraints on the reconstructed quantities, will be encountered in ptychographic reconstructions elaborated on in the next section.

5.4. STEM – DIFFERENTIAL PHASE CONTRAST AND PTYCHOGRAPHY

In this last section of this chapter, we discuss two closely related holographic techniques, namely STEM – Differential Phase Contrast (DPC) and Ptychography. They deviate from the previously discussed off-axis and inline schemes in that they do not seek the reconstruction of the electron's wave function or quantum state in STEM mode.[9] Instead they utilize the scanning mode of the TEM (STEM) to spatially resolve some integral information of the scattering process. In case of Differential Phase Contrast, this is the average scattering direction per scanning point (Rose, 1973; Dekkers & de Lang, 1974; Rose, 1976; Shibata et al., 2012) (Fig. 5.17). A ptychographic dataset consists of a complete 2D diffraction pattern per scanning point, hence is 4D in total (Rodenburg, 2008). Based on these data and using the axial scattering approximation, Ptychography (Rodenburg & Bates, 1992) inverts the averaging over the probe intensity distribution involved in the STEM-DPC scheme, thereby permitting the reconstruction of the complete mutual object transparency with a spatial resolution not limited by the point spread imposed by the STEM probe.

DPC studies are reported for, e.g., X-rays (Hornberger et al., 2008) or visible light (Stewart, 1976; Amos, Reichelt, Cattermole, & Laufer, 2003). A multitude of STEM-DPC investigations at medium resolution have been used to study electric (Lohr et al., 2012) and magnetic (Chapman, Batson, Waddell, & Ferrier, 1978; Chapman, 1984; Chapman, McFadyen, &

---

[9] STEM holography as discussed by Leuthner et al. (1989) employs an electron biprism as beamsplitter in the illumination system to generate two STEM probes on the sample, which are superimposed below the object by a second biprism. Hence, it is a special variant of Off-Axis Holography, discussed previously, and shall not be discussed in this section.

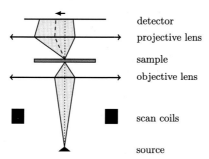

detector

projective lens

sample

objective lens

scan coils

source

Figure 5.17 Ptychography setup in one lateral dimension. The DPC signal, i.e., the shift of the momentum center of mass recorded in the far field of the specimen forming is indicated by an arrow.

McVitie, 1990) fields in solids. More recently, also atomic columns have been imaged with STEM-DPC at high-resolution (Shibata et al., 2012; Müller et al., 2014; Pennycook et al., 2015). It will be shown below that there is a close link between STEM-DPC and the previously discussed transport of intensity phase retrieval technique (cf. Section 5.2). These two techniques could be considered as bright field TEM and STEM implementations of lateral probability current measurements connected by a generalized notion of the principle of reciprocity (Cowley, 1969; Zeitler & Thomson, 1970). Contrary to TIE, however, the STEM-DPC setup does not permit a quantum state reconstruction of the electron beam itself.

Indeed, the generalization of STEM-DPC, i.e., Ptychography, represents a particularly elegant reconstruction of the phase space representation of the (mixed) object transparency (2.80) within the axial scattering regime (cf. Section 2.3). Acquisition, storage and processing of large 4D ptychographic datasets require advanced detector technology as well as considerable computational resources, which restricted the application of the technique for a long time, e.g., to periodic specimen (Hoppe, 1969a; Hoppe & Strube, 1969). With the recent advent of very fast pixel detectors, however, the acquisition of 4D ptychographic datasets became possible within reasonable time frames, leading to a revival of Ptychography, including the demonstration of atomic resolution (Pennycook et al., 2015).

In the following, we consider the reconstruction of the mixed object transparency first. Subsequently, the reconstruction of the transmission function, corresponding to the elastic object transparency, is treated as a special case. The algorithms employed for the latter bear a close resemblance to those used with focal series reconstructions, which allows us to recycle a number of results from the previous section.

5.4.1 Mutual Dynamic Object Transparency Reconstruction

The typical Ptychography setup is displayed in Fig. 5.17. Accordingly, a focused electron beam is scanned over the object to record one diffraction pattern $I\left(\mathbf{k}, \mathbf{r}_s\right)$ per scanning point $\mathbf{r}_s$. The (not normalized) STEM-DPC signal per scanning point $\mathbf{r}_s$,

$$\mathbf{s}\left(\mathbf{r}_s\right) = \int_{-\infty}^{\infty} \mathbf{k} I\left(\mathbf{k}, \mathbf{r}_s\right) \mathrm{d}^2 k, \tag{5.69}$$

then corresponds to the center of mass of the electron beam in the far-field of the object plane (equipped with coordinates $\mathbf{k}$). The experimental STEM-DPC signal may be obtained from pixel detectors as a discrete approximation

$$\mathbf{s}\left(\mathbf{r}_s\right) \approx \sum_p \mathbf{k}_p I\left(\mathbf{k}_p, \mathbf{r}_s\right) \tag{5.70}$$

to the above quantum mechanical momentum expectation value (5.69). In practice, dedicated STEM-DPC detectors have been developed (Chapman et al., 1990; Cowley, 1993; Feser et al., 2006; Hornberger et al., 2008; Shibata et al., 2010; Lohr et al., 2012; Müller et al., 2014; Majert & Kohl, 2015) (Fig. 5.18), where the number of pixels p is restricted in order to keep the scanning times short (dwell time < milliseconds). Due to its role as a weighting factor for the intensity, the $\mathbf{k}_p$ term is occasionally referred to as detector response function $\mathbf{d}\left(\mathbf{k}_p\right)$. Improved read-out speeds of detectors will permit an increasing number of pixels (or sectors) and hence more elaborate sampling schemes in future applications. In anticipation of that development, only the ideal DPC signal, as recorded by a continuous detector in the far-field, is considered below. Imaging theories for particular discrete detectors, in particular the four quadrant detector, can be found in literature (Majert & Kohl, 2015).

Inserting a generic mixed quantum state into the formula for the DPC signal (5.69), applying Parseval's theorem and integrating by parts (p.I.), one obtains

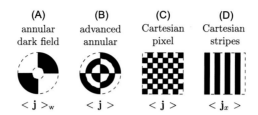

Figure 5.18 Non-exhaustive overview of detector geometries employed in STEM DPC: (A) conventional dark field annular four quadrant detector, (B) advanced segmented annular detector, (C) Cartesian pixel detector and (D) speculative Cartesian stripe detector illustrating the binning degree of freedom of (C). The lower line indicates the type of averaged probability currents measured with this detector.

$$
\begin{aligned}
\mathbf{s}(\mathbf{r}_s) &= \sum_m \int_{-\infty}^{\infty} \tilde{\Psi}_m^*(\mathbf{k}, \mathbf{r}_s)\, \mathbf{k}\, \tilde{\Psi}_m(\mathbf{k}, \mathbf{r}_s)\, \mathrm{d}^2 k \qquad (5.71) \\
&\overset{\text{Parseval}}{=} -i \sum_m \int_{-\infty}^{\infty} \Psi_m^*(\mathbf{r}, \mathbf{r}_s)\, \nabla \Psi_m(\mathbf{r}, \mathbf{r}_s)\, \mathrm{d}^2 r \\
&\overset{\text{p.I.}}{=} \int_{-\infty}^{\infty} \mathbf{j}(\mathbf{r}, \mathbf{r}_s)\, \mathrm{d}^2 r,
\end{aligned}
$$

which shows that the DPC signal corresponds to the spatially averaged probability current of the scattered electron beam in the object plane for each scanning point. Thus, referring to **s** as Differential Phase Contrast is somewhat misleading, because a pure state wave's phase cannot be sensibly defined for a mixed quantum state. The above notion of a probability current contrast, on the other hand, is well defined for beams of arbitrary coherence. Note, however, that a spectral (i.e., energy resolved) analysis of the probability current pertaining to inelastically scattered electrons is missing to date of this work.

Relying on the probability current interpretation, the previously discussed astigmatic TIE (cf. Section 5.2) can be considered as the broad beam TEM technique corresponding to STEM DPC. This analogy could be interpreted as an extension of the fundamental reciprocity principle (Cowley, 1969; Zeitler & Thomson, 1970), relating conventional TEM and STEM techniques by spatial inversion (i.e., reversing the beam path and swapping source and detector in Figs. 5.17 and 5.7). Here, the DPC-TIE current measurement reflects this principle even on an analytic level, in so far as in DPC the integral and in TIE the derivative of **j** is recorded.

Similar to TIE reconstructions, projected electric and magnetic fields are proportional to the DPC signal (i.e., the probability current) irrespective of the coherence of the beam (2.55) in the axial scattering regime. By inserting the axial scattering expression for the probability current (2.55), the following relationship between the DPC signal and electrostatic and magnetostatic fields may be derived

$$\mathbf{s}(\mathbf{r}_s) = \int_{-\infty}^{\infty} I(\mathbf{r}, \mathbf{r}_s) \int_{-\infty}^{\infty} -\frac{e}{\nu} \begin{pmatrix} E_x(\mathbf{r}, z) \\ E_y(\mathbf{r}, z) \end{pmatrix} + \begin{pmatrix} B_y(\mathbf{r}, z) \\ -B_x(\mathbf{r}, z) \end{pmatrix} dz \, d^2 r. \quad (5.72)$$

In the above derivation, it was assumed that the undisturbed beam has been aligned along z in a way to yield no DPC signal in vacuum (Lohr et al., 2012). Thus, under axial scattering conditions, the STEM-DPC is proportional to projected fields averaged over the (possibly attenuated) probe intensity distribution in the object exit plane. As discussed in Section 2.3, the axial scattering approximation holds under a wide range of medium resolution imaging conditions. In STEM, these conditions are obtained by reducing the convergence angle of the STEM probe and orienting the object such to avert channeling along atomic columns.

In order to interpret that signal directly in terms of projected fields, the latter must vary slowly over the spatial extension of the probe distribution $I(\mathbf{r}, \mathbf{r}_s)$ in the object exit plane, allowing to drag the electric and magnetic fields outside of the above spatial integration. If additionally normalizing the corresponding STEM-DPC signal with the total probe intensity, one obtains

$$\mathbf{s}(\mathbf{r}_s) = \int_{-\infty}^{\infty} -\frac{e}{\nu} \begin{pmatrix} E_x(\mathbf{r}, z) \\ E_y(\mathbf{r}, z) \end{pmatrix} + \begin{pmatrix} B_y(\mathbf{r}, z) \\ -B_x(\mathbf{r}, z) \end{pmatrix} dz, \quad (5.73)$$

which has been used in various electric and magnetic field STEM-DPC studies (Chapman, 1984; Chapman et al., 1990; Sandweg et al., 2008; Lohr et al., 2012).

The basic idea behind Ptychography is to utilize the incoming quantum state of the STEM beam as an invariant probe for the object transparency facilitating a direct reconstruction of the latter without taking a detour over the scattered beam quantum state as done by the previously discussed holographic methods. Note, however, that the validity of the single axial scattering approximation is severely limited toward atomic resolution (cf. Section 2.3), where dynamical scattering effects eventually invalidate the following ptychographic reconstruction principle.

We start by writing the far field intensity in the axial approximation (cf. Section 2.3)

$$I(\mathbf{r}_s, \mathbf{k}) = \iint_{-\infty}^{\infty} d^2r d^2r' \langle \mathbf{k}|\mathbf{r}\rangle \langle \mathbf{r}|\hat{\rho}_s|\mathbf{r}'\rangle \langle \mathbf{r}'|\mathbf{k}\rangle \tag{5.74}$$

$$= \frac{1}{4\pi^2} \iint_{-\infty}^{\infty} d^2r d^2r' \, e^{-i\mathbf{k}\mathbf{r}} \, T\left(\mathbf{r}, \mathbf{r}'\right) \rho_s\left(\mathbf{r} - \mathbf{r}_s, \mathbf{r}' - \mathbf{r}_s\right) e^{i\mathbf{k}\mathbf{r}'} .$$

Inserting the Wigner transform of the mutual object transparency and the probe state into (5.74), one obtains an alternative expression of the reciprocal space intensity at a particular probe position in terms of Wigner transforms

$$I\left(\mathbf{r}_s, \mathbf{k}\right) = \iint_{-\infty}^{\infty} d^2r d^2k' \, T_W(\mathbf{r}, \mathbf{k}') W_s(\mathbf{r} - \mathbf{r}_s, \mathbf{k} - \mathbf{k}') . \tag{5.75}$$

Consequently, the complete set of diffraction patterns depending on the probe position $\mathbf{r}$

$$I\left(\mathbf{r}, \mathbf{k}\right) = T_W(\mathbf{r}, \mathbf{k}) * W_s(-\mathbf{r}, \mathbf{k}) \tag{5.76}$$

corresponds to a convolution of the transparencies Wigner transform and the probe's Wigner function in phase space (Fig. 5.19), which reduces to simple multiplication of ambiguity functions in reciprocal phase space

$$\tilde{I}\left(\mathbf{k}', \mathbf{r}'\right) = -\tilde{T}_W(\mathbf{k}', \mathbf{r}') \tilde{W}_s(-\mathbf{k}', \mathbf{r}') . \tag{5.77}$$

Ptychography seeks an inversion (deconvolution) of the above phase space convolution. Several deconvolution strategies, eventually incorporating additional constraints, have been developed until now. The original approach, referred to as Wigner Distribution Deconvolution Method (WDDM), consists of using (regularized) deconvolution algorithms in combination with separately determined probe quantum states (Rodenburg & Bates, 1992; McCallum & Rodenburg, 1992; Chapman, 1996). Typically, the deconvolution is limited at the zeros of the probe's ambiguity function, defining a (band) limit for the quantum state reconstruction (Fig. 5.19). Consequently, a strictly positive probe's ambiguity function with an extension as large as possible, which is a (squeezed) coherent state (purity $\zeta = 1$), provides the optimal probe function for this type of reconstruction (Li, Edo, & Rodenburg, 2014). Therefore, Ptychography in the TEM may greatly profit from advanced electron sources, taking into account that current field emission

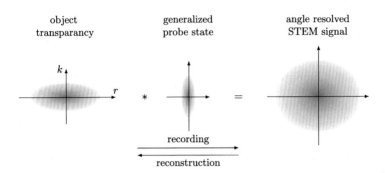

Figure 5.19 Ptychography as (de)convolution of the Wigner transformed axial scattering operator. Accordingly, the smaller (i.e. purer) the probe's Wigner function phase space volume, the larger the corresponding low-pass truncating the corresponding ambiguity function of the axial scattering operator (Fourier transform of Wigner transform).

electron guns at best provide purities[10] in the range of 10^{-2}–10^{-3} (Lubk et al., 2015).

To date of this work no ptychographic reconstruction of a mixed object transparency, e.g., to study inelastic low-loss scattering at plasmons, has been reported within the scope of TEM. However, the phase space perspective allows us to derive a number of important properties of the elastic transmission function reconstructions obtained from adapted pure state ptychographic reconstruction schemes. For instance, one may directly observe that the extension of the detector in momentum space should be large enough to facilitate a complete sampling of the object transparency's Wigner function (Fig. 5.19). Similarly, the scanning grid must be sufficiently dense to sample the localized STEM probe free from aliasing. These details are important when discussing the ptychographic reconstruction of pure states in the next section.

5.4.2 Transmission Function Reconstruction

In response to the large computational costs of a deconvolution in 4D phase space, iterative reconstruction methods (e.g., Ptychographic Iterative Engine; Rodenburg & Faulkner, 2004) incorporating additional constraints on both the purity of the probe quantum state and the mutual object transparency have been developed (Hesse, Luke, Sabach, & Tam, 2014; Horstmeyer et al., 2015). Owing to the required purity, the phase space

---

[10] Note that our purity values provided in Lubk et al. (2015) need to be multiplied with an additional factor of 2π to be consistent with the definition of the purity used here.

deconvolution may be performed within the wave function representation, cutting the number of dimensions of the problem in half. In combination with the arrival of novel fast pixelated detectors and advanced computational resources facilitating the recording and processing of huge 4D spatially resolved diffraction data, these transmission function reconstruction methods have been recently employed in an increasing number of ptychographic studies in the TEM.

To consider Ptychography as a wave reconstruction technique, it is instructive to first exhibit the relation between the DPC signal (5.69), i.e., the (spatially averaged) probability current and the phase of the wave function in the object plane

$$\mathbf{s}(\mathbf{r}_s) = \int_{-\infty}^{\infty} \mathbf{j}(\mathbf{r}, \mathbf{r}_s)\, d^2r = \int_{-\infty}^{\infty} I(\mathbf{r}, \mathbf{r}_s)\, \nabla \varphi(\mathbf{r}, \mathbf{r}_s)\, d^2r. \qquad (5.78)$$

This simple relation facilitates a straightforward discussion of various aspects of STEM-DPC, such as the influence of coherent and incoherent aberrations as well as the relation to the TIE reconstruction. Using that notation, a normalized DPC signal reads

$$\mathbf{s}_n(\mathbf{r}_s) = \frac{\int_{-\infty}^{\infty} I(\mathbf{r}, \mathbf{r}_s)\, \nabla \varphi(\mathbf{r}, \mathbf{r}_s)\, d^2r}{\int_{-\infty}^{\infty} I(\mathbf{r}, \mathbf{r}_s)\, d^2r}, \qquad (5.79)$$

and one can identify suitable conditions under which the above expression permits a reconstruction of the phase gradient.

Since both the intensity as well as the phase do vary across a sharply focused TEM probe, it is not possible to reconstruct the phase (gradient) from the above expression in the general case. If, however, a phase gradient with negligible variations within the STEM probe diameter is considered, the normalized STEM-DPC signal can be approximated by the object's phase (Cowley, 1993), i.e.,

$$\mathbf{s}_n(\mathbf{r}_s) \approx \nabla \varphi(\mathbf{r}_s). \qquad (5.80)$$

Within the axial approximation, the above phase gradient is given by projected electric and magnetic fields (Stewart, 1976; Rose, 1976; Waddell & Chapman, 1979; Cowley, 1993)

$$\mathbf{s}_n(\mathbf{r}_s) = -\frac{e}{v} \int_{-\infty}^{\infty} \mathbf{E}(\mathbf{r}_s, z)\, dz + \int_{-\infty}^{\infty} \begin{pmatrix} B_y(\mathbf{r}, z) \\ -B_x(\mathbf{r}, z) \end{pmatrix} dz, \qquad (5.81)$$

as noted previously.

Coherent aberrations affect the probe via a multiplication of a phase plate in the spatial frequency domain (condenser plane)

$$\tilde{\Psi}_a (\mathbf{k}) = \tilde{\Psi}_s (\mathbf{k}) \, e^{i\chi(\mathbf{k})} . \qquad (5.82)$$

This affects the DPC signal under axial scattering conditions (5.72) by modulating the probe intensity $I(\mathbf{r}, \mathbf{r}_s)$, which is integrated together with the projected electromagnetic fields. Consequently, the averaged projected electromagnetic fields are reduced in spatial resolution (Cowley, 1993). A similar effect is imposed by partial coherence, also smearing out probe intensity. Thus, partial spatial coherence is only dampening higher spatial frequencies but not the DPC signal as a whole, which again highlights our previous finding that not the differential of the phase but the probability current is the fundamental quantity behind the DPC signal.

Yet another interesting aspect of the pure state reconstruction pertains to the use of the conventional dark field four quadrant detector, where the detector contains a hole in the middle, which permits electrons, scattered into small scattering angles, to pass without interacting with the detector (Fig. 5.18). If the wave function on the detector is

$$\Psi_d = \Psi \times \mathrm{supp}_d, \quad \text{with } \mathrm{supp}_d(\mathbf{k}) = \begin{cases} 1| & \mathbf{k} \text{ on detector,} \\ 0| & \text{otherwise,} \end{cases} \qquad (5.83)$$

the normalized DPC signal reads

$$\mathbf{s} = \frac{\int_{\mathrm{supp}_d} \mathbf{k} I(\mathbf{k}) \mathrm{d}^2 k}{\int_{\mathrm{supp}_d} I(\mathbf{r}) \, \mathrm{d}^2 r} = \frac{\left\langle \Psi_d \left| \hat{\mathbf{k}} \right| \Psi \right\rangle}{\langle \Psi_d, \Psi \rangle} . \qquad (5.84)$$

Accordingly, the dark-field DPC signal can be considerably amplified with respect to the conventional one ($\mathrm{supp}_d = 1$), because the denominator in (5.84) decreases faster than the numerator containing $\mathbf{k}$ as argument, when reducing the overlap between wave function and detector. In particular in the limit of a very small overlap, where the amplification is largest, the influence of the measurement on the wave function is *weak*, and, following Aharonov, Albert, and Vaidman (1988), measurements described by expressions similar to (5.84) are referred to as *weak* quantum measurements (Vaidman, 2009; Tamir & Cohen, 2013). The amplification effect has been deliberately exploited in the dark-field DPC setup (Shibata et al., 2012; Lohr et al., 2012), where the collection angle has been chosen such to amplify minute scattering angles. Similar to other *weak* quantum measurements (Vaidman, 2009; Tamir & Cohen, 2013), however, one must be very careful with interpreting these amplified results as common *strong* expectation values (Aharonov et al., 1988; Hosten & Kwiat, 2008; Dixon, Starling, Jordan, & Howell, 2009), that is averaged momentums, or electromagnetic fields in our case.

In order to generalize STEM–DPC to Ptychography, we first rewrite the STEM–DPC signal in the axial scattering regime

$$\mathbf{s}\left(\mathbf{r}_s\right) = \int_{-\infty}^{\infty} d^2 r I\left(\mathbf{r} - \mathbf{r}_s\right) e^{-\int_{-\infty}^{\infty} \mu_{\mathrm{el}}(\mathbf{r}, z) dz} \tag{5.85}$$

$$\times \int_{-\infty}^{\infty} -\frac{e}{v} \left(\begin{matrix} E_x\left(\mathbf{r}, z\right) \\ E_y\left(\mathbf{r}, z\right) \end{matrix} \right) + \left(\begin{matrix} B_y\left(\mathbf{r}, z\right) \\ -B_x\left(\mathbf{r}, z\right) \end{matrix} \right) dz,$$

where we explicitly note the elastic attenuation and assume an invariant probe intensity at the object entrance face.

In the previous section, we have shown that Ptychography recovers the complex object transmission function (containing both the projected potential and attenuation). Thus, Ptychography may be alternatively considered as removing the convolution of the probe intensity from the reconstructed projected potentials, thereby effectuating an increase in spatial resolution, frequently referred to as superresolution (Humphry, Kraus, Hurst, Maiden, & Rodenburg, 2012). On top of that also the projected attenuation coefficient is recovered as part of the reconstructed transmission function. Unfortunately, the applicability of this superresolution scheme is somewhat limited toward high (i.e., atomic) resolution in the TEM, because of the violation of the axial scattering assumption forming the basis of the ptychographic principle. Therefore, similar to the above noted limits of the axial scattering interpretation of STEM–DPC, quantitative atomic resolution ptychographic transmission function reconstructions are restricted to very thin samples with thicknesses below a few nanometers only, depending on the atomic number of the constituents. Strong scatterers, i.e., scatterers containing a sufficiently large number of atoms in the interaction volume oriented along a low-index zone axis and scanned with an angstrom probe (high-resolution conditions), do not permit a description of the scattered wave function by multiplying an invariant (i.e., beam position independent) object transparency due to dynamical scattering (cf. Section 2.3; Rose, 1976; Kirkland, 1998; Vulovic, Voortman, van Vliet, & Rieger, 2014). Therefore, under atomic resolution conditions the interpretation of the ptychographic dataset becomes complicated and typically requires accurate scattering simulations similar to all other high-resolution (S)TEM techniques (Maiden, Humphry, & Rodenburg, 2012; Chen et al., 2016; Müller-Caspary et al., 2016).

Ptychographic transmission function reconstruction methods resemble the previously discussed Gerchberg–Saxton algorithm employed in focal series reconstructions in that iterative projections of the diffraction data obtained from complex valued probe and transmission functions are used to update the former in each iteration step (Fig. 5.20) (Hesse et al., 2014; Horstmeyer et al., 2015). To discuss the properties of these algorithms, we start with considering infinitesimal shifts of the probe, facilitating an analysis similar to that pertaining to infinitesimal defocus steps in focal series reconstructions. Within the scope of the axial scattering approximation the object exit wave in the image plane reads

$$\Psi\left(\mathbf{r}, \mathbf{r}_s\right) = T\left(\mathbf{r}\right)\left(\psi\left(\mathbf{r}\right) * \delta\left(\mathbf{r} - \mathbf{r}_s\right)\right), \tag{5.86}$$

where $\mathbf{r}_s$ denotes the scanning position of the probe. The corresponding wave function in the far field (detector plane)

$$\tilde{\Psi}\left(\mathbf{k}, \mathbf{r}_s\right) = \tilde{T}\left(\mathbf{k}\right) * \left(\tilde{\psi}\left(\mathbf{r}\right) e^{i\mathbf{k}\mathbf{r}_s}\right) \tag{5.87}$$

is then obtained as a convolution of the Fourier transform of the transmission function and the Fourier transformed probe wave function. Accordingly, a small shift $\delta\mathbf{r}_s$ slightly perturbs the convolution in the following way

$$\tilde{\Psi}_n\left(\mathbf{k}, \mathbf{r}_s + \delta\mathbf{r}_s\right) = \tilde{T}_n\left(\mathbf{k}\right) * \left(\tilde{\psi}_n\left(\mathbf{r}\right) e^{i\mathbf{k}\mathbf{r}_s}\left(1 + i\mathbf{k}\delta\mathbf{r}_s\right)\right). \tag{5.88}$$

Now, the iterative projection consists of updating the amplitude $A(\mathbf{k}, \mathbf{r}_s + \delta\mathbf{r}_s) = \sqrt{I(\mathbf{k}, \mathbf{r}_s + \delta\mathbf{r}_s)}$ at the shifted position with the experimental diffraction pattern, i.e.,

$$A\left(\mathbf{k}, \mathbf{r}_s + \delta\mathbf{r}_s\right) e^{i\tilde{\varphi}_n(\mathbf{k}, \mathbf{r}_s + \delta\mathbf{r}_s)} = \tilde{T}_{n+1}\left(\mathbf{k}\right) * \left(\tilde{\psi}_n\left(\mathbf{r}\right) e^{i\mathbf{k}\mathbf{r}_s}\left(1 + i\mathbf{k}\delta\mathbf{r}_s\right)\right), \tag{5.89}$$

followed by a back-propagation to the object exit plane

$$\Psi_{n+1}\left(\mathbf{r}, \mathbf{r}_s + \delta\mathbf{r}_s\right) = \mathcal{F}^{-1}\left\{A\left(\mathbf{k}, \mathbf{r}_s + \delta\mathbf{r}_s\right) e^{i\tilde{\varphi}_n(\mathbf{k}, \mathbf{r}_s + \delta\mathbf{r}_s)}\right\} \tag{5.90}$$

$$= T_{n+1}\left(\mathbf{r}\right)\left(\psi_n\left(\mathbf{r}\right) * \delta\left(\mathbf{r} - \mathbf{r}_s\right) + \psi_n\left(\mathbf{r}\right) * \delta\mathbf{r}_s \nabla \delta\left(\mathbf{r} - \mathbf{r}_s\right)\right)$$

$$\overset{\text{sym}}{\approx} T_{n+1}\left(\mathbf{r}\right)\left(\psi_n\left(\mathbf{r}\right) * \delta\left(\mathbf{r} - \mathbf{r}_s\right)\right).$$

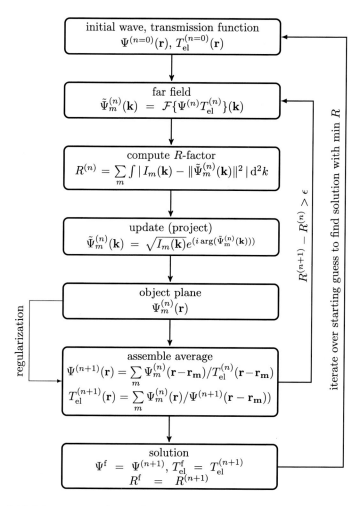

Figure 5.20 Schematic ptychographic reconstruction algorithm. The regularization involved in updating the transmission function is required to dampen artifacts, introduced when dividing the localized probe wave functions. One possibility is restricting the division to the domain, where the probe amplitude is sufficiently large. The iteration stops, if some convergence criteria (e.g., R-factor limit ϵ or maximal iteration number) are reached.

In the last two steps we exploited the symmetry of the probe and the smallness of the probe shift $\delta \mathbf{r}_s$. After solving for T, we may directly obtain the transmission function in the shifted interval.

To perform the next iteration, i.e., to jump to the next scanning position, we have to reevaluate (5.88), which requires an extrapolation of the

transmission function to the new probe interval. This may be achieved by
computing the Taylor decomposition around the old scanning position **s**

$$T_{n+1}(\mathbf{r}) = \sum_{m=0} \partial_m T_n(\mathbf{r}_s) \frac{(\mathbf{r}-\mathbf{r}_s)^m}{m!}, \tag{5.91}$$

which converges homogeneously according to the Weierstraß approximation theorem. Extrapolation to next probe interval is now straightforward:

$$T_{n+1}(\mathbf{r}) = \sum_{m=0} \partial_m T_n(\mathbf{r}_s + \delta\mathbf{r}_s) \frac{(\mathbf{r}-\mathbf{r}_s-\delta\mathbf{r}_s)^m}{m!} \tag{5.92}$$

$$= \sum_{m=0} \partial_m T_n(\mathbf{r}_s) \frac{(\delta\mathbf{r}_s)^{m-1}}{(m-1)!} \frac{(\mathbf{r}-\mathbf{r}_s-\delta\mathbf{r}_s)^m}{m!}.$$

Ptychography with small scanning shifts thus corresponds to a numerical integration of the transmission function from local Taylor expansions
facilitated by the ptychographic dataset. Correct transmission functions
corresponds to those solutions, which are self-consistent under any loop
integration.

   Under experimental conditions, the scan step is not infinitesimal, although it has been noted that a certain overlap of adjacent probe functions
is necessary for convergence. Similar to focal series reconstructions, an
equivalence between a numerical integration and the ptychographic reconstruction may not be established anymore. Due to the non-convexity of
the set of complex transmission functions yielding the observed diffraction
intensities, the emerging convergence behavior is generally complex and
difficult to characterize. For instance, one observes that the results depend
on the starting conditions and, in contrast to the focal series reconstruction,
little is known about the restricting role of the topology in this case. The
similarity of iterated projection algorithms employed in Ptychography and
focal series reconstructions, however, allows us to transfer a large number
of the previously discussed properties of such algorithms. For instance, an
improved convergence behavior may be obtained by performing the projections on the diffraction data pertaining to different scanning positions in
parallel (Fig. 5.20).

   It has been pointed out that this scheme even permits the concomitant reconstruction of the probe wave function (Thibault, Dierolf, Bunk,
Menzel, & Pfeiffer, 2009; Maiden & Rodenburg, 2009). Note, however,
that such an approach amounts to a blind deconvolution in the reciprocal phase space setting, i.e., the retrieval of both the Wigner transformed

transmission function and the Wigner function of the probe from one ptychographic dataset

$$\tilde{I}(\mathbf{q}, \mathbf{p}) = -\tilde{T}_W(\mathbf{q}, \mathbf{p}) \, \tilde{W}_s(-\mathbf{q}, \mathbf{p}) \,. \tag{5.93}$$

Blind deconvolution, being essentially ill-defined, has been extensively studied in image processing, and a number of additional conditions, such as support or shape constraints, stabilizing the deconvolution could be identified (Ahmed, Recht, & Romberg, 2014; Choudhary & Mitra, 2014). These have been transferred partially to Ptychography, e.g., by restricting the support of the probe function or the shape of the probe wave function. Other strategies are the pre-iteration of the transmission function with a sufficiently good guess for the probe function, before iterating both in parallel, and the restriction of the transmission function to a pure phase object (with positive phase shift). In particular in combination with the ill-posedness of the blind deconvolution, one therefore has to take into account that unique ptychographic reconstructions may not be generally possible, when interpreting the obtained results.

REFERENCES

Aharonov, Y., Albert, D. Z., & Vaidman, L. (1988). How the result of a measurement of a component of the spin of a spin-1/2 particle can turn out to be 100. *Physical Review Letters, 60*(14), 1351–1354.

Ahmed, A., Recht, B., & Romberg, J. (2014). Blind deconvolution using convex programming. *IEEE Transactions on Information Theory, 60*(3), 1711–1732.

Akhtari-Zavareh, A., Carignan, L. P., Yelon, A., Ménard, D., Kasama, T., Herring, R., ... Kavanagh, K. L. (2014). Off-axis electron holography of ferromagnetic multilayer nanowires. *Journal of Applied Physics, 116*(2), 023902.

Allen, L. J., Faulkner, H. M. L., Nugent, K. A., Oxley, M. P., & Paganin, D. (2001). Phase retrieval from images in the presence of first-order vortices. *Physical Review E, 63*(3), 037602.

Allen, L. J., Faulkner, H. M. L., Oxley, M. P., & Paganin, D. (2001). Phase retrieval and aberration correction in the presence of vortices in high-resolution transmission electron microscopy. *Ultramicroscopy, 88*, 85–97.

Allen, L. J., McBride, W., O'Leary, N. L., & Oxley, M. P. (2004). Exit wave reconstruction at atomic resolution. *Ultramicroscopy, 100*, 91–104.

Allen, L. J., & Oxley, M. P. (2001). Phase retrieval from series of images obtained by defocus variation. *Optics Communications, 199*, 65.

Allen, L. J., Oxley, M. P., & Paganin, D. (2001). Computational aberration correction for an arbitrary linear imaging system. *Physical Review Letters, 87*, 123902.

Allman, B. E., McMahon, P. J., Nugent, K. A., Paganin, D., Jacobson, D. L., Arif, M., & Werner, S. A. (2000). Imaging: Phase radiography with neutrons. *Nature, 408*(6809), 158–159.

Amos, W. B., Reichelt, S., Cattermole, D. M., & Laufer, J. (2003). Re-evaluation of differential phase contrast (DPC) in a scanning laser microscope using a split detector as an alternative to differential interference contrast (DIC) optics. *Journal of Microscopy, 210*(2), 166–175.

Atakishiyev, N. M., Chumakov, S., Rivera, A., & Wolf, K. (1996). On the phase space description of quantum nonlinear dynamics. *Physics Letters A, 215*, 128–134.

Baba, N., & Mutoh, K. (2001). Measurement of telescope aberrations through atmospheric turbulence by use of phase diversity. *Applied Optics, 40*(4), 544–552.

Band, W., & Park, J. L. (1971). A general method of empirical state determination in quantum physics: Part II. *Foundations of Physics, 1*(4), 339–357.

Barty, A., Nugent, K. A., Paganin, D., & Roberts, A. (1998). Quantitative optical phase microscopy. *Optics Letters, 23*(11), 817–819.

Bastiaans, M. J. (1978). The Wigner distribution function applied to optical signals and systems. *Optics Communications, 25*(1), 26–30.

Bastiaans, M. J. (1979). Wigner distribution function and its application to first-order optics. *Journal of the Optical Society of America, 69*(12), 1710–1716.

Bates, R. H. T., & Rodenburg, J. M. (1989). Sub-angstrom transmission microscopy: A Fourier transform algorithm for microdiffraction plane intensity information. *Ultramicroscopy, 31*(3), 303–307.

Bauschke, H. H., Combettes, P. L., & Luke, D. R. (2002). Phase retrieval, error reduction algorithm, and Fienup variants: A view from convex optimization. *Journal of the Optical Society of America A, Optics and Image Science, 19*(7), 1334–1345.

Bayh, W. (1962). Messung der kontinuierlichen Phasenschiebung von Elektronenwellen im kraftfeldfreien Raum durch das magnetische Vektorpotential einer Wolfram-Wendel. *Zeitschrift für Physik, 169*(4), 492–510.

Béché, A., Rouvière, J., Barnes, J., & Cooper, D. (2011). Dark field electron holography for strain measurement. *Ultramicroscopy, 111*(3), 227–238.

Beleggia, M., Kasama, T., Dunin-Borkowski, R. E., & Pozzi, G. (2010). Measurement of the charge distribution along an electrically biased carbon nanotube using electron holography. *Microscopy and Microanalysis, 16*(Suppl. 2), 562–563.

Beleggia, M., Kasama, T., & Dunin-Borkowski, R. E. (2010). The quantitative measurement of magnetic moments from phase images of nanoparticles and nanostructures – I. Fundamentals. *Ultramicroscopy, 110*(5), 425–432.

Bergmann, L., Niedrig, H., & Eichler, H. J. (2004). *Optik: Wellen- und Teilchenoptik.* De Gruyter.

Berriman, J., Bryan, R., Freeman, R., & Leonard, K. (1984). Methods for specimen thickness determination in electron microscopy. *Ultramicroscopy, 13*(4), 351–364.

Bethe, H. (1928). Theorie der Beugung von Elektronen an Kristallen. *Annalen der Physik, 392*(17), 55–129.

Bonevich, J. E., Harada, K., Matsuda, T., Kasai, H., Yoshida, T., Pozzi, G., & Tonomura, A. (1993). Electron holography observation of vortex lattices in a superconductor. *Physical Review Letters, 70*, 2952–2955.

Boucher, R. H. (1980). Convergence of algorithms for phase retrieval from two intensity distributions. *Proceedings of SPIE, 0231*, 130–141.

Breitenbach, G., Schiller, S., & Mlynek, J. (1997). Measurement of the quantum states of squeezed light. *Nature, 387*(6632), 471–475.

Bromwich, T. J., Kasama, T., Chong, R. K. K., Dunin-Borkowski, R. E., Petford-Long, A. K., Heinonen, O. G., & Ross, C. A. (2006). Remanent magnetic states and interactions in nano-pillars. *Nanotechnology, 17*(17), 4367.

Bruck, Y. M., & Sodin, L. G. (1979). On the ambiguity of the image reconstruction problem. *Optics Communications*, *30*(3), 304–308.

Bryngdahl, O., & Lohmann, A. W. (1968). Interferograms are image holograms. *Journal of the Optical Society of America*, *58*, 141–142.

Burrows, C. J. (1991). Hubble Space Telescope optics status. *Proceedings of SPIE*, *1567*, 284–293.

Burrows, C. J., Holtzman, J. A., Faber, S. M., Bely, P. Y., Hasan, H., Lynds, C. R., & Schroeder, D. (1991). The imaging performance of the Hubble Space Telescope. *The Astrophysical Journal*, *369*, L21–L25.

Cámara, A. (2015). *Optical beam characterization via phase-space tomography*. Switzerland: Springer International Publishing.

Cámara, A., Alieva, T., Rodrigo, J. A., & Calvo, M. L. (2009). Phase space tomography reconstruction of the Wigner distribution for optical beams separable in Cartesian coordinates. *Journal of the Optical Society of America A, Optics and Image Science*, *26*(6), 1301–1306.

Candès, E. J., Eldar, Y. C., Strohmer, T., & Voroninski, V. (2013). Phase retrieval via matrix completion. *SIAM Journal on Imaging Sciences*, *6*(1), 199–225.

Cederquist, J. N., Fienup, J. R., Wackerman, C. C., Robinson, S. R., & Kryskowski, D. (1989). Wave-front phase estimation from Fourier intensity measurements. *Journal of the Optical Society of America A, Optics and Image Science*, *6*(7), 1020–1026.

Chapman, H. N. (1996). Phase-retrieval X-ray microscopy by Wigner-distribution deconvolution. *Ultramicroscopy*, *66*, 153–172.

Chapman, J. N. (1984). The investigation of magnetic domain structures in thin foils by electron microscopy. *Journal of Physics D: Applied Physics*, *17*(4), 623.

Chapman, J. N., Batson, P., Waddell, E., & Ferrier, R. (1978). The direct determination of magnetic domain wall profiles by differential phase contrast electron microscopy. *Ultramicroscopy*, *3*, 203–214.

Chapman, J. N., McFadyen, I., & McVitie, S. (1990). Modified differential phase contrast Lorentz microscopy for improved imaging of magnetic structures. *IEEE Transactions on Magnetics*, *26*(5), 1506–1511.

Chen, J. W., Matteucci, G., Migliori, A., Missiroli, G. F., Nichelatti, E., Pozzi, G., & Vanzi, M. (1989). Mapping of microelectrostatic fields by means of electron holography: Theoretical and experimental results. *Physical Review A*, *40*(6), 3136–3146.

Chen, Z., Weyland, M., Ercius, P., Cistond, J., Zhenga, C., Fuhrera, M. S., . . . Findlay, S. D. (2016). Practical aspects of diffractive imaging using an atomic-scale coherent electron probe. *Ultramicroscopy*.

Choudhary, S., & Mitra, U. (2014). Sparse blind deconvolution: What cannot be done. In *2014 IEEE International Symposium on Information Theory (ISIT)* (pp. 3002–3006).

Clark, L., Guzzinati, G., Béché, A., Lubk, A., & Verbeeck, A. (2016). Symmetry-constrained electron vortex propagation. *Physical Review A*, *93*, 063840.

Coene, W., Janssen, G., Op de Beeck, M., & Van Dyck, D. (1992). Phase retrieval through focus variation for ultra-resolution in field-emission transmission electron microscopy. *Physical Review Letters*, *69*(26), 3743–3746.

Coene, W., Thust, A., Op de Beeck, M., & Van Dyck, D. (1996). Maximum-likelihood method for focus-variation image reconstruction in high resolution transmission electron microscopy. *Ultramicroscopy*, *64*(1–4), 109–135.

Combettes, P. L., & Trussell, H. J. (1990). Method of successive projections for finding a common point of sets in metric spaces. *Journal of Optimization Theory and Applications*, *67*(3), 487–507.

Cooper, D., Barnes, J. P., Hartmann, J. M., Béché, A., & Rouvière, J. (2009). Dark field electron holography for quantitative strain measurements with nanometer-scale spatial resolution. *Applied Physics Letters, 95*, 053501.

Cooper, D., de la Peña, F., Béché, A., Rouvirè, J.-L., Servanton, G., Pantel, R., & Morin, P. (2011). Field mapping with nanometer-scale resolution for the next generation of electronic devices. *Nano Letters, 11*(11), 4585–4590.

Cooper, D., Twitchett, A. C., Somodi, P. K., Midgley, P. A., Dunin-Borkowski, R. E., Farrer, I., & Ritchie, D. A. (2006). Improvement in electron holographic phase images of focused-ion-beam milled GaAs and Si p–n junctions by in situ annealing. *Applied Physics Letters, 88*, 63510.

Cowley, J. M. (1969). Image contrast in a transmission scanning electron microscope. *Applied Physics Letters, 15*(2), 58–59.

Cowley, J. M. (1992). Twenty forms of electron holography. *Ultramicroscopy, 41*(4), 335–348.

Cowley, J. M. (1993). Configured detectors for stem imaging of thin specimens. *Ultramicroscopy, 49*, 4–13.

Cowley, J. M. (1995). *Diffraction physics. North-Holland personal library.* Amsterdam: Elsevier Science.

Cumings, J., Zettl, A., McCartney, M. R., & Spence, J. C. H. (2002). Electron holography of field-emitting carbon nanotubes. *Physical Review Letters, 88*(5), 056804.

De Caro, L., Carlino, E., Caputo, G., Cozzoli, P. D., & Giannini, C. (2010). Electron diffractive imaging of oxygen atoms in nanocrystals at sub-angstrom resolution. *Nature Nanotechnology, 5*(5), 360–365.

de Knoop, L., Houdellier, F., Gatel, C., Masseboeuf, A., Monthioux, M., & Hÿtch, M. (2014). Determining the work function of a carbon-cone cold-field emitter by in situ electron holography. *Micron, 63*, 2–8.

Dekkers, N. H., & de Lang, H. (1974). Differential phase contrast in a STEM. *Optik, 41*, 452–456.

den Hertog, M. I., Schmid, H., Cooper, D., Rouviere, J.-L., Björk, M. T., Riel, H., ... Riess, W. (2009). Mapping active dopants in single-silicon nanowires using off-axis electron holography. *Nano Letters, 9*(11), 3837–3843.

Dietrich, J., Abou-Ras, D., Schmidt, S. S., Rissom, T., Unold, T., Cojocaru-Mirédin, O., ... Boit, C. (2014). Origins of electrostatic potential wells at dislocations in polycrystalline $Cu(In, Ga)Se_2$ thin films. *Journal of Applied Physics, 115*(10), 103507.

Dixon, P. B., Starling, D. J., Jordan, A. N., & Howell, J. C. (2009). Ultrasensitive beam deflection measurement via interferometric weak value amplification. *Physical Review Letters, 102*, 173601.

Dong, B.-Z., Zhang, Y., Gu, B.-Y., & Yang, G.-Z. (1997). Numerical investigation of phase retrieval in a fractional Fourier transform. *Journal of the Optical Society of America A, Optics and Image Science, 14*(10), 2709–2714.

Dücker, H., & Möllenstedt, G. (1955). Fresnelscher interferenzversuch mit einem biprisma für elektronenwellen. *Zeitschrift für Physik, 42*, 41.

Dunin-Borkowski, R. E., Kasama, T., Wei, A., & Tripp, S. L. (2003). Magnetic flux closure states in self-assembled nanoparticle rings. In *Proceedings of EMC 2003 (European Microscopy Society): Vol. 42.*

Dunin-Borkowski, R. E., Kasama, T., Wei, A., Tripp, S. L., Hÿtch, M. J., Snoeck, E., ... Putnis, A. (2004). Off-axis electron holography of magnetic nanowires and chains, rings, and planar arrays of magnetic nanoparticles. *Microscopy Research and Technique, 64*(5–6), 390–402.

Dunin-Borkowski, R. E., McCartney, M. R., Frankel, R. B., Bazylinski, D. A., Pósfai, M., & Buseck, P. R. (1998). Magnetic microstructure of magnetotactic bacteria by electron holography. *Science, 282*(5395), 1868–1870.

Dunin-Borkowski, R. E., McCartney, M. R., Kardynal, B., Parkin, S. S. P., Scheinfein, M. R., & Smith, D. J. (2000). Off-axis electron holography of patterned magnetic nanostructures. *Journal of Microscopy, 200*(3), 187–205.

Dunin-Borkowski, R. E., McCartney, M. R., Kardynal, B., Scheinfein, M. R., Smith, D. J., & Parkin, S. S. P. (2001). Off-axis electron holography of exchange-biased CoFe/FeMn patterned nanostructures. *Journal of Applied Physics, 90*(6), 2899–2902.

Dunin-Borkowski, R. E., McCartney, M. R., Kardynal, B., & Smith, D. J. (1998). Magnetic interactions within patterned cobalt nanostructures using off-axis electron holography. *Journal of Applied Physics, 84*(1), 374–378.

Dunin-Borkowski, R. E., McCartney, M. R., Smith, D. J., & Parkin, S. S. P. (1998). Towards quantitative electron holography of magnetic thin films using in situ magnetization reversal. *Ultramicroscopy, 74*(1–2), 61–73.

Edström, A., Lubk, A., & Rusz, J. (2016). Elastic scattering of electron vortex beams in magnetic matter. *Physical Review Letters, 116*(12), 127203.

Egerton, R. F. (1996). *Electron energy-loss spectroscopy in the electron microscope.* Plenum Press.

Einsle, J. F., Gatel, C., Masseboeuf, A., Cours, R., Bashir, M., Gubbins, M., . . . Snoeck, E. (2015). In situ electron holography of the dynamic magnetic field emanating from a hard-disk drive writer. *Nano Research, 8*(4), 1241–1249.

Evans, L. C. (1998). *Partial differential equations. Graduate studies in mathematics.* American Mathematical Society.

Feser, M., Hornberger, B., Jacobsen, C., Geronimo, G. D., Rehak, P., Holl, P., & Sträder, L. (2006). Integrating silicon detector with segmentation for scanning transmission X-ray microscopy. *Nuclear Instruments & Methods in Physics Research Section A: Accelerators, Spectrometers, Detectors and Associated Equipment, 565*(2), 841–854.

Fick, E. (1988). *Einführung in die Grundlagen der Quantentheorie.* Aula.

Fienup, J. R. (1982). Phase retrieval algorithms – a comparison. *Applied Optics, 21,* 2758–2769.

Fienup, J. R. (1987). Reconstruction of a complex-valued object from the modulus of its Fourier-transform using a support constrain. *Journal of the Optical Society of America A, Optics and Image Science, 4*(1), 118–123.

Fienup, J. R., Marron, J. C., Schulz, T. J., & Seldin, J. H. (1993). Hubble Space Telescope characterized by using phase-retrieval algorithms. *Applied Optics, 32*(10), 1747–1767.

Fienup, J. R., & Wackerman, C. C. (1986). Phase-retrieval stagnation problems and solutions. *Journal of the Optical Society of America A, Optics and Image Science, 3*(11), 1897–1907.

Fox, P. J., Mackin, T. R., Turner, L. D., Colton, I., Nugent, K. A., & Scholten, R. E. (2002). Noninterferometric phase imaging of a neutral atomic beam. *Journal of the Optical Society of America B: Optical Physics, 19*(8), 1773–1776.

Frabboni, S., Gabrielli, A., Carlo Gazzadi, G., Giorgi, F., Matteucci, G., Pozzi, G., . . . Zoccoli, A. (2012). The Young–Feynman two-slits experiment with single electrons: Build-up of the interference pattern and arrival-time distribution using a fast-readout pixel detector. *Ultramicroscopy, 116,* 73–76.

Frabboni, S., Matteucci, G., & Pozzi, G. (1985). Electron holographic observations of the electrostatic field associated with thin reverse-biased p–n junctions. *Physical Review Letters, 55,* 2196.

Frabboni, S., Matteucci, G., & Pozzi, G. (1987). Observation of electrostatic fields by electron holography: The case of reverse biased p–n junctions. *Ultramicroscopy, 23,* 29–38.

Freund, I. (1999). Critical point explosions in two-dimensional wave fields. *Optics Communications, 159*(1), 99–117.

Fujita, T., Chen, M., Wang, X., Xu, B., Inoke, K., & Yamamoto, K. (2007). Electron holography of single-crystal iron nanorods encapsulated in carbon nanotubes. *Journal of Applied Physics, 101*(1), 014323.

Fujita, T., Hayashi, Y., Tokunaga, T., & Yamamoto, K. (2006). Cobalt nanorods fully encapsulated in carbon nanotube and magnetization measurements by off-axis electron holography. *Applied Physics Letters, 88*(24), 243118.

Gabor, D. (1948). A new microscopic principle. *Nature, 161,* 777.

Gabor, D., & Goss, W. P. (1966). Interference microscope with total wavefront reconstruction. *Journal of the Optical Society of America, 56*(7), 849–858.

Gao, Y., Shindo, D., Bao, Y., & Krishnan, K. (2007). Electron holography of core–shell Co/CoO spherical nanocrystals. *Applied Physics Letters, 90*(23), 233105.

Gatel, C., Lubk, A., Pozzi, G., Snoeck, E., & Hÿtch, M. (2013). Counting elementary charges on nanoparticles by electron holography. *Physical Review Letters, 111*(2), 025501.

Genz, F., Niermann, T., Buijsse, B., Freitag, B., & Lehmann, M. (2014). Advanced double-biprism holography with atomic resolution. *Ultramicroscopy, 147,* 33–43.

Gerchberg, R. W., & Saxton, W. O. (1972). A practical algorithm for the determination of phase from image and diffraction plane pictures. *Optik, 35*(2), 237–246.

Ghiglia, D. C., & Pritt, M. D. (1998). *Two-dimensional phase unwrapping.* Wiley.

Gilbarg, D., & Trudinger, N. S. (1983). *Elliptic partial differential equations of second order.* Springer.

Gonsalves, R. A. (1976). Phase retrieval from modulus data. *Journal of the Optical Society of America, 66*(9), 961–964.

Gribelyuk, M., McCartney, M., Li, J., Murthy, C., Ronsheim, P., Doris, B., . . . Smith, D. (2002). Mapping of electrostatic potential in deep submicron CMOS devices by electron holography. *Physical Review Letters, 89*(2), 25502.

Grillo, V., Carlo Gazzadi, G., Karimi, E., Mafakheri, E., Boyd, R. W., & Frabboni, S. (2014). Highly efficient electron vortex beams generated by nanofabricated phase holograms. *Applied Physics Letters, 104*(4), 043109.

Gureyev, T. E. (2003). Composite techniques for phase retrieval in the Fresnel region. *Optics Communications, 220,* 49–58.

Gureyev, T. E., Roberts, A., & Nugent, K. A. (1995a). Partially coherent fields, the transport-of-intensity equation, and phase uniqueness. *Journal of the Optical Society of America A, Optics and Image Science, 12,* 1942–1946.

Gureyev, T. E., Roberts, A., & Nugent, K. A. (1995b). Phase retrieval with the transport-of-intensity equation: Matrix solution with use of Zernike polynomials. *Journal of the Optical Society of America A, Optics and Image Science, 12*(9), 1932–1941.

Gureyev, T. E., & Wilkins, S. W. (1998). On X-ray phase imaging with a point source. *Journal of the Optical Society of America A, Optics and Image Science, 15*(3), 579–585.

Haigh, S. J., Jiang, B., Alloyeau, D., Kisielowski, C., & Kirkland, A. (2013). Recording low and high spatial frequencies in exit wave reconstructions. *Ultramicroscopy, 133,* 26–34.

Haine, M. E., & Mulvey, T. (1952). The formation of the diffraction image with electrons in the Gabor diffraction microscope. *Journal of the Optical Society of America, 42,* 763–773.

Hanszen, K.-J. (1986). Method of off-axis electron holography and investigations of the phase structure in crystals. *Journal of Physics D: Applied Physics, 19,* 373–395.

Harada, K., Togawa, Y., Akashi, T., Matsuda, T., & Tonomura, A. (2004). Double-biprism electron holography and its applications. *Microscopy and Microanalysis*, *10*(Suppl. 2), 986–987.

Harscher, A. (1999). *Elektronenholographie biologischer Objekte: Grundlagen und Anwendungsbeispiele* (PhD thesis).

Harscher, A., Lichte, H., & Mayer, J. (1997). Interference experiments with energy filtered electrons. *Ultramicroscopy*, *69*, 201–209.

He, K., Cho, J.-H., Jung, Y., Picraux, S. T., & Cumings, J. (2013). Silicon nanowires: Electron holography studies of doped p–n junctions and biased Schottky barriers. *Nanotechnology*, *24*(11), 115703.

He, K., Smith, D. J., & McCartney, M. R. (2009a). Direct visualization of three-step magnetization reversal of nanopatterned spin-valve elements using off-axis electron holography. *Applied Physics Letters*, *94*(17), 172503.

He, K., Smith, D. J., & McCartney, M. R. (2009b). Observation of asymmetrical pinning of domain walls in notched permalloy nanowires using electron holography. *Applied Physics Letters*, *95*(18), 182507.

Helgason, S. (2011). *Integral geometry and Radon transforms*. New York: Springer.

Henderson, C. A., Williams, G. J., Peele, A. G., Quiney, H. M., & Nugent, K. A. (2009). Astigmatic phase retrieval: An experimental demonstration. *Optics Express*, *17*(14), 11905–11915.

Herring, R. A., Pozzi, G., Tanji, T., & Tonomura, A. (1993). Realization of a mixed type of interferometry using convergent-beam electron diffraction and an electron biprism. *Ultramicroscopy*, *50*(1), 94–100.

Hesse, R., Luke, D. R., Sabach, S., & Tam, M. K. (2014). Proximal heterogeneous block input-output method and application to blind ptychographic diffraction imaging. ArXiv e-prints.

Hoppe, W. (1969a). Beugung im inhomogenen Primärstrahlwellenfeld. I. Prinzip einer Phasenmessung von Elektronenbeugungsinterferenzen. *Acta Crystallographica Section A*, *25*(4), 495–501.

Hoppe, W. (1969b). Beugung im inhomogenen Primärstrahlwellenfeld. III. Amplituden- und Phasenbestimmung bei unperiodischen Objekten. *Acta Crystallographica Section A*, *25*(4), 508–514.

Hoppe, W., & Strube, G. (1969). Beugung in inhomogenen Primärstrahlenwellenfeld. II. Lichtoptische Analogieversuche zur Phasenmessung von Gitterinterferenzen. *Acta Crystallographica Section A*, *25*(4), 502–507.

Hornberger, B., de Jonge, M. D., Feser, M., Holl, P., Holzner, C., Jacobsen, C., ... Vogt, S. (2008). Differential phase contrast with a segmented detector in a scanning X-ray microprobe. *Journal of Synchrotron Radiation*, *15*(4), 355–362.

Horstmeyer, R., Chen, R. Y., Ou, X., Ames, B., Tropp, J. A., & Yang, C. (2015). Solving ptychography with a convex relaxation. *New Journal of Physics*, *17*(5), 053044.

Hosten, O., & Kwiat, P. (2008). Observation of the spin hall effect of light via weak measurements. *Science*, *319*(5864), 787–790.

Houdellier, F., & Hÿtch, M. J. (2008). Diffracted phase and amplitude measurements by energy-filtered convergent-beam holography (CHEF). *Ultramicroscopy*, *108*(3), 285–294.

Hsieh, W.-K., Chen, F.-R., Kai, J.-J., & Kirkland, A. I. (2004). Resolution extension and exit wave reconstruction in complex HREM. *Ultramicroscopy*, *98*, 99–114.

Huiser, A. M. J., Drenth, A. J. J., & Ferwerda, H. A. (1976). On phase retrieval in electron microscopy from image and diffraction pattern. *Optik*, *45*, 303.

Huiser, A. M. J., & Ferwerda, H. A. (1976). On the problem of phase retrieval in electron microscopy from image and diffraction pattern. II: On the uniqueness and stability. *Optik*, *46*, 407.

Huiser, A. M. J., van Toorn, P., & Ferwerda, H. A. (1977). On the problem of phase retrieval in electron microscopy from image and diffraction pattern. III: The development of an algorithm. *Optik*, *47*, 1.

Humphry, M. J., Kraus, B., Hurst, A. C., Maiden, A. M., & Rodenburg, J. M. (2012). Ptychographic electron microscopy using high-angle dark-field scattering for sub-nanometre resolution imaging. *Nature Communications*, *3*, 730.

Hurt, N. E. (2001). *Phase retrieval and zero crossings: Mathematical methods in image reconstruction. Mathematics and its applications*. Springer.

Hÿtch, M., Houdellier, F., Hue, F., & Snoeck, E. (2008). Nanoscale holographic interferometry for strain measurements in electronic devices. *Nature*, *453*(7198), 1086–1089.

Hÿtch, M., Houdellier, F., Hüe, F., & Snoeck, E. (2011). Dark-field electron holography for the measurement of geometric phase. *Ultramicroscopy*, *111*(8), 1328–1337.

Ishizuka, K., & Allman, B. (2005). Phase measurement of atomic resolution image using transport of intensity equation. *Journal of Electron Microscopy*, *54*(3), 191–197.

Jaming, P. (2014). Uniqueness results in an extension of Pauli's phase retrieval problem. *Applied and Computational Harmonic Analysis*, *37*(3), 413–441.

Javon, E., Gatel, C., Masseboeuf, A., & Snoeck, E. (2010). Electron holography study of the local magnetic switching process in magnetic tunnel junctions. *Journal of Applied Physics*, *107*(9), 09D310.

Javon, E., Lubk, A., Cours, R., Reboh, S., Cherkashin, N., Houdellier, F., . . . Hÿtch, M. (2014). Dynamical effects in strain measurements by dark-field electron holography. *Ultramicroscopy*, *147*, 70–85.

Jönsson, C. (1974). Electron diffraction at multiple slits. *American Journal of Physics*, *42*(1), 4–11.

Juchtmans, R., Béché, A., Abakumov, A., Batuk, M., & Verbeeck, J. (2015). Using electron vortex beams to determine chirality of crystals in transmission electron microscopy. *Physical Review B*, *91*(9), 094112.

Junginger, F., Kläui, M., Backes, D., Krzyk, S., Rüdiger, U., Kasama, T., . . . Heyderman, L. J. (2008). Quantitative determination of vortex core dimensions in head-to-head domain walls using off-axis electron holography. *Applied Physics Letters*, *92*(11), 112502.

Junginger, F., Kläui, M., Backes, D., Rüdiger, U., Kasama, T., Dunin-Borkowski, R. E., . . . Bland, J. A. C. (2007). Spin torque and heating effects in current-induced domain wall motion probed by transmission electron microscopy. *Applied Physics Letters*, *90*(13), 132506.

Kasama, T., Barpanda, P., Dunin-Borkowski, R. E., Newcomb, S. B., McCartney, M. R., Castano, F. J., & Ross, C. A. (2005). Off-axis electron holography of pseudo-spin-valve thin-film magnetic elements. *Journal of Applied Physics*, *98*, 013903.

Kasama, T., Dunin-Borkowski, R. E., & Beleggia, M. (2011). Electron holography of magnetic materials. In F. Monroy (Ed.), *Holography – different fields of application*. InTech.

Kasama, T., Dunin-Borkowski, R. E., & Eerenstein, W. (2006). Off-axis electron holography observation of magnetic microstructure in a magnetite (001) thin film containing antiphase domains. *Physical Review B*, *73*, 104432.

Kasama, T., Dunin-Borkowski, R. E., Scheinfein, M. R., Tripp, S. L., Liu, J., & Wei, A. (2008). Reversal of flux closure states in cobalt nanoparticle rings with coaxial magnetic pulses. *Advanced Materials*, *20*(22), 4248–4252.

Kasama, T. M., Pósfai, M., Chong, R. K. K., Finlayson, A. P., Dunin-Borkowski, R. E., & Frankel, R. B. (2006). Magnetic microstructure of iron sulfide crystals in magnetotactic bacteria from off-axis electron holography. *Physica B: Condensed Matter*, *384*, 249–252.

Kawasaki, T., Takai, Y., Ikuta, T., & Shimizu, R. (2001). Wave field restoration using three-dimensional Fourier filtering method. *Ultramicroscopy*, *90*, 47–59.

Keinert, F. (1989). Inversion of k-plane transforms and applications in computer tomography. *SIAM Review*, *31*(2), 273–298.

Kienle, S. H. (2000). *Matter wave reconstruction: Interferometric methods* (PhD thesis). University of Ulm.

Kim, J. J., Hirata, K., Ishida, Y., Shindo, D., Takahashi, M., & Tonomura, A. (2008). Magnetic domain observation in writer pole tip for perpendicular recording head by electron holography. *Applied Physics Letters*, *92*(16), 162501.

Kirkland, A. I., & Meyer, R. R. (2004). "Indirect" high-resolution transmission electron microscopy: Aberration measurement and wavefunction reconstruction. *Microscopy and Microanalysis*, *10*(04), 401–413.

Kirkland, E. J. (1984). Improved high resolution image processing of bright field electron micrographs. *Ultramicroscopy*, *15*, 151–172.

Kirkland, E. J. (1998). *Advanced computing in electron microscopy*. New York: Plenum Press.

Koch, C. T. (2008). A flux-preserving non-linear inline holography reconstruction algorithm for partially coherent electrons. *Ultramicroscopy*, *108*(2), 141–150.

Koch, C. T. (2014). Towards full-resolution inline electron holography. *Micron*, *63*, 69–75.

Koch, C. T., & Lubk, A. (2010). Off-axis and inline electron holography: A quantitative comparison. *Ultramicroscopy*, *110*(5), 460–471.

Koch, C. T., Özdöl, V. B., & van Aken, P. A. (2010). An efficient, simple, and precise way to map strain with nanometer resolution in semiconductor devices. *Applied Physics Letters*, *96*(9), 091901.

Kohn, A., Habibi, A., & Mayo, M. (2016). Experimental evaluation of the "transport-of-intensity" equation for magnetic phase reconstruction in Lorentz transmission electron microscopy. *Ultramicroscopy*, *160*, 44–56.

Komrska, J. (1971). *Advances in Electronics and Electron Physics*, *30*, 139.

Kruse, P., Rosenauer, A., & Gerthsen, D. (2003). Determination of the mean inner potential in III–V semiconductors by electron holography. *Ultramicroscopy*, *96*(1), 11–16.

Kruse, P., Schowalter, M., Lamoen, D., Rosenauer, A., & Gerthsen, D. (2006). Determination of the mean inner potential in III–V semiconductors, Si and Ge by density functional theory and electron holography. *Ultramicroscopy*, *106*(2), 105–113.

Kübel, C., & Thust, A. (2006). TrueImage. In T. E. Weirich, J. L. Labar, & X. Zou (Eds.), *Electron crystallography: Novel approaches for structure determination of nanosized materials* (pp. 373–392). Dordrecht: Springer Netherlands.

Latychevskaia, T., & Fink, H.-W. (2007). Solution to the twin image in holography. *Physical Review Letters*, *98*, 233901.

Latychevskaia, T., Formánek, P., Koch, C., & Lubk, A. (2010). Off-axis and inline electron holography: Experimental comparison. *Ultramicroscopy*, *110*(5), 472–482.

Lau, B., & Pozzi, G. (1978). Off-axis electron microholography of magnetic domain walls. *Optik*, *51*, 287–296.

Lehmann, M., & Lichte, H. (2002). Tutorial on off-axis electron holography. *Microscopy and Microanalysis*, *8*(06), 447–466.

Leibfried, D., Pfau, T., & Monroe, C. (1998). Shadows and mirrors: Reconstructing quantum states of atom motion. *Physics Today*, *51*, 22–28.

Leith, E. N., & Upatnieks, J. (1962). Reconstructed wavefronts and communication theory. *Journal of the Optical Society of America*, *52*, 1123–1130.

Leith, E. N., & Upatnieks, J. (1963). Wavefront reconstruction with continuous-tone objects. *Journal of the Optical Society of America*, *53*(12), 37–43.

Lenz, F. (1954). Zur Streuung mittelschneller Elektronen in kleinste Winkel. *Zeitschrift für Naturforschung*, *9A*, 185–204.

Lenz, F. (1988). Statistics of phase and contrast determination in electron holograms. *Optik*, *79*, 13–14.

Leuthner, T., Lichte, H., & Herrman, K. H. (1989). STEM-holography using the electron biprism. *Physica Status Solidi A*, *116*, 113–121.

Levi, A., & Stark, H. (1984). Image restoration by the method of generalized projections with application to restoration from magnitude. *Journal of the Optical Society of America A, Optics and Image Science*, *1*(9), 932–943.

Li, P., Edo, T. B., & Rodenburg, J. M. (2014). Ptychographic inversion via Wigner distribution deconvolution: Noise suppression and probe design. *Ultramicroscopy*, *147*, 106–113.

Lichte, H. (1996). Electron holography: Optimum position of the biprism in the electron microscope. *Ultramicroscopy*, *64*, 79–86.

Lichte, H. (2008). Performance limits of electron holography. *Ultramicroscopy*, *108*(3), 256–262.

Lichte, H., Börrnert, F., Lenk, A., Lubk, A., Röder, F., Sickmann, J., ... Wolf, D. (2013). Electron holography for fields in solids: Problems and progress. *Ultramicroscopy*, *134*, 126–134.

Lichte, H., & Freitag, B. (2000). Inelastic electron holography. *Ultramicroscopy*, *81*, 177–186.

Lichte, H., & Möllenstedt, G. (1979). Measurement of the roughness of supersmooth surfaces using an electron mirror interference microscope. *Journal of Physics E: Scientific Instruments*, *12*, 941–944.

Lin, F., Chen, F. R., Chen, Q., Tang, D., & Peng, L.-M. (2006). The wrap-around problem and optimal padding in the exit wave reconstruction using HRTEM images. *Journal of Electron Microscopy*, *55*(4), 191–200.

Lohmann, A. W., Testorf, M. E., & Ojeda-Castañeda, J. (2004). Holography and the Wigner function. In H. J. Caulfield (Ed.), *The art and science of holography* (pp. 129–144). SPIE Press.

Lohr, M., Schregle, R., Jetter, M., Wächter, C., Wunderer, T., Scholz, F., & Zweck, J. (2012). Differential phase contrast 2.0: Opening new fields for an established technique. *Ultramicroscopy*, *117*, 7–14.

Lubk, A., Béché, A., & Verbeeck, J. (2015). Electron microscopy of probability currents at atomic resolution. *Physical Review Letters*, *115*(17), 176101.

Lubk, A., Clark, L., Guzzinati, G., & Verbeeck, J. (2013). Topological analysis of paraxially scattered electron vortex beams. *Physical Review A*, *87*(3), 033834.

Lubk, A., Guzzinati, G., Börrnert, F., & Verbeeck, J. (2013). Transport of intensity phase retrieval of arbitrary wave fields including vortices. *Physical Review Letters*, *111*(17), 173902.

Lubk, A., Javon, E., Cherkashin, N., Reboh, S., Gatel, C., & Hÿtch, M. J. (2014). Dynamic scattering theory for dark-field electron holography of 3D strain fields. *Ultramicroscopy*, *136*, 42–49.

Lubk, A., & Röder, F. (2015a). Phase-space foundations of electron holography. *Physical Review A*, *92*(3), 033844.

Lubk, A., & Röder, F. (2015b). Semiclassical TEM image formation in phase space. *Ultramicroscopy*, *151*, 136–149.

Lubk, A., Röder, F., Niermann, T., Gatel, C., Joulie, S., Houdellier, F., ... Hÿtch, M. J. (2012). A new linear transfer theory and characterization method for image detectors. Part II: Experiment. *Ultramicroscopy*, *115*, 78–87.

Lubk, A., Vogel, K., Röder, F., Wolf, D., Clark, L., & Verbeeck, J. (2016). Fundamentals of focal series inline electron holography. *Advances in Imaging and Electron Physics*, *197*, 105–147.

Lubk, A., & Zweck, J. (2015). Differential phase contrast: An integral perspective. *Physical Review A*, *91*(2), 023805.

Luke, D., Burke, J., & Lyon, R. (2002). Optical wavefront reconstruction: Theory and numerical methods. *SIAM Review*, *44*(2), 169–224.

Lvovsky, A. I., & Raymer, M. G. (2009). Continuous-variable optical quantum-state tomography. *Reviews of Modern Physics*, *81*(1), 299–332.

Lyon, R. G., Dorband, J. E., & Hollis, J. M. (1997). Hubble Space Telescope faint object camera calculated point-spread functions. *Applied Optics*, *36*(8), 1752–1765.

Lyon, R. G., Miller, P. E., & Gruszczak, A. (1991). HST phase retrieval: A parameter estimation. *Proceedings of SPIE*, *1567*, 317–326.

Maiden, A. M., Humphry, M. J., & Rodenburg, J. M. (2012). Ptychographic transmission microscopy in three dimensions using a multi-slice approach. *Journal of the Optical Society of America A, Optics and Image Science*, *29*(8), 1606–1614.

Maiden, A. M., & Rodenburg, J. M. (2009). An improved ptychographical phase retrieval algorithm for diffractive imaging. *Ultramicroscopy*, *109*(10), 1256–1262.

Majert, S., & Kohl, H. (2015). High-resolution stem imaging with a quadrant detector – conditions for differential phase contrast microscopy in the weak phase object approximation. *Ultramicroscopy*, *148*, 81–86.

Mankos, M., Cowley, J. M., & Scheinfein, M. R. (1996). Quantitative micromagnetics at high spatial resolution using far-out-of-focus STEM electron holography. *Physica Status Solidi A*, *154*, 469.

Marín, L., Rodríguez, L. A., Magén, C., Snoeck, E., Arras, R., Lucas, I., ... Ibarra, M. R. (2015). Observation of the strain induced magnetic phase segregation in manganite thin films. *Nano Letters*, *15*(1), 492–497.

Martin, A. V., & Allen, L. J. (2007). Phase imaging from a diffraction pattern in the presence of vortices. *Optics Communications*, *277*(2), 288–294.

Martin, A. V., Chen, F.-R., Hsieh, W.-K., Kai, J.-J., Findley, S. D., & Allen, L. J. (2006). Spatial incoherence in phase retrieval based on focus variation. *Ultramicroscopy*, *106*, 914–924.

Masseboeuf, A., Marty, A., Bayle-Guillemaud, P., Gatel, C., & Snoeck, E. (2009). Quantitative observation of magnetic flux distribution in new magnetic films for future high density recording media. *Nano Letters*, *9*(8), 2803–2806.

Matsuda, T., Hasegawa, S., Igarashi, M., Kobayashi, T., Naito, M., Kajiyama, H., ... Aoki, R. (1989). Magnetic field observation of a single flux quantum by electron-holographic interferometry. *Physical Review Letters*, *62*(21), 2519–2522.

McCallum, B. C., & Rodenburg, J. M. (1992). Two-dimensional demonstration of Wigner phase-retrieval microscopy in the STEM configuration. *Ultramicroscopy*, *45*, 371–380.

McCartney, M. R. (2005). Characterization of charging in semiconductor device materials by electron holography. *Journal of Electron Microscopy, 54,* 239–242.

McCartney, M. R., Agarwal, N., Chung, S., Cullen, D. A., Han, M.-G., He, K., . . . Smith, D. J. (2010). Quantitative phase imaging of nanoscale electrostatic and magnetic fields using off-axis electron holography. *Ultramicroscopy, 110*(5), 375–382.

McCartney, M. R., Ponce, F. A., & Cai, J. (2000). Mapping electrostatic potential across an AlGaN/InGaN/AlGaN diode by electron holography. *Applied Physics Letters, 76,* 3055–3057.

McCartney, M. R., & Smith, D. J. (1994). Direct observation of potential distribution across Si/Si p–n junctions using off-axis electron holography. *Applied Physics Letters, 65*(20), 2603.

McCartney, M. R., & Smith, D. J. (2007). Electron holography: Phase imaging with nanometer resolution. *Annual Review of Materials Research, 37*(1), 729–767.

McCartney, M. R., & Zhu, Y. (1998). Induction mapping of Nd2Fe14B magnetic domains by electron holography. *Applied Physics Letters, 72*(11), 1380–1382.

McMorran, B. J., Agrawal, A., Anderson, I. M., Herzing, A. A., Lezec, H. J., McClelland, J. J., & Unguris, J. (2011). Electron vortex beams with high quanta of orbital angular momentum. *Science, 331*(6014), 192–195.

Mecklenbräuker, W., & Hlawatsch, F. (Eds.). (1997). *The Wigner distribution – theory and applications in signal processing* (pp. 375–426). Elsevier.

Merli, P. G., Missiroli, G. F., & Pozzi, G. (1976). On the statistical aspect of electron interference phenomena. *American Journal of Physics, 44*(3), 306–307.

Meyer, R. (2002). *Quantitative automated object wave restoration in high-resolution electron microscopy* (PhD thesis). Technische Universität Dresden.

Meyer, R. R., Kirkland, A. I., & Saxton, W. O. (2002). A new method for the determination of the wave aberration function for high resolution TEM. 1. Measurement of the symmetric aberrations. *Ultramicroscopy, 92,* 89–109.

Meyer, R. R., Kirkland, A. I., & Saxton, W. O. (2004). A new method for the determination of the wave aberration function for high-resolution TEM. 2. Measurement of the antisymmetric aberrations. *Ultramicroscopy, 99,* 115–123.

Millane, R. P. (1990). Phase retrieval in crystallography and optics. *Journal of the Optical Society of America A, Optics and Image Science, 7*(3), 394–411.

Misell, D. L. (1973). An examination of an iterative method for the solution of the phase problem in optics and electron optics: I. Test calculations. *Journal of Physics D: Applied Physics, 6*(18), 2200.

Möllenstedt, G., & Düker, H. (1955). Interferenzversuch mit einem Biprisma für Elektronenwellen. *Naturwissenschaften, 42,* 41.

Möllenstedt, G., & Wahl, H. (1968). Elektronenholographie und Rekonstruktion mit Laserlicht. *Naturwissenschaften, 55,* 340–341.

Morishita, S., Yamasaki, J., Nakamura, K., Kato, T., & Tanaka, N. (2008). Diffractive imaging of the dumbbell structure in silicon by spherical-aberration-corrected electron diffraction. *Applied Physics Letters, 93*(18), 183103.

Müller, K., Krause, F. F., Béché, A., Schowalter, M., Galioit, V., Löffler, S., . . . Rosenauer, A. (2014). Atomic electric fields revealed by a quantum mechanical approach to electron picodiffraction. *Nature Communications, 5,* 5653.

Müller-Caspary, K., Krause, F. F., Grieb, T., Löffler, S., Schowalter, M., Béché, A., . . . Rosenauer, A. (2016). Measurement of atomic electric fields and charge densities from average momentum transfers using scanning transmission electron microscopy. *Ultramicroscopy, 178,* 62–80.

Murakami, Y., Shindo, D., Oikawa, K., Kainuma, R., & Ishida, K. (2004). Microstructural change near the martensitic transformation in a ferromagnetic shape memory alloy Ni51Fe22Ga27 studied by electron holography. *Applied Physics Letters, 85,* 6170–6172.

Murakami, Y., Yanagisawa, K., Niitsu, K., Park, H., Matsuda, T., Kainuma, R., . . . Tonomura, A. (2013). Determination of magnetic flux density at the nanometer-scale antiphase boundary in Heusler alloy Ni50Mn25Al12.5Ga12.5. *Acta Materialia, 61*(6), 2095–2101.

Murakami, Y., Yano, T., Shindo, D., Kainuma, R., Oikawa, K., & Ishida, K. (2010). Elucidation of microstructures produced in Ni51Fe22Ga27 ferromagnetic shape memory alloy. *Journal of Applied Physics, 107*(4), 043904.

Nakajima, N. (1999). Reconstruction of a wave function from the Q function using a phase-retrieval method in quantum-state measurements of light. *Physical Review A, 59*(6), 4164–4171.

Niermann, T., & Lehmann, M. (2016). Holographic focal series: Differences between in-line and off-axis electron holography at atomic resolution. *Journal of Physics D: Applied Physics, 49*(19), 194002.

Niermann, T., Lubk, A., & Röder, F. (2012). A new linear transfer theory and characterization method for image detectors. Part I: Theory. *Ultramicroscopy, 115,* 68–77.

Nugent, K. A. (2007). X-ray noninterferometric phase imaging: a unified picture. *Journal of the Optical Society of America A, Optics and Image Science, 24*(2), 536–547.

Nugent, K. A. (2010). Coherent methods in the X-ray sciences. *Advances in Physics, 59*(1), 1–99.

Nugent, K. A., & Paganin, D. (2000). Matter-wave phase measurement: A noninterferometric approach. *Physical Review A, 61*(6), 063614.

Nye, J. F., & Berry, M. V. (1974). Dislocations in wave trains. *Proceedings of the Royal Society A: Mathematical, Physical and Engineering Sciences, 336*(1605), 165–190.

Ohneda, Y., Baba, N., Miura, N., & Sakurai, T. (2001). Multiresolution approach to image reconstruction with phase-diversity technique. *Optical Review, 8*(1), 32–36.

O'Holleran, K., Padgett, M. J., & Dennis, M. R. (2006). Topology of optical vortex lines formed by the interference of three, four, and five plane waves. *Optics Express, 14*(7), 3039–3044.

Ophus, C., & Ewalds, T. (2012). Guidelines for quantitative reconstruction of complex exit waves in HRTEM. *Ultramicroscopy, 113,* 88–95.

Paganin, D., & Nugent, K. A. (1998). Noninterferometric phase imaging with partially coherent light. *Physical Review Letters, 80*(12), 2586–2589.

Pantzer, A., Vakahy, A., Eliyahou, Z., Levi, G., Horvitz, D., & Kohn, A. (2014). Dopant mapping in thin FIB prepared silicon samples by off-axis electron holography. *Ultramicroscopy, 138,* 36–45.

Park, H. S., Hirata, K., Yanagisawa, K., Ishida, Y., Matsuda, T., Shindo, D., & Tonomura, A. (2012). Nanoscale magnetic characterization of tunneling magnetoresistance spin valve head by electron holography. *Small, 8*(23), 3640–3646.

Park, H. S., Murakami, Y., Shindo, D., Chernenko, V. A., & Kanomata, T. (2003). Behavior of magnetic domains during structural transformations in Ni2MnGa ferromagnetic shape memory alloy. *Applied Physics Letters, 83*(18), 3752–3754.

Park, H. S., Murakami, Y., Yanagisawa, K., Matsuda, T., Kainuma, R., Shindo, D., & Tonomura, A. (2012). Electron holography studies on narrow magnetic domain walls observed in a Heusler alloy Ni50Mn25Al12.5Ga12.5. *Advanced Functional Materials, 22*(16), 3434–3437.

Park, J. B., Niermann, T., Berger, D., Knauer, A., Koslow, I., Weyers, M., ... Lehmann, M. (2014). Impact of electron irradiation on electron holographic potentiometry. *Applied Physics Letters, 105*(9), 094102.

Park, J. L., & Band, W. (1971). A general theory of empirical state determination in quantum physics: Part I. *Foundations of Physics, 1*(3), 211–226.

Parvizi, A., Van den Broek, W., & Koch, C. T. (2016). Recovering low spatial frequencies in wavefront sensing based on intensity measurements. *Advanced Structural and Chemical Imaging, 2*(1), 1–9.

Pauli, W. (1933). Die allgemeinen Prinzipien der Wellenmechanik. In H. Geiger, & K. Scheel (Eds.), *Handbuch der Physik: Vol. 24*. Berlin: Springer Verlag.

Peele, A. G., Nugent, K. A., Mancuso, A. P., Paterson, D., McNulty, I., & Hayes, J. P. (2004). X-ray phase vortices: Theory and experiment. *Journal of the Optical Society of America A, Optics and Image Science, 21*(8), 1575–1584.

Pennington, R. S., Boothroyd, C. B., & Dunin-Borkowski, R. E. (2015). Surface effects on mean inner potentials studied using density functional theory. *Ultramicroscopy, 159*(Part 1), 34–45.

Pennycook, T. J., Lupini, A. R., Yang, H., Murfitt, M. F., Jones, L., & Nellist, P. D. (2015). Efficient phase contrast imaging in STEM using a pixelated detector. Part 1: Experimental demonstration at atomic resolution. *Ultramicroscopy, 151*, 160–167.

Petersen, T. C., & Keast, V. J. (2007). Astigmatic intensity equation for electron microscopy based phase retrieval. *Ultramicroscopy, 107*, 635–643.

Petersen, T. C., Keast, V. J., Johnson, K., & Duvall, S. (2007). TEM-based phase retrieval of p–n junction wafers using the transport of intensity equation. *Philosophical Magazine, 87*(24), 3565–3578.

Petersen, T. C., Keast, V. J., & Paganin, D. M. (2008). Quantitative TEM-based phase retrieval of MgO nano-cubes using the transport of intensity equation. *Ultramicroscopy, 108*, 805–815.

Petersen, T. C., Paganin, D. M., Weyland, M., Simula, T. P., Eastwood, S. A., & Morgan, M. J. (2013). Measurement of the Gouy phase anomaly for electron waves. *Physical Review A, 88*(4), 043803.

Petford-Long, A. K., & De Graef, M. (2002). Lorentz microscopy. In *Characterization of materials*. John Wiley & Sons, Inc.

Petruccelli, J. C., Tian, L., & Barbastathis, G. (2013). The transport of intensity equation for optical path length recovery using partially coherent illumination. *Optics Express, 21*(12), 14430–14441.

Pinhasi, S. V., Alimi, R., Perelmutter, L., & Eliezer, S. (2010). Topography retrieval using different solutions of the transport intensity equation. *Journal of the Optical Society of America A, Optics and Image Science, 27*(10), 2285–2292.

Polking, M. J., Han, M.-G., Yourdkhani, A., Petkov, V., Kisielowski, C. F., Volkov, V. V., ... Ramesh, R. (2012). Ferroelectric order in individual nanometre-scale crystals. *Nature Materials, 11*(8), 700–709.

Potapov, P., Lichte, H., Verbeeck, J., & van Dyck, D. (2006). Experiments on inelastic electron holography. *Ultramicroscopy, 106*(11–12), 1012–1018.

Potapov, P. L., Lichte, H., Verbeeck, J., Van Dyck, D., & Schattschneider, P. (2003). Point-to-point coherence in inelastic scattering. In *Proceedings of EMC 2003 (European Microscopy Society): Vol. 81* (pp. 26–27).

Pozzi, G., Beleggia, M., Kasama, T., & Dunin-Borkowski, R. E. (2014). Interferometric methods for mapping static electric and magnetic fields. *Comptes Rendus. Physique, 15*, 126–139.

Quatieri, T., & Oppenheim, A. (1981). Iterative techniques for minimum phase signal reconstruction from phase or magnitude. *IEEE Transactions on Acoustics, Speech, and Signal Processing*, *29*(6), 1187–1193.

Rau, W. D., Schwander, P., Baumann, F. H., Höppner, W., & Ourmazd, A. (1999). Two-dimensional mapping of the electrostatic potential in transistors by electron holography. *Physical Review Letters*, *82*(12), 2614–2617.

Raymer, M. G., Beck, M., & McAlister, D. (1994). Complex wave-field reconstruction using phase-space tomography. *Physical Review Letters*, *72*(8), 1137–1140.

Rodenburg, J. M. (2008). Ptychography and related diffractive imaging methods. *Advances in Imaging and Electron Physics*, *150*, 87–184.

Rodenburg, J. M., & Bates, R. H. T. (1992). The theory of super-resolution electron microscopy via Wigner-distribution deconvolution. *Philosophical Transactions of the Royal Society of London. Series A: Physical and Engineering Sciences*, *339*(1655), 521–553.

Rodenburg, J. M., & Faulkner, H. M. L. (2004). A phase retrieval algorithm for shifting illumination. *Applied Physics Letters*, *85*(20), 4795–4797.

Röder, F., Hlawacek, G., Wintz, S., Hübner, R., Bischoff, L., Lichte, H., ... Bali, R. (2015). Direct depth- and lateral-imaging of nanoscale magnets generated by ion impact. *Scientific Reports*, *5*, 16786.

Röder, F., & Lichte, H. (2011). Inelastic electron holography – first results with surface plasmons. *Journal of Applied Physics*, *54*, 33504.

Röder, F., & Lubk, A. (2014). Transfer and reconstruction of the density matrix in off-axis electron holography. *Ultramicroscopy*, *146*, 103–116.

Röder, F., Lubk, A., Wolf, D., & Niermann, T. (2014). Noise estimation for off-axis electron holography. *Ultramicroscopy*, *144*, 32–42.

Rose, H. (1973). Phase contrast in scanning transmission electron microscopy. *Optik*, *39*, 416.

Rose, H. (1976). Nonstandard imaging methods in electron microscopy. *Ultramicroscopy*, *2*, 251–267.

Rother, A., Gemming, T., & Lichte, H. (2009). The statistics of the thermal motion of the atoms during imaging process in transmission electron microscopy and related techniques. *Ultramicroscopy*, *109*(2), 139–146.

Sandweg, C. W., Wiese, N., McGrouther, D., Hermsdoerfer, S. J., Schultheiss, H., Leven, B., ... Chapman, J. N. (2008). Direct observation of domain wall structures in curved permalloy wires containing an antinotch. *Journal of Applied Physics*, *103*(9), 093906.

Saxton, W. O. (1994). Accurate alignment of sets of images. *Journal of Microscopy*, *174*(2), 61–68.

Schattschneider, P., & Lichte, H. (2005). Correlation and the density-matrix approach to inelastic electron holography in solid state plasmas. *Physical Review B*, *71*(4), 045130.

Schattschneider, P., Stöger-Pollach, M., & Verbeeck, J. (2012). Novel vortex generator and mode converter for electron beams. *Physical Review Letters*, *109*(8), 1–5.

Schiske, P. (1968). Image reconstruction by means of focus series. In *Proceedings of the 4th European regional conference on electron microscopy I* (p. 145).

Schiske, P. (1973). In P. W. Hawkes (Ed.), *Image processing and computer-aided design in electron optics* (p. 82). London: Academic Press.

Schleich, W. P. (2001). *Quantum optics in phase space*. Berlin: Wiley VCH.

Schmalz, J. A., Gureyev, T. E., Paganin, D. M., & Pavlov, K. M. (2011). Phase retrieval using radiation and matter-wave fields: Validity of Teague's method for solution of the transport-of-intensity equation. *Physical Review A*, *84*(2), 023808.

Schofield, M., Beleggia, M., Zhu, Y., Guth, K., & Jooss, C. (2004). Direct evidence for negative grain boundary potential in Ca-doped and undoped YBa2Cu3O7-x. *Physical Review Letters*, *92*(20), 195502.

Schowalter, M., Rosenauer, A., Lamoen, D., Kruse, P., & Gerthsen, D. (2006). Ab initio computation of the mean inner coulomb potential of wurtzite-type semiconductors and gold. *Applied Physics Letters*, *88*(23), 232108.

Schowalter, M., Titantah, J. T., Lamoen, D., & Kruse, P. (2005). Ab initio computation of the mean inner coulomb potential of amorphous carbon structures. *Applied Physics Letters*, *86*(11), 112102.

Seldin, J. H., & Fienup, J. R. (1990). Numerical investigation of the uniqueness of phase retrieval. *Journal of the Optical Society of America A, Optics and Image Science*, *7*(3), 412–427.

Seraphin, S., Beeli, C., Bonard, J.-M., Jiao, J., Stadelmann, P. A., & Chatelain, A. (1999). Magnetization of carbon-coated ferromagnetic nanoclusters determined by electron holography. *Journal of Materials Research*, *14*(07), 2861–2870.

Shibata, N., Findlay, S. D., Kohno, Y., Sawada, H., Kondo, Y., & Ikuhara, Y. (2012). Differential phase-contrast microscopy at atomic resolution. *Nature Physics*, *8*(8), 611–615.

Shibata, N., Kohno, Y., Findlay, S. D., Sawada, H., Kondo, Y., & Ikuhara, Y. (2010). New area detector for atomic-resolution scanning transmission electron microscopy. *Journal of Electron Microscopy*, *59*(6), 473–479.

Simon, P., Huhle, R., Lehmann, M., Lichte, H., Mönter, D., Bieber, T., . . . Michler, G. H. (2002). Electron holography on beam sensitive materials: Organic polymers and mesoporous silica. *Chemistry of Materials*, *14*(4), 1505–1514.

Simon, P., Lichte, H., Drechsel, J., Formánek, P., Graff, A., Wahl, R., . . . Michler, G. H. (2003). Electron holography of organic and biological molecules. *Advanced Materials*, *15*(17), 1475–1481.

Simon, P., Lichte, H., Formánek, P., Lehmann, M., Huhle, R., Carrillo-Cabrera, W., . . . Ehrlich, H. (2008). Electron holography of biological samples. *Micron*, *39*, 229–256.

Simon, P., Zahn, D., Lichte, H., & Kniep, R. (2006). Intrinsic electric dipole fields and the induction of hierarchical form developments in fluorapatite–gelatine nanocomposites: A general principle for morphogenesis of biominerals? *Angewandte Chemie*, *118*(12), 1945–1949.

Smithey, D. T., Beck, M., Raymer, M. G., & Faridani, A. (1993). Measurement of the Wigner distribution and the density matrix of a light mode using optical homodyne tomography: Application to squeezed states and the vacuum. *Physical Review Letters*, *70*(9), 1244–1247.

Snoeck, E., Dunin-Borkowski, R. E., Dumestre, F., Renaud, P., Amiens, C., Chaudret, B., & Zurcher, P. (2003). Quantitative magnetization measurements on nanometer ferromagnetic cobalt wires using electron holography. *Applied Physics Letters*, *82*(1), 88–90.

Snoeck, E., Gatel, C., Lacroix, L. M., Blon, T., Lachaize, S., Carrey, J., . . . Chaudret, B. (2008). Magnetic configurations of 30 nm iron nanocubes studied by electron holography. *Nano Letters*, *8*(12), 4293–4298.

Song, K., Shin, G.-Y., Kim, J. K., Oh, S. H., & Koch, C. T. (2013). Strain mapping of LED devices by dark-field inline electron holography: Comparison between deterministic and iterative phase retrieval approaches. *Ultramicroscopy*, *127*, 119–125.

Sonnentag, P., & Hasselbach, F. (2007). Measurement of decoherence of electron waves and visualization of the quantum–classical transition. *Physical Review Letters*, *98*(20), 200402.

Stark, H., & Sezan, M. I. (1994). Image processing using projection methods. In B. Javidi, & J. L. Horner (Eds.), *Real-time optical information processing* (pp. 185–232). London, UK: Academic Press.

Stewart, W. C. (1976). On differential phase contrast with an extended illumination source. *Journal of the Optical Society of America*, *66*(8), 813–818.

Strutt, J. W. (1892). On the interference bands of approximately homogeneous light; in a letter to Prof. A. Michelson. *Philosophical Magazine*, *34*, 407–411.

Sugawara, A., Fukunaga, K-i., Scheinfein, M. R., Kobayashi, H., Kitagawa, H., & Tonomura, A. (2007). Electron holography study of the temperature variation of the magnetic order parameter within circularly chained nickel nanoparticle rings. *Applied Physics Letters*, *91*(26), 262513.

Takahashi, Y., Yajima, Y., Ichikawa, M., & Kuroda, K. (1994). Observation of magnetic induction distribution by scanning interference electron microscopy. *Japanese Journal of Applied Physics*, *33*, 1352–1354.

Takajo, H., Takahashi, T., Itoh, K., & Fujisaki, T. (2002). Reconstruction of an object from its Fourier modulus: Development of the combination algorithm composed of the hybrid input–output algorithm and its converging part. *Applied Optics*, *41*(29), 6143–6153.

Tamir, B., & Cohen, E. (2013). Introduction to weak measurements and weak values. *Quanta*, *2*(1), 7–17.

Tavabi, A. H., Migunov, V., Savenko, A., & Dunin-Borkowski, R. E. (2014). Electron holography of the magnetic phase shift of a current-carrying wire. *Microscopy and Microanalysis*, *20*(Suppl. S3), 278–279.

Teague, M. R. (1983). Deterministic phase retrieval: A Green's function solution. *Journal of the Optical Society of America*, *73*, 1434–1441.

Testorf, M., & Lohmann, A. W. (2008). Holography in phase space. *Applied Optics*, *47*(4), A70–A77.

Thibault, P., Dierolf, M., Bunk, O., Menzel, A., & Pfeiffer, F. (2009). Probe retrieval in ptychographic coherent diffractive imaging. *Ultramicroscopy*, *109*(4), 338–343.

Thibault, P., & Menzel, A. (2013). Reconstructing state mixtures from diffraction measurements. *Nature*, *494*(7435), 68–71.

Thomas, J. M., Simpson, E. T., Kasama, T., & Dunin-Borkowski, R. E. (2008). Electron holography for the study of magnetic nanomaterials. *Accounts of Chemical Research*, *41*(5), 665–674.

Thust, A., Coene, W., de Beeck, M. O., & Van Dyck, D. (1996). Focal-series reconstruction in HRTEM: Simulation studies on non-periodic objects. *Ultramicroscopy*, *64*, 211–230.

Tillmann, K., Houben, L., Thust, A., & Urban, K. (2006). Spherical-aberration correction in tandem with the restoration of the exit-plane wavefunction: Synergetic tools for the imaging of lattice imperfections in crystalline solids at atomic resolution. *Journal of Materials Science*, *41*(14), 4420–4433.

Tonomura, A. (1972). The electron interference method for magnetization measurement of thin films. *Japanese Journal of Applied Physics*, *11*, 493–502.

Tonomura, A. (1987). Applications of electron holography. *Reviews of Modern Physics*, *59*, 639.

Tonomura, A., Allard, L. F., Pozzi, G., Joy, D. C., & Ono, Y. A. (Eds.). (1995). *Electron holography. Proceedings of the international workshop on electron holography. North-Holland delta series* (pp. 267–276).

Tonomura, A., Matsuda, T., Endo, J., Arii, T., & Mihama, K. (1986). Holographic interference electron microscopy for determining specimen magnetic structure and thickness distribution. *Physical Review B*, *34*(5), 3397–3402.

Trahan, C. J., & Wyatt, R. E. (2006). *Quantum dynamics with trajectories: Introduction to quantum hydrodynamics. Interdisciplinary applied mathematics*. New York: Springer.

Twitchett, A. C., Dunin-Borkowski, R. E., & Midgley, P. A. (2002). Quantitative electron holography of biased semiconductor devices. *Physical Review Letters, 88*, 238302.

Twitchett, A. C., Dunin-Borkowski, R. E., & Midgley, P. A. (2006). Comparison of off-axis and in-line electron holography as quantitative dopant-profiling techniques. *Philosophical Magazine, 86*(36), 5805–5823.

Uchida, M., & Tonomura, A. (2010). Generation of electron beams carrying orbital angular momentum. *Nature, 464*(7289), 737–739.

Vaidman, L. (2009). Weak value and weak measurements. In D. Greenberger, K. Hentschel, & F. Weinert (Eds.), *Compendium of quantum physics* (pp. 840–842). Springer.

Van Dyck, D., & Coene, W. (1987). A new procedure for wave function restoration in high resolution electron microscopy. *Optik, 77*, 125–128.

Varón, M., Beleggia, M., Kasama, T., Harrison, R. J., Dunin-Borkowski, R. E., Puntes, V. F., & Frandsen, C. (2013). Dipolar magnetism in ordered and disordered low-dimensional nanoparticle assemblies. *Scientific Reports, 3*, 1234.

Verbeeck, J., Bertoni, G., & Lichte, H. (2011). A holographic biprism as a perfect energy filter? *Ultramicroscopy, 111*(7), 887–893.

Verbeeck, J., Bertoni, G., & Schattschneider, P. (2008). The Fresnel effect of a defocused biprism on the fringes in inelastic holography. *Ultramicroscopy, 108*(3), 263–269.

Verbeeck, J., Van Dyck, D., Lichte, H., Potapov, P., & Schattschneider, P. (2005). Plasmon holographic experiments: Theoretical framework. *Ultramicroscopy, 102*, 239–255.

Verbeeck, J., Tian, H., & Schattschneider, P. (2010). Production and application of electron vortex beams. *Nature, 467*(7313), 301–304.

Völkl, E., Allard, L. F., & Joy, D. C. (Eds.). (1999). *Introduction to electron holography*. Kluwer Academic/Plenum Publishers.

Volkov, V. V., Zhu, Y., & De Graef, M. (2002). A new symmetrized solution for phase retrieval using the transport of intensity equation. *Micron, 33*(5), 411–416.

Vulovic, M., Voortman, L. M., van Vliet, L. J., & Rieger, B. (2014). When to use the projection assumption and the weak-phase object approximation in phase contrast cryo-EM. *Ultramicroscopy, 136*, 61–66.

Waddell, E. M., & Chapman, J. N. (1979). *Optik, 54*, 83–96.

Wang, Z., Hirayama, T., Sasaki, K., Saka, H., & Kato, N. (2002). Electron holographic characterization of electrostatic potential distributions in a transistor sample fabricated by FIB. *Applied Physics Letters, 80*, 246–248.

Wang, Y. Y., Li, J., Domenicucci, A., & Bruley, J. (2013). Variable magnification dual lens electron holography for semiconductor junction profiling and strain mapping. *Ultramicroscopy, 124*, 117–129.

Wei, K. (2015). Solving systems of phaseless equations via Kaczmarz methods: A proof of concept study. *Inverse Problems, 31*(12), 125008.

Wolf, K. B. (1996). Wigner distribution function for paraxial polychromatic optics. *Optics Communications, 132*, 343–352.

Wolf, K. B., & Rivera, A. L. (1997). Holographic information in the Wigner function. *Optics Communications, 144*, 36–42.

Wolfke, M. (1920). Über die Möglichkeit der optischen Abbildung vom Molekulargittern. *Physikalische Zeitschrift, 21*, 495–497.

Xu, Q. Y., Wang, Y. G., You, B., Du, J., Hu, A., & Zhang, Z. (2004). Electron holography study on the microstructure of magnetic tunnelling junctions. *Ultramicroscopy, 98*, 297.

Yamamoto, K., Hogg, C. R., Yamamuro, S., Hirayama, T., & Majetich, S. A. (2011). Dipolar ferromagnetic phase transition in Fe3O4 nanoparticle arrays observed by Lorentz microscopy and electron holography. *Applied Physics Letters, 98*(7), 072509.

Yamamoto, K., Majetich, S. A., McCartney, M. R., Sachan, M., Yamamuro, S., & Hirayama, T. (2008). Direct visualization of dipolar ferromagnetic domain structures in Co nanoparticle monolayers by electron holography. *Applied Physics Letters*, *93*(8), 082502.

Yang, G. Z., Dong, B. Z., Gu, B. Y., Zhuang, J. Y., & Ersoy, O. K. (1994). Gerchberg–Saxton and Yang–Gu algorithms for phase retrieval in a nonunitary transform system: A comparison. *Applied Optics*, *33*(2), 209–218.

Zeitler, E., & Thomson, M. G. R. (1970). Scanning transmission electron microscopy. *Optik*, *31*, 258.

Zhang, D., Ray, N. M., Petuskey, W. T., Smith, D. J., & McCartney, M. R. (2014). Magnetic domain structure in nanocrystalline Ni–Zn–Co spinel ferrite thin films using off-axis electron holography. *Journal of Applied Physics*, *116*(8), 083901.

Zhang, H., Egerton, R. F., & Malac, M. (2010). Local thickness measurement in TEM. *Microscopy and Microanalysis*, *16*(Suppl. S2), 344–345.

Zhang, H.-R., Egerton, R. F., & Malac, M. (2012). Local thickness measurement through scattering contrast and electron energy-loss spectroscopy. *Micron*, *43*(1), 8–15.

Zheng, S., Xue, B., Xue, W., Bai, X., & Zhou, F. (2012). Transport of intensity phase imaging from multiple noisy intensities measured in unequally-spaced planes. *Optics Express*, *20*(2), 972–985.

Zou, M.-Y., & Unbehauen, R. (1997). Methods for reconstruction of 2-D sequences from Fourier transform magnitude. *IEEE Transactions on Image Processing*, *6*(2), 222–233.

Zuo, C., Chen, Q., & Asundi, A. (2014). Boundary-artifact-free phase retrieval with the transport of intensity equation: Fast solution with use of discrete cosine transform. *Optics Express*, *22*(8), 9220–9244.

Zuo, J. M., Vartanyants, I., Gao, M., Zhang, R., & Nagahara, L. A. (2003). Atomic resolution imaging of a carbon nanotube from diffraction intensities. *Science*, *300*, 1419–1421.

Zysk, A. M., Schoonover, R. W., Carney, P. S., & Anastasio, M. A. (2010). Transport of intensity and spectrum for partially coherent fields. *Optics Letters*, *35*(13), 2239–2241.

CHAPTER SIX

Electron Holographic Tomography

Axel Lubk

Institute for Structure Physics, Physics Department, Faculty of Mathematics and Natural Sciences, Technical University of Dresden, Dresden, Germany
e-mail address: axel.lubk@tu-dresden.de

Contents

"The light microscope opened the first gate to microcosm. The electron microscope opened the second gate to microcosm. What will we find opening the third gate?"

(Ernst Ruska)

This chapter contains case studies, which exemplify how the previous considerations lead to working experimental procedures and finally materials properties in terms of three-dimensional physical fields. We put stronger focus on methodology at the expense of interpretation of the obtained results. Thus, solid state properties obtained in the course of the experiments, such as the three-dimensional distribution of the electrostatic potential within a

231

solid, will be discussed in a compact form only and the reader is referred to the literature for a more comprehensive overview on their significance for materials science.

The presented tomographic techniques are of highly varying states of maturity. As a result of the outstanding efforts of D. Wolf, off-axis Electron Holographic Tomography (EHT) has been developed to an automated, reproducible and flexible TEM technique enabling a variety of applications. That, and the favorable properties of Off-Axis Holography lead to its predominant use for reconstructing mostly electric fields in 3D with nanometer resolution. We discuss this technique and its applications at length in the first section. More recently, and partly in this volume, a widening of the scope of off-axis Electron Holographic Tomography to other fields such as magnetic, strain and attenuation fields has been proposed and in part experimentally demonstrated. We will take a peek into these emerging fields and discuss the first results, properties, prospects and current limits. Note, however, that other holographic setups, notably those discussed in Chapter 5, are occasionally better suited for tomography, e.g., because of stringent coherency or optical restrictions.

The following physical quantities are reconstructed in three dimensions:

1. Mean Inner Potentials, i.e., nanoscale volume averages of electrostatic potentials in solids, at the example of a GaAs–$Al_{0.33}Ga_{0.67}As$ core–shell nanowire. This nanowire contains sharp phase boundaries, e.g., between the core and the shell, and long range fields, e.g., due to compositional gradients, which makes it an ideal showcase for discussing basic principles and limits of Holographic Tomography. We will discuss spatial and signal resolution, the use of additional constraints in the reconstruction, such as imposed by symmetry and the influence of the ubiquitous regularization.

2. Atomic potentials in a Au nanoparticle. This purely computational study will serve to investigate some limits encountered when reconstructing the potential of the nanowire and to explore future routes toward holographic tomography at atomic resolution. Here, the main issue is the violation of the linear projection due to dynamical electron scattering, which produces artifacts in the standard tomographic reconstruction scheme.

3. Magnetostatic flux density at the example of a Co nanowire. In contrast to the electrostatic potential, the magnetic potential is a vector field, which requires adapted holographic and tomographic reconstruction strategies. Additionally, the magnetic phase shift is typically one or-

der of magnitude smaller than the electric one, necessitating a careful separation of both phase shifts in combination with a stronger regularization.

4. Mean free path lengths or linear attenuation coefficients at the example of the GaAs–$Al_{0.33}Ga_{0.67}As$ core–shell nanowire. It is shown, how off-axis Electron Holographic Tomography can be used to separately reconstruct the distribution of elastic and inelastic attenuation coefficients in 3D by effectively combining bright field tomography with zero-loss energy filtered tomography in one experiment. This data are closely related to the chemical composition and the dielectric losses in the material, which complements the electric and magnetic field reconstruction; thereby yielding a more comprehensive picture of the investigated material without additional experimental effort.

5. Strain fields in strain-engineered semiconductors. This section contains a derivation of the nonlinear projection law pertaining to dark field electron holography and a discussion of tomographic reconstruction schemes adapted to this nonlinear projection. The feasibility of the latter is discussed at the example of a strained Si–Ge multilayer structure.

The following section on holographic tomography of Mean Inner Potentials contains a large number of detailed explanations on the experimental setup. It is therefore advisable to read it first. The order of the following sections is somewhat arbitrary and might be chosen by the reader according to specific interests.

Moreover, it is well understood that electron tomography is applied to a much wider class of signals collected in the TEM than those obtained from holography. Prominent examples are bright-field TEM tomography (Glaeser, 1985; Spontak, Williams, & Agard, 1988; Koster, Ziese, Verkleij, Janssen, & de Jong, 2000; Nickell, 2001; Kuebel et al., 2005; Lucic et al., 2005; Friedrich, McCartney, & Buseck, 2005; McIntosh, Nicastro, & Mastronarde, 2005; Friedrich et al., 2009; Loos et al., 2009; Jinnai & Spontak, 2009), dark field TEM tomography (Barnard, Sharp, Tong, & Midgley, 2006; Hata et al., 2008), bright field phase contrast TEM tomography (Glaeser, 1985; Toyoshima & Unwin, 1988; Lucic, Foerster, & Baumeister, 2005; McIntosh et al., 2005; Barton, Joos, & Schröder, 2008; Danev, Kanamaru, Marko, & Nagayama, 2010), energy-filtered TEM tomography (Midgley & Weyland, 2003; Möbus, Doole, & Inkson, 2003; Leapman, Kocsis, Zhang, Talbot, & Laquerriere, 2004; Yurtsever, Weyland, & Muller, 2006; Jin-Phillipp, Koch, & van Aken, 2011; Goris et al., 2011; Friedrich et al., 2013), high-angular annular dark field STEM tomography (Weyland, 2001; Weyland, Midgley, & Thomas, 2001; Ziese, Kübel, Verkleij, & Koster, 2002; Midgley & Weyland, 2003; Friedrich, McCartney, & Buseck, 2005; Midgley et al., 2006; Bals, Van Tendeloo, & Kisielowski, 2006; Midgley, Ward, Hungría, & Thomas, 2007; Friedrich et al., 2009; Biermans, Molina, Batenburg, Bals, & Van Tendeloo, 2010; Perassi et al., 2010; Hindson, Saghi, Hernandez-Garrido, Midgley, & Greenham, 2011; Van den Broek et al., 2012) (see Frank, 2008 for a detailed introduction into the field of electron tomography). While the recording and projection properties of these signals might differ significantly, the tomographic principle exploited in the various techniques is often similar. Thus, a considerable portion of the methodology presented below builds upon work done in these fields and parts of the newly developed procedures might be favorably exploited there in return.

6.1. MEAN INNER POTENTIALS

In this section we discuss the reconstruction of 3D electrostatic potentials in solids by means of off-axis Electron Holographic Tomography. These potentials contain a wealth of information as has been discussed in more detail in Section 5.1. Most importantly, the Mean Inner Potential reflects the chemical composition of the sample, where the general trend is that solids composed of light elements have a lower averaged electrostatic potential than heavy ones. Space charge regions, such as introduced by dopant variations in semiconducting materials, may be identified by their corresponding drift-diffusion potentials. Other potential sources pertain to surface dipoles or charging (e.g., due to the electron beam). It has been noted in Section 5.1 that electron holographic studies have been mainly concerned with revealing these electrostatic potentials and hence crucial properties of materials, such as the dopant distribution in semiconductors. One important challenge in these studies has been the projection of the electrostatic potential into the holographically reconstructed phase as given by (5.20)

$$\varphi(\mathbf{r}) = \frac{e}{v} \int \Phi(\mathbf{r}, z)\, dz, \qquad (6.1)$$

which prevents discerning 3D configuration changes (e.g., thickness variations) from potential variations in lateral direction (Lenk, Lichte, & Muehle, 2005; Cooper, de la Peña, et al., 2011; Lichte & Lehmann, 2008).

As early as 1994, Lai, Hirayama, Ishizuka, and Tonomura (1994) showed how to generally solve that problem by performing the first semiquantitative Electron Holographic Tomography experiment. In spite of the promising results, the experimental hurdles prevented a widespread use of the technique at that time. In particular, dedicated tomography specimen holders allowing large tilt angles, computer-controlled goniometers with nanometer precision, as well as powerful computers were not available yet. Furthermore, Electron Holographic Tomography is a comprehensive and time consuming technique requiring a large amount of automation (similar to X-ray tomography), which still had to be developed. Therefore, more quantitative reconstructions have been delayed for over ten years (Twitchett-Harrison, Yates, Newcomb, Dunin-Borkowski, & Midgley, 2007; Fujita & Chen, 2009; Wolf, Lubk, Lichte, & Friedrich, 2010) and the strife for improved experimental conditions is still continuing. Below we will discuss the crucial steps in state-of-the-art holographic tomography, beginning with the tilt series acquisition and ending with the reconstruction of the electrostatic potential. A GaAs–Al$_{0.33}$Ga$_{0.67}$As core–shell nanowire will serve as a showcase for the various procedures, because its sharp and smooth compositional gradients are ideally suited to gauge the limits of the technique.

Such nanowires exhibit a number of intriguing properties (e.g., quantum confinement) distinct from the bulk state, which are currently explored for a number of applications in (opto-)electronic and photovoltaic devices, photon ballistic wave guides, sensors and tribological additives (e.g., Appenzeller et al., 2008; Mongillo, Spathis, Katsaros, Gentile, & De Franceschi, 2012). By providing a unique non-destructive nanoscale 3D characterization for nanowires (but also nanoparticles, etc.), Electron Holographic Tomography fosters the progress in these fields as has been demonstrated by a number of recent studies (Wolf, Lichte, Pozzi, Prete, & Lovergine, 2011; Wolf, Lubk, Lenk, Sturm, & Lichte, 2013; Wolf, Lubk, Röder, & Lichte, 2013; Lubk, Wolf, Prete, et al., 2014).

6.1.1 Experimental Implementation

In the following, we will elaborate on the experimental details of Electron Holographic Tomography. We will begin with a small schematic of the workflow containing the most important processing steps, illustrated with the help of the GaAs–Al$_{0.33}$Ga$_{0.67}$As core–shell nanowire. Based on that, we discuss the most important experimental issues and corresponding remedies in greater detail in order to highlight the current pitfalls and artifacts of the technique. The general workflow of EHT is given in Fig. 6.1. Each step

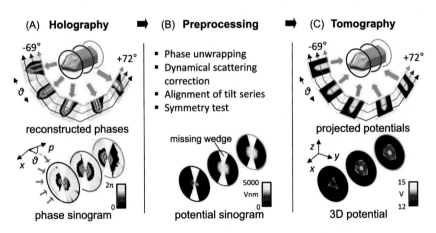

Figure 6.1 Holographic tomographic principle: (A) Acquisition and reconstruction of holographic tilt series (top) with selected sinograms (bottom). (B) Processing of reconstructed phases to obtain aligned projected potentials. The 3-fold symmetry of the nanowire is revealed by the potential sinograms. (C) Tomographic reconstruction (top) and reconstructed potential (bottom). Throughout the figure, the five green arrows denote a set of angles ($-66°$, $-33°$, $0°$, $33°$, $66°$). Three deliberately chosen slices at $x = 100, 400, 700$ nm are highlighted by black, dark gray and light gray lines and frames.

contains of a number of subroutines, which are shortly explained in the following.

(A) The acquisition of the holographic tilt series consists of recording one object and one empty hologram at each tilt angle. The latter is required to remove distortions such as introduced by the biprism or the projective system from the reconstructed waves by subtracting the reference phase from the object phase. The recordings need to be taken immediately after each other under the same imaging conditions. To reduce the noise in the reconstructed phase it can be advantageous to average a multitude reference holograms (and object holograms). Each hologram acquisition is preceded by centering the object on the holographic field of view, because tilting the specimen almost inevitably introduces a small shift proportional to the distance between the specimen and the tilt axis of the goniometer. Furthermore, an adaption of the imaging conditions might be required because of focus variations, aberration drift, and other instabilities occurring during the tilt series. To control this large number of parameters, while recording the tilt series, a high degree of automation is generally very helpful. To this end, a dedicated tilt series acquisition tool, Tomographic and HOlographic Microscope Acquisition Software (THOMAS), performing the adaptations noted above, has been developed (Wolf et al., 2010). The use of that tool decreases the acquisition time from previously $\mathcal{O}(10\ h)$ to now $\mathcal{O}(1\ h)$ for a typical holographic tilt series (Wolf et al., 2010), while increasing the reproducibility and quality of the obtained data. All Electron Holographic Tomography studies in this work acquired the tilt series with THOMAS. A persisting experimental obstacle is the limited tilt range of typical tomographic sample holders in the TEM, which renders tomography blind to the spatial frequencies laying in the missing tilt interval (Fig. 3.8). We will discuss this issue in more detail below.

(B) The holographic reconstruction of the object and reference hologram is performed by applying Fourier techniques detailed in Section 5.1. Here, most of the holographic reconstruction steps are performed for each tilt independently, albeit with the same parameters (e.g., holographic mask size). In the particular case of phase unwrapping, however, a combined approach using the data from many tilt angles may be used to improve the unwrapping, in particular in the presence of strong phase gradients such as present at the edge of the nanowire. This 3D phase unwrapping is detailed further below.

(C) Before performing a tomographic reconstruction, a tilt series has to be aligned such that the tilt axis is located in the same place in each projection. Due to imperfections of the goniometer, a (statistical) shift is introduced upon tilting. This needs to be removed. A short overview of pertinent alignment procedures is given below.

(D) A small overview of tomographic reconstruction algorithms has been given in Section 3.6. Here, we employ the polar sampling of the reconstruction domain, described in detail in Section 3.3, in combination with a Conjugate Gradient solver (cf. Section 3.6) as implemented in the LSQR algorithm (Paige & Saunders, 1982). That allows us to adapt the azimuthal sampling in the reconstruction to the azimuthal bandwidth observed experimentally, thereby saving computation time and memory. Even more importantly, optimizing the number of reconstructed pixels helps to regularize the reconstruction (cf. Section 3.5). This is the most crucial step in the tomographic reconstruction, because experimentally acquired projection data (from the electrostatic potential in this case) are typically corrupted by non-projective features, such as dynamical scattering effects and detector noise in our case. These errors amplify upon reconstruction because of the (mildly) ill-conditioned nature of the Radon transform (cf. Section 3.5). It is therefore crucial to reduce these non-projective errors as much as possible by applying regularization schemes as discussed in Section 3.5. We discuss these things separately subsequently. Before turning to the explicit example of a GaAs–Al$_{0.33}$Ga$_{0.67}$As core–shell nanowire we proceed with a discussion of the most important issues affecting the reconstruction in more detail.

Missing Wedge

One important experimental issue in almost all electron-microscopic tomographic techniques is the limited tilt range. Due to the small space in between the pole pieces of the objective lens[1] and the mechanical fixing of the specimen, the available tilt interval is usually restricted to ±70° instead of ±90°. Owing to the Fourier slice theorem (cf. Section 3.2.3), spatial frequencies of the object laying in the corresponding double wedge in Fourier space cannot be reconstructed from the data recorded in the tilt series. Accordingly, this leads to an anisotropic reduction of spatial resolution in the direction of the missing wedge and additional artifacts by the non-local propagation of inconsistencies in the reconstruction.

Several strategies allow to mitigate the missing wedge problem, some of which will be employed in the following. First, novel 360° tilt holders (e.g., Fischione Model 2050 on-axis 360° tomography holder) permit straight access to all tilt angles. Unfortunately, these holders currently require a complicated specimen handling and mounting, which is partly related to the missing β-tilt capability. We employ such a holder in Section 6.3. Second, dedicated tomographic rotation holders allow for recording incomplete (tilt range < 180°) tilt series around multiple tilt axes. That reduces the missing wedge to a missing pyramid (two perpendicular tilt series) or missing cone (many tilt series). This method, however, is only capable of reducing the missing wedge while increasing both experimental and reconstruction efforts significantly. It is most favorably employed under circumstances, where several tilt series are required anyway (cf. Section 6.3). Last but not least it is frequently possible to infer some of the information inside of the missing wedge from additional knowledge about the object. For instance, if the object features a (discrete) rotational symmetry, the missing information could be "copy-pasted" from other regions of Fourier space (cf. Section 3.5). Partial remedy is also possible by imposing a support constraint or a constraint on the range of reconstructed potential values (Batenburg et al., 2009; Biermans et al., 2010).

---

[1] See Börrnert et al. (2015) for a radical instrumental solution to that problem.

Alignment

A further experimental issue, affecting TEM tomography, is the accurate alignment of the tomographic tilt series around the tilt axis. Misaligned projections are inconsistent and lead to characteristic artifacts (e.g., "arc" distortion; Kobayashi, Fujigaya, Itoh, Taguchi, & Takano, 2009) depending on the regularization level. A variety of alignment procedures has been developed which can be coarsely divided into three groups (Frank & McEwen, 1992; Mastronarde, 2006; Kobayashi et al., 2009; Houben & Bar Sadan, 2011):

1. Cross-correlation methods seek to maximize the cross-correlation between subsequent projections by registering them appropriately (Dengler, 1989; Liu, Penczek, McEwen, & Frank, 1995). They necessitate small tilt increments to ensure a maximization of the cross-correlation, if residual displacements are removed. Thus, the accuracy of these methods crucially depends on the tilt increment and the contrast of the images to be aligned. Strong contrasts, generating a pronounced cross-correlation peak, may be registered more accurately (Tonomura, 1987; Tonomura, Allard, Pozzi, Joy, & Ono, 1995; Völkl, Allard, & Joy, 1999; McCartney et al., 2010; Kasama, Dunin-Borkowski, & Beleggia, 2011; Lichte et al., 2013; Pozzi, Beleggia, Kasama, & Dunin-Borkowski, 2014; Pozzi, 2016). A drawback is that alignment errors may sum up throughout the tilt series. Consequently, cross-correlation alignment is often used for an initial coarse alignment only and the following two methods are used for refinement.

2. Point-tracking methods monitor the evolution of distinguished points throughout the tilt series and compare the observed trajectories with the analytic solution, i.e., circles in the polar representation of the sinogram introduced in Section 3.2.2. Residual displacements are identified as deviations from the best fitting circle and removed accordingly. Introducing such point features artificially in the form of fiducial markers is commonly employed for the alignment of bright field TEM tomographic tilt series of biological (low contrast) specimen since they allow a precise global alignment of the complete tilt series without error amplification (Jing & Sachs, 1991). When sufficiently fine fiducial markers are not available, other point features such as the center-of-mass or edges and other characteristic features, which can be tracked throughout the tilt series, may be employed (Brandt, Heikkonen, & Engelhardt, 2001; Brandt, 2006). Since point-tracking alignment is predominantly used in the course of this work, we further expound on it below.

3. Iterative reconstruction-alignment methods rely on cross-correlating the projections of reconstructed 3D data with the experimental projection data (Winkler & Taylor, 2006; Houben & Bar Sadan, 2011). That information is used for removing residual shifts and performing an improved reconstruction, which then serves as input for the next iteration cycle. These methods permit a very accurate alignment provided that the starting guess (e.g., from the above two methods) is sufficiently good and the reconstruction quality is high. The latter requirement is often violated in Electron Holographic Tomography (e.g., due to the missing wedge). Therefore, the correction of found displacements may reduce the quality of the reconstruction, thereby decreasing the misfit between the reprojected and experimental data. A further improvement of the method is required to overcome these obstacles.

In the following, we elaborate on the point-tracking alignment. The biggest obstacle to this method is the identification and tracking of suitable point features, with the most important ones being

- Fiducial markers, such as small particles made of some heavy material (e.g., gold) dispersed on purpose on the specimen during preparation. Fiducial markers have the advantage of providing a large number of point features building up a statistic of traceable trajectories with very precise mean values. The drawback is a certain modification of the specimen and a certain resolution limit imposed by the size of the nanoparticles. The latter may be mitigated by tracking the center of mass of the markers.

- Center of mass of the specimen. The advantage is that the centroid may be determined very precisely from an integral over the whole image without requiring additional preparation steps. The disadvantage is that the center of mass may be not-well defined because the amount of specimen visible within the field of view may vary during the tilt series. For the needle-shaped specimen used in this work the problem could be reduced by aligning the experimental tilt axis with the nanowire. Moreover, it turned out useful to multiply a uniform mask with the shape of the projected density (e.g., generated using a threshold value) and thereby removing the background disturbing the centroid computation. The center of mass alignment is frequently used in this work.

- Edges and other characteristic features, which appear localized in the first derivative of the projected data. This method also needs no additional specimen preparation. However, the visibility of the derivative strongly depends on the orientation of the specimen. An edge, which appears sharp along one direction, may completely disappear in another. Consequently, one feature serves well for a certain angular interval only, and a larger set of features needs to be tracked to cover the whole tilt interval.

Once projected positions of point features are available throughout the tilt series, one can proceed as follows. First, a projection model for the points, while rotating around the tilt axis, is generated. Besides the rotation itself, such a model may include magnification changes, image rotations, as well as distortions (Mastronarde, 2006), which, fortunately, can be neglected for the imaging conditions and samples used in this work.

Consequently, the motion of a fixed point with coordinates $\mathbf{r}_j$, while rotating around an arbitrary tilt axis with a normal segment $\mathbf{r}_t$ (smallest distance from tilt axis to the origin) reads

$$\mathbf{r}_{ij} = \mathbf{R}_i\left(\mathbf{r}_j - \mathbf{r}_t\right) + \mathbf{r}_t, \tag{6.2}$$

with the rotation matrix $\mathbf{R}_i$ being parameterized by the rotation angle θ in the axis-angle representation

$$\mathbf{R}_i = \exp\left(\theta_i \mathbf{K}\right) = \mathbf{I} + \sin(\theta_i)\mathbf{K} + \left(1 - \cos(\theta_i)\right)\mathbf{K}^2. \tag{6.3}$$

Here, the skew symmetric matrix $\mathbf{K}$ (i.e., the infinitesimal generator of the rotation group) is obtained from the rotation axis direction unit vector ω as follows

$$\mathbf{K} = \begin{pmatrix} 0 & -\omega_z & \omega_y \\ \omega_z & 0 & -\omega_x \\ -\omega_y & \omega_x & 0 \end{pmatrix} \tag{6.4}$$

and consequently

$$\mathbf{K}^2 = \begin{pmatrix} -\omega_y^2 - \omega_z^2 & \omega_x\omega_y & \omega_x\omega_z \\ \omega_x\omega_y & -\omega_x^2 - \omega_z^2 & \omega_y\omega_z \\ \omega_x\omega_z & \omega_y\omega_z & -\omega_x^2 - \omega_y^2 \end{pmatrix}. \tag{6.5}$$

Such a transformation may be considered within the setting of an affine space, i.e., as affine transformation. In that setup, only differences between marker points j in the reconstruction domain (the affine space)

$$\mathbf{d}_j = \mathbf{r}_{j+1} - \mathbf{r}_j \tag{6.6}$$

are considered, which transform according to

$$\mathbf{d}_{ij} = \mathbf{R}_i \mathbf{d}_j .$$ (6.7)

The advantage of considering only the transformation of relative distances of marker points instead of absolute marker coordinates is that the orientation of the tilt axis suffices to describe the rotation and the normal segment may be disregarded.

To determine the orientation of the tilt axis, we write the projection matrix of the point distances in the detector plane (xy-plane), describing their trajectories upon rotation, as the first two rows of the rotation matrix $\mathbf{R}_i$ (rotation axis situated in xy-plane, i.e., $\omega_z = 0$)

$$\mathbf{R}_i^{\perp} = \left(\begin{array}{ccc} 1 + (1 - \cos(\theta_i))\omega_y^2 & (1 - \cos(\theta_i))\omega_x\omega_y & \sin(\theta_i)\omega_y \\ (1 - \cos(\theta_i))\omega_x\omega_y & 1 + (1 - \cos(\theta_i))\omega_x^2 & -\sin(\theta_i)\omega_x \end{array} \right)$$ (6.8)

and solve the following nonlinear fit problem

$$(\boldsymbol{\omega}_{\perp}, \mathbf{d}_{j\perp}) = \arg\min \sum_{ij} \left\| \mathbf{d}_{ij}^{(\exp)} - \mathbf{R}_i \mathbf{d}_j \right\|$$ (6.9)

with the experimental distances $\mathbf{d}_{ij}^{(\exp)}$ slightly deviating from the ideal ones $\mathbf{d}_{ij}$ due to instabilities of the goniometer and point-feature tracking errors. Consequently, the orientation of the tilt axis may be determined from distances or even mere orientation vectors identified in each projection.

The latter observation is important for the alignment of nanowires representing one typical specimen geometry investigated by means of Electron Holographic Tomography. Here, it is often much more convenient and accurate to determine a certain orientation, typically the main axis of the nanowire, whereas the identification of single point features may be difficult in the absence of fiducial markers. For instance a common center of mass within the projections may be corrupted by the varying amount of specimen visible in each projection. In particular the latter problem is mitigated once the orientation of the tilt axis has been found. In this case one can deliberately align the rotation axis along the, say, y-coordinate of the reconstruction domain, thereby avoiding that different parts of the nanowire move into the reconstruction volume. The corresponding reconstruction domain is then a cylinder and the center of mass within the cylinder and the projection is well-defined. Finally, the determination of the normal segment of the tilt axis is typically incorporated into fixing that position in the middle of the field of view containing the specimen at its center. This is achieved by introducing displacements such that the center of mass is situated on the tilt axis.

Phase Unwrapping

The phase, reconstructed by means of electron holography, is only defined modulo 2π (5.20). This so-called wrapping needs to be inverted in order to obtain projected potentials from reconstructed phases. Phase unwrapping problems are common to a large variety of phase retrieval techniques and a multitude of unwrapping techniques has been developed to date of this work (see Ghiglia & Pritt, 1998 for an overview). Typically, they consist of two steps. First, loci with phase jumps larger than some predefined threshold (e.g., π) are identified as phase wrapping loci. In the second step, the phase on one side of the jump is corrected by an appropriate multiple of 2π.

Both steps can be problematic for various reasons. However, problems related to the removal of correctly identified phase jumps are more severe, because phase vortices render the removal fundamentally non-unique. Additional assumptions in the form of boundary conditions need to be introduced to allow a unique solution to the unwrapping problem (i.e., to fix the topology of the unwrapped phase). These additional boundary conditions often represent the main difference between the various phase unwrapping algorithms. Unfortunately, phase vortices are an intrinsic feature of electron waves scattered at solids (Lubk, Guzzinati, Börrnert, & Verbeeck, 2013) and are readily visible in atomic resolution off-axis holographic reconstructions. Fortunately, they average out under medium resolution imaging conditions used here (the topology of the incoming wave is trivial). Consequently, the second part of the unwrapping procedure is typically not critical in our case. Detecting the phase wrap loci, on the other hand, is difficult in a digital representation of the phase, if the true phase difference between two adjacent reconstructed pixels exceeds π (Ghiglia & Pritt, 1998), a situation often present in noisy data with sharp thickness jumps at specimen edges. In this case the phase unwrapping algorithm may not detect the phase wrap or erroneously detects only one 2π-jump, where there have been two. We show below that this problem can be mitigated by extending the original 2D phase unwrapping performed in each individual holographic reconstruction independently to a 3D procedure working on the complete tilt series. Here, the continuous dependency of the phase on the tilt angle is crucial.

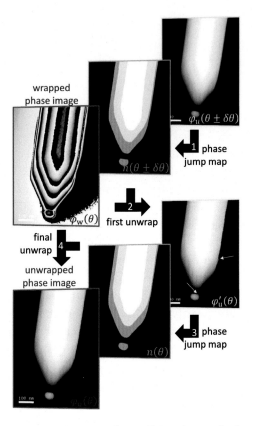

Figure 6.2 3D phase unwrapping procedure utilizing the continuity of the phase as a function of the tilt angle θ.

The basic principle of the 3D phase unwrapping procedure is shown in Fig. 6.2. Accordingly, we first identify those tilt angles, for which the standard 2D unwrapping failed. In those cases, the "unwrapping information" of a properly unwrapped phase image $\varphi_u \left(\theta \pm \delta\theta \right)$ at one of the adjacent tilt angles is used in the following way. First, a preliminary phase jump map

$$n \left(\theta \pm \delta\theta \right) = \lfloor \varphi_u \left(\theta \pm \delta\theta \right) / 2\pi \rfloor \times 2\pi \qquad (6.10)$$

containing integer multiples of 2π is calculated. Second, this map is added to the wrapped phase image $\varphi_w \left(\theta \right)$ yielding a preliminary unwrapped phase

$$\varphi_u' \left(\theta \right) = \varphi_w \left(\theta \right) + n \left(\theta \pm \delta\theta \right) . \qquad (6.11)$$

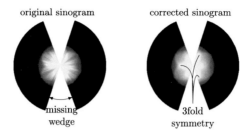

Figure 6.3 Suppression of dynamical scattering artifacts in the projection data (sinogram). The depicted sinograms pertain to a central cross-section of the nanowire. After the correction, the 3-fold symmetry of the central nanowire is readily visible.

By virtue of the smooth dependency of the phase on the tilt angle, this preliminary version is close to the true unwrapped $\varphi_u(\theta)$. However, it also contains some artifacts indicated by yellow arrows in Fig. 6.2, because the added phase jumps do not exactly match the true ones. Exploiting the smoothness of the phase again, these remaining discrepancies are removed by calculating a new phase jump map $n(\theta)$ from a smoothed $\varphi_u'(\theta)$ in a third step. Finally, this phase jump map $n(\theta)$ is added to the original wrapped phase image $\varphi_w(\theta)$ resulting in the final unwrapped phase image $\varphi_u(\theta)$. Steps 3 and 4 may be repeated if the first iteration was not satisfactory. This algorithm is particularly suited to remove unwrapping ambiguities at sharp edges in projection.

Regularization

Besides phase unwrapping errors, misalignment and the missing wedge, dynamical scattering effects and shot noise represent the most important errors in the projection data. While the impact of noise may be reduced by generic regularization methods such as discussed in Section 3.5, fluctuation of the reconstructed phase due to dynamical scattering can be mitigated with the help of a dynamical correction factor approach (Lubk, 2010) (cf. Section 2.3 and Appendix A). Accordingly, the phase of the axially scattered beam is modulated by a tilt-dependent dynamical scattering term under axial scattering conditions. This offset can be largely removed, however, by normalizing the projected potential average at all angles to the same value. The corresponding improvement of the projection data pertaining to the GaAs–Al$_{0.33}$Ga$_{0.67}$As core–shell nanowire is illustrated in Fig. 6.3.

We further reiterate that the conjugate gradient method, employed in the following, exhibits a Tikhonov-like regularization behavior in that larger spatial frequency features are reconstructed at higher iteration steps. However, because we typically do not know the projection errors, we cannot sensibly define an optimal iteration number (cf. Section 3.5). A correct representation and interpretation of the reconstructed data therefore requires the inspection of reconstructions obtained at different iteration numbers (regularization strengths). This inspection of differently regularized results is frequently ignored in contemporary TEM tomography.

One main issue in the tomographic reconstruction is the non-local nature of the inverse Radon transformation (cf. Section 3.2.3). Accordingly, each projection including its corresponding noise, misalignment, phase unwrapping, dynamical scattering, and other non-projective artifacts, is contributing to the reconstructed value at each coordinate. In combination with the error amplification inherent to the inverse Radon transformation this results in artificial stray fields originating from the non-projective artifacts inside the object.

When only seeking the reconstruction of the stray field in vacuum, a restriction of the tomographic reconstruction to the region outside of the nanostructure can largely circumvent some of these issues. In the course of proving the support theorem in Section 3.2.4, it was shown that the reconstruction of some function (here, electric potential) outside a convex domain (here nanowire) from projections outside the domain, has a unique solution. Because the projection artifacts within that domain are invisible in this approach, this unique solution does not suffer from the above issues. Furthermore, the small azimuthal band width of the stray fields allows largely mitigating missing wedge artifacts because large parts of the missing interval may be padded from outside due to the smoothness of the field in azimuthal direction. In practice this is achieved by increasing the azimuthal width of the reconstructed circular pixels in the polar sampling scheme (cf. Section 3.3). Below we will apply this regularization strategy to reveal electrostatic stray fields due to charging of the nanowire in the electron beam and to reveal magnetic stray fields around a magnetic nanowire in Section 6.3. Note that even though the interior is completely disregarded in this approach, some integral knowledge about the object may be obtained by virtue of the laws of electro-(magneto)statics, which relate exterior potentials to charges, currents and dipoles in the interior (Beleggia, Kasama, & Dunin-Borkowski, 2010; Gatel, Lubk, Pozzi, Snoeck, & Hÿtch, 2013) (cf. Section 5.1.2). In spite of the ease of the numerical implementation, we additionally note that hardware approaches to the reconstruction of stray fields, e.g., using a hollow cone illumination to shadow certain parts of the object (Williams & Carter, 1996), could help to reduce electron beam induced specimen modifications (charging, radiation damage) in future applications.

Summary

To give a vivid picture of the experimental details noted above, the EHT workflow, as realized for the GaAs–$Al_{0.33}Ga_{0.67}As$ nanowire, is shown in Fig. 6.1. That nanowire has been grown by metal–organic vapor phase epitaxy (MOVPE) using Au nanoparticles as metal catalyst (Prete et al., 2008; Wolf et al., 2011). The nanowire has a total diameter of 300 nm (80 nm core, 110 nm shell). The holographic tilt series (Fig. 6.1A) covers a range from $-69°$ to $+72°$ sampled in $3°$ steps. Thus, the missing wedge encompassed an interval of $39°$. The tilt series has been recorded on a FEI Titan 80-300 Berlin Holography Special TEM operated at 300 kV acceleration voltage in aberration corrected Lorentz mode, providing a resolution of about 2 nm. We used a double biprism setup to adjust a field of view of 1 μm and a fringe spacing of 3 nm. The object exit wave tilt series has been reconstructed from the object holograms and corresponding empty holograms applying a Butterworth filter of 0.13 nm^{-1} full-width-half-maximum (FWHM) to separate the side band (cf. Section 5.1 for further details of the holographic method).

Subsequently, the tilt axis has been coarsely oriented along the x-axis by means of the alignment procedures discussed above. We then applied the 3D phase unwrapping discussed in Section 6.1.1, suppressed dynamic scattering artifacts as illustrated in Fig. 6.3, and precisely aligned the center of mass to the image x-axis to facilitate a polar sampling of the reconstruction domain with minimal azimuthal band width (cf. Section 3.3). The final tilt series used in the tomographic reconstruction is depicted in Fig. 6.1B. Besides the substantial improvement of the data, one observes a 6-fold symmetry in the core–shell region reducing to a 3-fold symmetry in the tapered section below the Au nanoparticle on top used as catalyst in the growth process. The 3-fold symmetry has been used in the subsequent tomographic reconstruction to remove missing wedge artifacts and to improve regularization. Ten conjugate gradient LSQR iterations (Paige & Saunders, 1982) on a 13800 × 11700 polar grid Radon matrix have been found sufficient to ensure a good reconstruction quality (good spatial resolution) containing all important potential features, while keeping the noise level small. Note, however, that the regularization strength ($\cong$ the number of iterations) is a free parameter. We therefore also show reconstructions with 5 and 15 iterations and a reconstruction without symmetry constraint in Fig. 6.6 to assess the regularization error.

6.1.2 Results

Using the optimized procedures described above, we now reconstruct the 3D potential of a $\langle 111 \rangle$-oriented GaAs–Al$_{0.33}$Ga$_{0.67}$As core–shell nanowire. It has been shown by high-angle annular dark field (HAADF) STEM of cross-sections that the AlGaAs shell contains a self-assembled Al segregation due to the different Ga and Al ad-atom mobilities on the nanowire's surface during growth (Rudolph et al., 2013; Heiss et al., 2013). Such compositional and structural fluctuation can severely alter the electronic, optoelectronic, and photovoltaic properties of the NW in future electronic, optoelectronic, and photovoltaic devices (Bryllert, Wernersson, Löwgren, & Samuelson, 2006; Gallo et al., 2011; Krogstrup et al., 2013).

We first note that a small charging of the NW in the electron beam was present throughout the tilt series. A detailed reconstruction and assessment of this field is shown further below. For the moment, we note that its influence on the reconstruction is around 0.1 V, which was tested by reconstructing with and without constraining the vacuum to 0 V. We now turn to the general features of the reconstructed 3D potential of the GaAs–Al$_{0.33}$Ga$_{0.67}$As nanowire (Fig. 6.4B). It exhibits a slightly deformed 6-fold symmetry in the core–shell region corroborating the symmetry test performed on the projection data. Missing wedge artifacts are completely absent (see Fig. 6.7 for comparison), which is corroborated by the sharp nanowire boundaries in all directions. Consequently, an *isotropic* spatial resolution of 6 nm (determined from the FWHM of the potential gradient at the NW boundary) is observed throughout. It is mainly limited by the holographic imaging process and not by the tomographic reconstruction. Similarly, the signal precision of 0.2 V, computed from the standard deviations of the corresponding homogeneous region, reflects the accuracy of the measured potential throughout the reconstruction volume, which is in contrast to the non-symmetric reconstruction suffering from missing wedge artifacts. Last but not least, we note a small influence of the Fresnel propagation at the boundary of the GaAs–Al$_{0.33}$Ga$_{0.67}$As nanowire (Fig. 6.4G), leading to a loss of resolution as well as small oscillation artifacts. Thus, regions close to the boundary must be interpreted with care. Based on the achieved spatial and signal resolution, the following features can be detected *in 3D*.

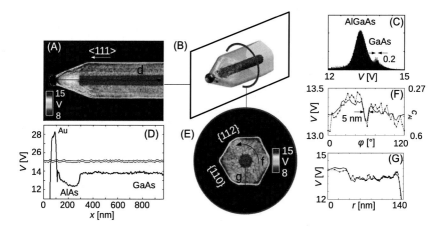

Figure 6.4 Reconstructed 3D potential of GaAs–$Al_{0.33}Ga_{0.67}$As core shell NW: The iso-surfaces (B) at 9 V and 13.75 V illustrate the core–shell morphology corroborated by the double peaks in the corresponding histogram (C). The zx-plane (A) with corresponding linescan (D) on the left shows the sequence Au, AlAs, and GaAs unraveled through their different MIPs. The projected yz-plane (E) on the right with corresponding linescans in azimuthal (F) and radial (G) direction shows the core–shell structure with local Al accumulation identified through characteristic potential reduction in the azimuthal scan. Here, the corresponding Al concentration c_{Al} determined from Vegard's Law is shown on the right y-scale. The thick lines in (F) and (G) indicate averaged values from the whole hexagonal region of the nanowire.

The Au nanoparticle at the nanowire tip has an average potential of 27.8 ± 0.2 V (Fig. 6.4A, D) that agrees well with the theoretical value of 28 V ($\Phi_0^{(\text{theo})} = 30.1$ V; Schowalter, Rosenauer, Lamoen, Kruse, & Gerthsen, 2006) taking into account the error pertaining to ab-initio computations of Mean Inner Potentials and the impact of dynamical scattering as noted previously. The average potentials of the GaAs core, $\Phi_0^{(\text{exp})} = 13.9 \pm 0.1$ V, and the $Al_{0.33}Ga_{0.67}$As shell, $\Phi_0^{(\text{exp})} = 13.4 \pm 0.2$, show a similar level of agreement to the theoretic values of $\Phi_0^{(\text{theo})} = 14.19$ V (Kruse, Schowalter, Lamoen, Rosenauer, & Gerthsen, 2006) and

$$\Phi^{(\text{theo})} = 0.33 \cdot \underbrace{12.34 \text{ V}}_{\Phi_0^{(\text{AlAs})}} + 0.67 \cdot 14.19 \text{ V} = 13.58 \text{ V}, \qquad (6.12)$$

respectively. These results reflect the previously discussed correspondence between average potential and chemical composition. Taking into account the signal resolution of 0.2 V, we can infer a sensitivity to compositional changes of about 10%. Under such conditions, the small potential ($\Phi = 12.5 \pm 0.1$ V) region directly above the GaAs core (Fig. 6.4D) can be interpreted as a formation of AlAs alloy as reported for similar GaAs–Ga$_c$In$_{1-c}$P core–shell nanowires (Sköld et al., 2005).[2] The axial decay of the potential, starting at the Au–Al$_{0.33}$Ga$_{0.67}$As interface and extending over approximately 100 nm in direction of the NW (Fig. 6.4D), on the other hand, is of yet unresolved origin. It could indicate a Fermi-level-pinned metal–AlGaAs junction or a varying chemical composition. We will shed more light on this feature in Section 6.4.

The finest spatially resolved detail in the reconstruction is characteristic 3-fold symmetric lines of reduced potential along {112}-directions of the nanowire (Fig. 6.4E). These radial lines are due to facet-dependent Al-segregation, which are ascribed to a polarity-driven surface reconstruction during AlGaAs shell growth (Rudolph et al., 2013; Zheng et al., 2013). These studies determined an Al concentration enhancement to $c_{Al} \approx 0.67$, which agrees well with our value $c_{Al} = 0.55 \pm 0.5$ when taking into account a further increase toward higher conjugate gradient iterations (Fig. 6.6).

A further inspection of the 3D potential distribution reveals small fluctuations of the Al-content along the x-direction. We ascribe them to modulations of the local core facets as shown in Fig. 6.5. The depicted set of cross-sections reveals two effects. First, there is a change in facet length as observed in the sequence $x = 335$ nm to $x = 345.5$ nm. That effect has been reported by Johansson et al. (2007), where it has been ascribed to alternating {111}-nanofacets accumulating to the six-fold {110}-facets observed on a larger length scale. Secondly, we observe a complete rotation of about 30° between $x = 345.5$ nm to $x = 347.6$ possibly introduced by a twin boundary or a switch from {110}-facets to {112}-facets. These changes could be observed along the whole NW and are subjected to influence the growth of the shell structure in general and the Al-segregation in particular.

It has been noted above that there is no distinguished regularization strength, because the error in the projection data is not known. We therefore supplement reconstructions pertaining to different regularization strengths. Fig. 6.6 shows line scans corresponding to Fig. 6.4 obtained from reconstructions with 5, 10 (used in Fig. 6.4) and 15 LSQR iteration cycles as well as from a reconstruction with no symmetry constraint. Here, one readily observes an improvement of spatial resolution at the cost of signal

---

[2] The small discrepancy could be due to small imperfections of the generalized-gradient approximation of exchange-correlation in the theoretic calculations, the choice of the phase reference in the holographic experiment or the small charging field.

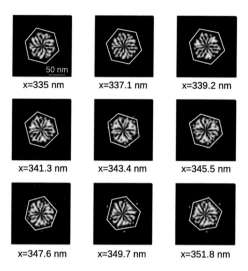

Figure 6.5 GaAs core showing a modulated facet structure from the unperturbed middle part (i.e. far away from the tapered region) of the nanowire. White frames are drawn around the core to indicate the core facets. The color range was restricted to an interval between 13.5 V and 14.5 V to highlight the core region.

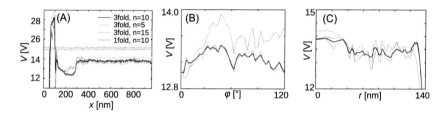

Figure 6.6 Comparison of different regularization parameter values. Potential linescans corresponding to Fig. 6.4 along x (A), azimuth φ (B), and r (C). The results previously displayed in Fig. 6.4 are plotted with a black line.

resolution (noise) with growing iteration number. For instance the potential drop due to Al segregation at the nanofacets becomes slightly deeper at 15 LSQR iterations (Fig. 6.6) with the value itself being less precise due to the increased noise. In detail, the results obtained from 5 iterations revealed a resolution around 8 nm determined from the FWHM of the potential gradient at the NW edge, which subsequently improved to 6 nm at 10 and 15 iterations. At the same time the reconstructed noise started to significantly increase at around 15 iterations (Fig. 6.6C), which is why 10 iterations have been judged a good compromise.

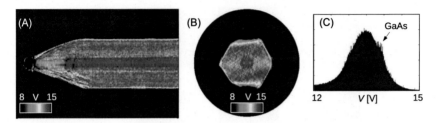

Figure 6.7 Reconstructed potential without symmetry constraint. The cross-sections in the *xz*-plane (A) and *yz*-plane (B) correspond to the ones shown in Fig. 6.4. The missing wedge shows up as horizontal stripes in (B). The histogram of the potential values is shown in (C).

To analyze the impact of the symmetry constraint, we performed a reconstruction without symmetry constraint using 10 LSQR iterations (see Fig. 6.7 for cross-section corresponding to Fig. 6.4). We obtain large missing wedge artifacts most prominently visible as horizontal lines within the NW in Fig. 6.7B and an anisotropic reduction of resolution. These missing wedge artifacts significantly reduce the signal resolution (Fig. 6.7C) and obscure fine potential details such as the Al segregation lines or the precision of the Mean Inner Potential values (Figs. 6.6 and 6.7C). Note, however, that the Al segregation is faintly present in Fig. 6.6B underlining that they cannot be considered an artifact of the symmetry-constrained reconstruction.

We finally reconstruct the weak charging field around the GaAs–AlGaAs nanowire by restricting the reconstruction to the exterior of the nanowire (cf. Section 3.2.4) by blocking the interior above a predefined threshold value. In Fig. 6.8, the solution of the exterior field reconstruction is compared to the standard one containing the interior of the nanowire. The beam induced charging on the GaAs–$Al_{0.33}Ga_{0.67}As$ nanowire is the result of a dynamical balance between secondary electron emission, induced by beam electrons, and replenishing electrons from the holder, hence is determined by the secondary electron yield and the conductivity of the sample. It typically represents an unintended artifact, affecting in particular all sorts of holographic investigations including EHT (Lubk, Javon, et al., 2014). In the case of the GaAs–$Al_{0.33}Ga_{0.67}As$ nanowire, only a weak, nearly radially symmetric, projected charging potential could be observed outside the NW (Fig. 6.8A), which is furthermore artificially modulated during the tilt series acquisition by various effects. The non-continuous variations in the sinogram (Fig. 6.8A) indicate a varying amount of charging as well

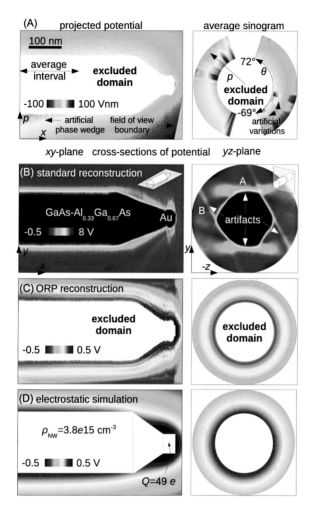

Figure 6.8 GaAs–Al$_{0.33}$Ga$_{0.67}$As NW charging field characterization: (A) holographically reconstructed projected potential (artifacts indicated), (B) standard (interior) tomographic reconstruction, (C) restricted tomographic reconstruction of the exterior, and (D) electrostatic finite element model. The indicated x-interval (= 150 nm) in (A) is used for computing the averaged sections shown in the second column. The small insets in (B) show the position of the cross-sections.

as varying phase wedges. The latter effect could stem from reference waves tilted by the weak stray fields or the numerical procedure determining the holographic sideband center (Lehmann, 1992).

The standard (interior) reconstruction (Fig. 6.8B) of the stray field largely fails, mainly due to non-local artifacts from the missing wedge:

One observes the typical anisotropic loss of resolution in wedge direction (artifact A in Fig. 6.8B) as well as sharp streaks (artifact B in Fig. 6.8B) originating from the missing wedge convolution kernel interacting with sharp object boundaries. As discussed above, the exterior reconstruction does not suffer from object features leaking erroneously into vacuum. However, the missing wedge still introduces artificial modulations in azimuthal direction, surmounting those of the almost radially symmetric charging field. To suppress them, we restrict the reconstruction to the radial symmetric part in Fig. 6.8C. A comparison to a finite element electrostatic simulation (Fig. 6.8D, cf. Lubk, Wolf, Simon, et al., 2014 for details) shows good agreement with the stray field of a homogeneous positively charged "nanobottle" with an increased charge at the Au tip ($\rho_{NW} = 3.8e15$ cm^{-3}, $\rho_{Au} = 1.3e17$ cm^{-3}). The latter can be explained by the higher yield of ejected secondary electrons at the Au tip. The remaining differences to that model are due to a violation of the model's homogeneous charge density assumption as well as the tilt series inconsistencies noted above.

6.2. ATOMIC POTENTIALS

In the previous chapter we observed some of the artifacts, in particular at the edges of the GaAs–Al$_{0.33}$Ga$_{0.67}$As nanowire, which may be related to a violation of the linear projection law, i.e., the phase object approximation (cf. Section 2.3), forming the basis of EHT. As discussed in Section 2.3, the POA is largely valid under imaging conditions used for medium resolution holography, i.e., small collection apertures and weakly diffracting crystal orientations (Matteucci, Missiroli, & Pozzi, 2002). However, when approaching atomic resolution imaging conditions, the influence of dynamical scattering, i.e., the intermixing of propagation and scattering (see Section 2.2), increasingly violates the POA (Lubk, 2010; Vulovic, Voortman, van Vliet, & Rieger, 2014). When further pushing EHT toward atomic resolution, these effects are getting more pronounced, rendering it futile to neglect them (see Section 2.3 for a discussion of the scope of the POA). Therefore, in order to reconstruct atomic potentials, dynamical scattering has to be incorporated into the EHT reconstruction. To do so is a formidable task, because a linear relationship between the potential and some reconstructed signal (phase, amplitude, etc.) cannot be found anymore. Different strategies to overcome this problem are conceivable. In the following, some of them are introduced by reviewing attempts on atomic resolution tomography reported based on other TEM techniques.

Extending tomographic reconstructions to atomic-resolution represents one of the holy grails in TEM. The successful implementation of this goal would allow tremendous insight into the properties of materials, e.g., by resolving point defect structures in crystals. Thus, atomic resolution tomography is a very active field of research accompanied by buzzing claim-staking typical for the modern academic publishing practice (Conrad, 2015). Previously reported results are therefore subject to considerable controversy, as will be detailed in the following.

First of all, it has to be noted that all of the results, reported to date of this work, were obtained in regimes, where the nonlinear effects introduced by dynamical scattering were not dominant. Current studies focus on metallic nanoparticles (Van Aert, Batenburg, Rossell, Erni, & Van Tendeloo, 2011; Scott et al., 2012; Goris, Van den Broek, et al., 2012; Chen et al., 2013; Chen, Kisielowski, & Van Dyck, 2015) or shallow tips of crystals (Jia et al., 2014; Xu et al., 2015), or almost 2D materials such as two layers of graphene (Van Dyck, Jinschek, & Chen, 2012), where the problem of 3D reconstruction is reduced to determining the elevation of a certain atom. This restriction guarantees a partially intact linearity, which is crucial for the existing tomographic approaches. The solution to the full problem of reconstructing atomic data within the fully dynamical regime is still beyond reach. Second, all current strategies involve some kind of a priori atomicity constraint, assuming that the material consists of a discrete set of scatterers. This assumption takes into account the dominant influence of the atom core on electron scattering, in particular into large angles. The implementation of an atomicity constraint, however, is highly non-trivial with different strategies being proposed to date of this work.

One can categorize the current approaches as (A) tomographic approaches containing some kind of atomicity constraint (Van den Broek, Van Aert, & Van Dyck, 2009; Van Aert et al., 2011; Gris, Van den Broek, et al., 2012; Chen et al., 2013; Xu et al., 2015) and (B) model based approaches seeking for a best fit between a dynamical scattering simulation and the measured result (Saghi, Xu, & Moebus, 2009; Van den Broek & Koch, 2012; Van Dyck et al., 2012; Van den Broek & Koch, 2013; Chen et al., 2013; Jia et al., 2014).

The idea behind category (A) is to use linear tomographic approaches and suppress artifacts due to nonlinear effects phenomenologically by means of a regularizing constraint limiting the number of scatterers. The projection data are typically acquired by High-Angle Annular Dark-Field (HAADF) Scanning TEM. This is an imaging mode, where a focused electron beam is scanned over the sample, while recording only those electrons, which have been scattered into very large angles (>50 mrad). When traversing the crystal lattice, the focused electron beam undergoes a transformation, exhibiting similar oscillations as those visible in Fig. 2.3, which are frequently referred to as channeling in the context of (S)TEM. However, the few electrons scattered at each atom into large angles sum up incoherently, since any phase relation is destroyed by the thermal motion (Howie, 1979). Thus, an annular detector integrating over large scattering angles intercepts an electron flux given by the product of an incoming flux, the atomic density and the atomic large-angle scattering cross-section. The latter is approximately proportional to Z^2 (2.53), which confers the atomic resolution HAADF-STEM signal its characteristic peaked contrast at atomic column positions proportional to the atomic number ("Z-contrast"; Pennycook & Nellist, 2011).

Van Aert et al. exploit that property in a column-wise atom-counting scheme (Van Aert et al., 2011), effectively corresponding to some sort of discrete space tomography. Goris et al. slightly adapted this approach by reconstructing atomistic data pertaining to atomic columns imaged along four low-index zone axes (Goris, Bals, et al., 2012). Extensive Fourier and atomic contrast filtering in the course of the reconstruction was applied by Chen et al. (2013) and Xu et al. (2015), respectively. From a general point of view all these constraints could be considered as regularization, which suppress reconstruction errors but also introduce regularization errors (cf. Section 3.5). The latter have stirred considerable discussion about the validity of the various regularization strategies within the scope of atomic resolution tomography (Rez & Treacy, 2013; Wang, Yu, Verbridge, & Sun, 2014). In spite of the progress made, little is known about the magnitude of the regularization error in the above noted results. For instance, the accuracy of atomically resolved deformation fields (Goris, Bals, et al., 2012; Xu et al., 2015) remains essentially speculative in the absence of a better understanding of the pertaining reconstruction and regularization errors.

In response to these problems, the approaches from category (B) focus on implementing the proper projection of the electron beam as given by the laws of dynamical scattering into the reconstruction procedure. By doing so the realm of tomography, i.e., the reconstruction from projections along submanifolds, is essentially abandoned and replaced by model-based approaches. The latter postulate a certain model for the contrast generation including dynamical scattering within the TEM and fit the free parameters of that model to the experimental results. While such approaches might be the only feasible strategy in the presence of the above noted nonlinear dependency between atomic potential and experimentally acquired data, model-based approaches are always threatened by wrong assumptions of the model, in addition to getting trapped in local minima of the error metric. For instance, "big bang" tomography (Van Dyck et al., 2012) (i.e., the fitting of the propagation distance from an atomic scatterer) quickly cools down beyond 2D materials featuring several atoms behind each other. Similarly, the reported sensitives for correlated fits of height, filling and position of atomic columns crucially depends on the correct modeling of input covariance given by the detection noise, object and beam instabilities as well as the correct modeling of the structural details of the object including surface layers (Jia et al., 2014; Chen et al., 2015).

In the following, we elaborate on incorporating the principles of dynamical scattering into a linear reconstruction scheme, with the goal to facilitate a well-behaved tomographic reconstruction without relying on model-based fitting approaches. In other words, we seek an approximate linear and continuous mapping between the electrostatic potential and a (possibly nonlinear) function of the projections in the tilt series, because such a mapping can be inverted in a straightforward manner. It is most beneficial to use holographic projection data in such an approach, because this offers the largest flexibility in choosing a suitable function permitting a linear projection. In order to identify that function and the corresponding linear mapping, we investigate the consequences of non-linearities in a standard tomographic reconstruction from atomic resolution phases under idealized conditions. The input data consist of phases pertaining to simulated noise-free waves, which imply that any artifacts in the reconstructed potential result from nonlinear effects. A characterization of these artifacts suggests possible modifications to the standard EHT reconstruction suited to achieve atomic resolution. One of these generalizations, containing the conventional tomographic reconstructions as a special case, is discussed in more detail in a second step.

6.2.1 Experimental Implementation

As a suitable model potential for the atomic resolution study we chose a gold nanoparticle because it represents a strong scatterer in spite of contain-

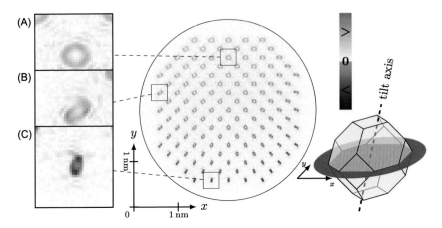

Figure 6.9 Cross-section of the reconstructed potential. Its position and orientation is indicated by means of a gray plane in the lower right corner.

ing a limited number of atoms only, which allows a tomographic reconstruction with reasonable computational resources. The gold nanoparticle has the form of a truncated octahedron (Fig. 6.9) with a diameter of approximately 3.6 nm, corresponding to a width of 8 unit cells and a total number of approximately 1300 atoms. In order to avoid low-index-zone-axis orientation during the tilt series, the initial tilt of the crystal was set about 13° away from the [100]-axis. To simulate the holographic tilt series we used a multislice algorithm (cf. Section 2.2). To avoid artifacts from the periodic continuation the simulated field of view was extended to 6 nm × 6 nm, which was sampled with 512 × 512 points to ensure a well-converged scattering simulation. The electron's acceleration voltage was set to 300 kV and atomic potentials in the parametrization of Weickenmeier and Kohl (1991) were assembled to generate the crystal potential. The tilt series comprises −90° to 89° with 1° tilt increments. The focus of each wave of the tilt series was set to the mid-plane of the nanoparticle. The tomographic reconstruction was performed along the same lines as expounded on in the previous section. Accordingly, the discrete Radon transformation is inverted by means of the conjugate gradient LSQR (Paige & Saunders, 1982) algorithm. The number of iterations was set to 20, which empirically yielded a good compromise between the clearness of the reconstructed features and amplification of artifacts. In the following, we will only discuss those features, which did not vary upon varying the regularization strength.

6.2.2 Results

The three-dimensional electric potential distribution reconstructed from the simulated tilt series is depicted in Fig. 6.9. The atomistic structure is clearly visible, with the centers of mass of the reconstructed potential features approximately corresponding to the underlying atomic positions used as input. However, one readily observes that the shapes of the reconstructed "atomic" potentials are not meaningful in a physical sense as they are deformed, contain multiple maxima and even minima. We will now elaborate on these artifacts.

First of all, there is a characteristic dependency of the reconstructed potential shape on the position in the nanoparticle. The most condensed peaks can be found on the rotation axis (C), with potential features getting spread out, the further the distance to the tilt axis. At a sufficient distance they exhibit a double maximum with a dip in the middle and two spiral arms. Additionally, the reconstructed potentials are rotated depending on their azimuthal placing with respect to the tilt axis. In particular the last observation suggests that the principal cause of the observed distortions in the reconstruction is the propagation distance between the atomic potentials (i.e., the scattering site) and the focal plane (midsection).

The atoms in the vicinity of the origin are imaged with nearly the same defocus in every projection direction, which produces a stable projection throughout the tilt series. The further away an atom from the tilt axis, the larger the differences (i.e., variation in spreading), which appear upon tilting. As these differences are essentially non-local projections (convolution with the Fresnel kernel), they cannot be reconstructed properly. The tomographic algorithm merely preserves the shape under different viewing angles leading to the observed deformations of the atomic potential. The details of these shapes are complicated because they are linear representations of nonlinear effects including the one of multiple scattering on strings of atoms appearing behind each other under a certain tilt angle.

We conclude that under ideal imaging conditions, EHT allows reconstructing atomistic data from high-resolution phases of a stable nanoparticle. The artifacts due to a violation of the linear projection law are well contained within the peaks and can be traced back to a dominant influence of the Fresnel propagation. A quantitative interpretation of the reconstructed data proves difficult as the reconstructed potentials bear little resemblance to real atomic potentials. Nevertheless, it is possible to extract information about the atomic number of the atoms (Krehl & Lubk, 2015b). Furthermore, the results show that a proper control of the defocus during tilting is crucial. This adds to the already long list of experimental prerequisites complicating atomic resolution EHT such as stable specimen or positional alignment.

6.2.3 Toward Atomic-Resolution Holographic Tomography

Following the findings of the previous section, EHT aiming for an accurate reconstruction of atomic potentials should take into account the Fresnel propagation effect in the first place. Moreover, it appears expedient to use the full information of the holographic tilt series, i.e., amplitude and phase, to extract the potential. Both observations are incorporated in the following approach, where we take advantage of the free-space propagator acting linearly on the wave function and of the potentials primarily affecting the phase of the scattered wave function (cf. Section 2.2). In other words, the leading order effect is a phase shift (i.e., axial scattering), which is modulated by propagation effects in the second order, before multiple scattering on various atoms eventually establish the complicated dynamical scattering behavior in its whole beauty as discussed in Section 2.2.

While it is generally impossible to reformulate the complete dynamical scattering as a linear projection problem of the potential, it is possible to approximately incorporate the propagation, and even some effects due to multiple scattering, into a linear tomographic reconstruction algorithm by reformulating the scattering problem with the following ansatz for the wave function

$$\Psi(\mathbf{r}, z) = e^{i\Lambda(\mathbf{r}, z)} . \tag{6.13}$$

Inserting this expression in the paraxial equation (2.21) leads to a nonlinear partial differential equation for the complex "phase" Λ

$$-\partial_z \Lambda(\mathbf{r}, z) = -\frac{i}{2k_0} \Delta \Lambda(\mathbf{r}, z) + \frac{1}{2k_0} (\nabla \Lambda(\mathbf{r}, z))^2 - \frac{e}{v} \Phi(\mathbf{r}, z) . \tag{6.14}$$

This equation is even less amenable to a solution due to the nonlinear $(\nabla \Lambda(\mathbf{r}, z))^2$. If, however, this nonlinear term can be neglected, the above equation transforms into a free-space paraxial equation for Λ

$$i\partial_z \Lambda(\mathbf{r}, z) + \frac{1}{2k_0} \Delta \Lambda(\mathbf{r}, z) \approx i\frac{e}{v} \Phi(\mathbf{r}, z) \tag{6.15}$$

containing an additional inhomogeneity proportional to the electrostatic potential Φ. This so-called Rytov approximation (Devaney, 1981) is related to other exponential series expansions such as the eikonal approximation mentioned in Section 2.2, and has the advantage to be valid for large phase shifts, where the first order Born approximation fails. Instead, the Rytov approximation breaks down, if large phase gradients are present in the wave field (Slaney, Kak, & Larsen, 1984; Kak & Slaney, 1988). We will see below, however, that the Rytov approximation contains the conventional projection along straight lines as limiting case, which facilitated a (qualitative) reconstruction of confined atomic potentials above. This observation will serve as heuristic argument for the validity of the Rytov approximation (6.15).

The above partial differential equation (6.15) can be transformed into an integral equation

$$\Lambda\left(\mathbf{r}, z\right) = i\frac{e}{v} \iint_{-\infty}^{\infty} d^2 r dz'\, G\left(\mathbf{r}, z, \mathbf{r}', z'\right) \Phi\left(\mathbf{r}', z'\right) \qquad (6.16)$$

by inserting the Green's function of the free paraxial equation

$$G\left(\mathbf{r}, z, \mathbf{r}', z'\right) = -i\Theta\left(z - z'\right) K_T\left(\mathbf{r}, z; \mathbf{r}', z'\right) \qquad (6.17)$$

$$= -\frac{\Theta(z - z')}{\lambda(z - z')} e^{i\frac{k}{2(z-z')}(\mathbf{r}-\mathbf{r}')^2}.$$

Accordingly, the complex "phase" Λ at the object exit plane $(z > z')$ reads

$$\Lambda\left(\mathbf{r}, z = t\right) = -\frac{ie}{\lambda v} \iint_{-\infty}^{\infty} d^2 r dz' \frac{\exp\left(i\frac{k}{2(z-z')} \left|\mathbf{r} - \mathbf{r}'\right|^2\right)}{z - z'} \Phi\left(\mathbf{r}', z'\right), \qquad (6.18)$$

and we readily verify that the "projection" law (6.18) contains the conventional projection along straight lines in the limit $z - z' \to 0$. Thus, taking the nonlinear complex logarithm of the wave function facilitates an approximate linear mapping between the potential and said logarithm. It is important to note that the complex logarithm is a multivalued function, and hence one has to perform a phase unwrapping prior to the reconstruction procedure (Chen & Stamnes, 1998), similar to that incorporated in the conventional EHT approach (cf. Section 6.1.1).

In order to build a tomographic reconstruction from this linear map, it remains to be shown that the integral equation (6.18) can be inverted provided a tilt series of reconstructed complex "phases". The pertaining inverse transformation, referred to as back-propagation (instead of back-projection), has been previously derived within the related theory

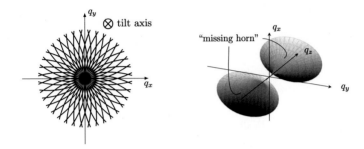

Figure 6.10 Fourier slice theorem of diffraction tomography. In the single tilt axis geometry (tilt axis ∥ q_z), the sampled planes in Fourier space correspond to paraboloids rotated around the tilt axis. Accordingly, a 360° tilt series is required to sample the $q_z = 0$ section completely (i.e., without missing regions) with the sampling depending only on $q_x^2 + q_y^2$ and not on the azimuth. Parallel to the tilt axis direction a "missing horn" of information may not be reconstructed in the single tilt axis setup.

of diffraction tomography (Kak & Slaney, 1988; Müller, Schürmann, & Guck, 2015). Here, we limit ourselves to a sketch of the reconstruction principle based on an adapted Fourier slice theorem (Fig. 6.10), which may be derived along the same lines as the original one (cf. Section 3.2.3). Accordingly, we write the Fourier transform of the complex "phase" in the detector plane z

$$\tilde{\Lambda}(\mathbf{q}) = \frac{e}{\nu} \int_{-\infty}^{\infty} dz' e^{-i\frac{z-z'}{2k_0}\mathbf{q}^2} \tilde{\Phi}(\mathbf{q}, z') \tag{6.19}$$

$$= \frac{e}{\nu} \int_{-\infty}^{\infty} dq_z \delta\left(-\frac{\mathbf{q}^2}{2k_0} - q_z\right) \tilde{\Phi}(\mathbf{q}, q_z) e^{iq_z z}$$

$$= \frac{e}{\nu} \tilde{\Phi}\left(\mathbf{q}, -\frac{\mathbf{q}^2}{2k_0}\right) e^{-i\frac{\mathbf{q}^2}{2k_0}z},$$

and note that the Fourier transform of the projection data now corresponds to paraboloid sections of the potential to be reconstructed multiplied with an additional Fresnel phase. The corresponding distribution of the single tilt series data, shown in Fig. 6.10, exhibits a missing spatial frequency interval along the tilt axis and a non-trivial sampling density. Moreover, the problem cannot be separated along the tilt axis anymore and a rotation around 360° is required in order to homogeneously sample Fourier space.

Based on the adapted Fourier slice theorem one may now derive a (weighted) back-propagation resembling the (weighted) back-projection of conventional tomography including the pertaining reconstruction algorithms (Müller et al., 2015). Currently available numerical pretests indicate that the back-projection algorithm in combination with the Rytov approximation indeed permits a largely improved tomographic reconstruction of atomic potentials from holographic tilt series (Krehl & Lubk, 2015a). This also suggests that higher-order Rytov-type expansions of the exponential of the wave function, such as the Magnus series and derivatives of which (Blanes, Casas, Oteo, & Ros, 2009), represent a promising route toward atomic potential retrieval in 3D. That includes the related eikonal approach consisting of integrating the potential along curved lines corresponding to classical paths (Li et al., 2006).

Finally, it is emphasized that the above theoretical considerations represent but one puzzle piece to the successful implementation of 3D atomic potential reconstruction. The experimental hurdles are at least as immense, e.g., because the requirements toward tilt axis alignment, goniometer precision, and specimen stability are significantly increased compared to the medium resolution setup discussed in the previous section. A further challenge is the realization of multiple 360° tilt series around several axes necessary to experimentally suppress missing horn and inhomogeneous sampling artifacts. Thus, it is only fair to say that atomic resolution Electron Holographic Tomography represents one of the rare goals in TEM, which are suited to trigger a particularly broad range of developments including a better understanding of dynamical scattering approximations, novel tomographic approaches and new instrumental possibilities.

6.3. MAGNETIC VECTOR FIELDS

Three-dimensional ferromagnetic nanoscale materials are currently investigated for a large number of applications in data storage (Parkin, Hayashi, & Thomas, 2008), logic circuits (Lavrijsen et al., 2013), and magnetic sensing (Liu, Nah, Varahramyan, & Tutuc, 2010). Thus, they constitute one important line of research in modern nanomagnetics and spintronics (Nasirpouri & Nogaret, 2011; Shinjo, 2013; Stamps et al., 2014). The fabrication of patterned magnetic structures, spin textures or nonplanar spintronic devices not only requires advanced fabrication techniques such as lithography, but also advanced imaging techniques that can resolve the magnetic configuration in 3D at the nanometer scale. Available local magnetic field probes including, e.g., Photoemission Electron Microscopy (PEEM) (Schönhense, 1999; Thomas, 2005), Magnetic Force Microscopy (MFM) (Shinjo, Okuno, Hassdorf, Shigeto, & Ono, 2000; Li et al., 2001), Spin Polarized Scanning Tunneling Microscopy (Wernsdorfer et al., 1997;

Wachowiak et al., 2002), or Lorentz (transmission electron) Microscopy (LM) (Petford-Long & De Graef, 2002) are, however, not suited to determine the 3D distribution of the magnetic field with nanometer spatial resolution.

Electron holography, on the other hand, probes projected magnetic fields with a lateral spatial resolution down to the angstrom regime, which has been proven a valuable characterization tool for ferromagnetic nanostructures such as nanoparticles, nanowires, and nanopillars (cf. Section 5.1). In spite of the similar projection law for magnetostatic and electrostatic fields

$$\partial_{x,y}\varphi_{\mathrm{mag}}\left(\mathbf{r}\right) = \int_{-\infty}^{\infty}\left(\begin{array}{c} B_y\left(\mathbf{r}, z\right) \\ -B_x\left(\mathbf{r}, z\right) \end{array}\right)\mathrm{d}z, \tag{6.20}$$

however, only a few tomographic reconstructions of magnetic fields have been reported so far. The main challenge for an electron holographic tomographic reconstruction of magnetic fields is the small magnetic phase shift, typically one order of magnitude smaller than the electric one, in combination with the large noise level due to the numerical derivative involved when computing the projected magnetic fields. Additionally, the experimental effort for reconstructing all three Cartesian components of the magnetic induction is considerably larger compared to the electric potential case. Each Cartesian component of the magnetic induction requires one 360° tilt series (i.e., two conventional 180° tilt series) around a tilt axis parallel to the respective component (cf. Section 5.1), where the two half series comprising 180° facilitate an unambiguous separation of electric and magnetic phase shifts (Lai, Hirayama, Fukuhara, et al., 1994; Wolf, Lubk, Röder, et al., 2013). Taking into account that one Cartesian component may be calculated from the other two by virtue of Gauss' law for magnetism, two 360° tilt series, i.e., four times the effort for the electrostatic potential EHT is necessary. This requires dedicated TEM specimen stages facilitating the acquisition of 360° tomographic tilt series around two perpendicular axes. The availability of such stages is pending to date of this work.[3] Current commercial designs (e.g., Fischione Model 2040) facilitate the recording of two perpendicular tilt series around some restricted tilt interval smaller than 180°. Outside of magnetic vector field tomography, they have been employed to reduce missing wedge artifacts occurring in single tilt axis

---

[3] Only one working implementation has been reported so far (Tanigaki et al., 2015).

tomographies to missing pyramids (Penczek, Marko, Buttle, & Frank, 1995; Mastronarde, 1997).

The proof-of-concept for applying EHT to magnetostatic fields was given by Lai, Hirayama, Fukuhara, et al. (1994), who reconstructed one component of the magnetic induction surrounding a micron-sized barium ferrite particle in a semiquantitative way (spatial resolution 100–150 nm, tilt interval of ±60° with 5° increments). Shortly afterward, Ferrier, Liu, Martin, and Arnoldussen (1995) combined Differential Phase Contrast (cf. Section 5.4) and tomography to reveal the magnetic stray field emerging from a magnetic recording head with a spatial resolution of about 1 μm. It is noteworthy that the DPC technique facilitated the reconstruction of the two in-plane components of the magnetic induction over several tens of microns here, which is beyond the capabilities of off-axis electron holography due to the limited coherence (cf. Chapter 5). Considering the reconstruction of a magnetic stray field as an exterior reconstruction (cf. Section 3.2.4), we reconstructed one axial component of the magnetic stray field of a free-standing Co_2FeGa nanowire by EHT (Lubk, Wolf, Simon, et al., 2014). The most recent EHT study revealed one component of the magnetic induction pertaining to Cobalt nanorods (Wolf et al., 2015), which will also serve as a showcase to illustrate the capabilities and properties of EHT applied to magnetic fields below.

To date of this work, only three EHT studies aiming at the reconstruction of all three Cartesian components of the magnetic induction have been reported. Stolojan and Dunin-Borkowski (2001) utilized EHT to visualize (tilting range was limited to ±40°) the magnetic induction of permalloy nanorods. Phatak, Petford-Long, and De Graef (2010) employed TIE phase retrieval and tomography to reveal the 3D magnetic induction distribution of a magnetic vortex state in a permalloy structure. Similar to the DPC study of Ferrier et al., a very large reconstruction volume with an extension of several tens of microns could be realized by employing TIE instead of Off-Axis Holography. However, the study suffered from a limited tilt range and the shortcomings of the TIE technique with respect to low spatial frequencies and unknown boundary conditions (cf. Section 5.2). An apparently perfect (i.e., artifact- and noise-free) reconstruction of all three Cartesian components of the magnetic induction pertaining to vortex cores in stacked ferromagnetic discs has been reported recently by Tanigaki et al. (2015). However, a number of issues such as the large tilt increment of 10°, the susceptibility of the vortex core to small stray fields from the Lorentz lens, the absence of any electric potential data as well as noise in the reconstruction in combination with a spatial and signal resolution significantly (i.e., one order of magnitude) better than in the previous studies, currently raise some concerns on the data treatment employed there.

In the following, the capability of EHT to reconstruct one component of the magnetic induction parallel to the tilt axis is explored. As discussed previously, the reconstruction of one component is the prerequisite for solving the full problem. The presented results pertain to a Co nanowire, which exhibits an internal multidomain structure at the tip formed in response to minimizing the demagnetization field. The nanowire (100 nm thick and a few microns long) was grown by focused electron beam induced deposition (FEBID) in a focused ion beam (FIB)/scanning electron microscopy (SEM) dual beam instrument on top of a copper tip of a dedicated 360° tomography sample holder. Micromagnetic simulations are shown for comparison. A more detailed account of the following results has been published in Wolf et al. (2015).

6.3.1 Experimental Implementation

In principle, two distinct experimental procedures may be employed to record a 360° tomographic tilt series required for an unambiguous reconstruction of one component of the magnetic induction. It is possible to acquire two distinct tilt series, ideally ranging from 0° to 180°, with the specimen manually flipped in between. This procedure works with standard tomography holders, which are, however, typically limited in their tilt range. An additional drawback of this method is that the manual flip can change the orientation of the specimen, its relative position to the optical axis and even the specimen surface (cf. Section 5.1), which complicates the accurate alignment of the two tilt series and hence the tomographic reconstruction. Moreover, modifications of the imaging conditions (e.g., aberrations, distortions) and the specimen lead to inconsistencies in the projected data, thereby further reducing the quality of the reconstruction.

In order to reduce these effects and to avoid any missing wedge artifacts, the Co nanowire was mounted on a dedicated 360° electron tomography holder (Model 2050, Fischione Instruments). Here, particular care must be taken to ensure a good alignment of the nanowire's symmetry axis and the tilt axis of the holder. In this study, tilt axis and longitudinal axis of the nanowire are almost parallel, comprising only a small angle of 3°. Therefore, the longitudinal B-field component and its projection onto the tilt axis differ only slightly by ca. 0.1% and can be considered identical when discussing the results of the ensuing tomographic reconstruction.

The 360° holographic tilt series (3° tilt increment) of the Co nanowire was recorded with the Qu-An-TEM (FEI Titan[3]) operated at 300 kV in non-Cs-corrected Lorentz mode (objective lens off) fitted with an electron biprism in the SA aperture plane. We employed a biprism voltage of 200 V to obtain a fringe spacing of 2.2 nm. Moreover, the diffraction lens current was maximal to facilitate a large field of view. The holograms were acquired with a US1000XP Gatan CCD camera and the whole acquisition process was controlled through THOMAS. The obtained resolution of the reconstructed object exit wave of 4.6 nm was determined by the size of the numerical mask used in the Fourier reconstruction.

Pairs of phase images were aligned by applying a linear affine transformation on the "flipped" phase images (cf. Section 6.1.1), which took into account displacements, rotation, and direction-dependent magnification changes. The final alignment of the tilt axis including corrections for residual sub-pixel displacements was performed by tracking the center of mass in the electric phase tilt series (center-of-mass method) and applying the obtained displacements to both the electric and magnetic phases. This approach exploits the higher signal-to-noise ratio of the electric phase shift and allows aligning both the electric and magnetic data consistently. Subsequently, the magnetic phase shifts were differentiated perpendicular to the tilt axis. To suppress the noise amplification due to the numerical derivative, the latter has been slightly regularized by applying a median filter. We furthermore note that, similar to the above GaAs–$Al_{0.33}Ga_{0.67}As$ nanowire, a small and constant phase gradient due to charging was observed throughout the tilt series. Thus, artifacts in the magnetic phase shift from varying charging can be largely excluded.

The tomographic reconstruction of the interior electric potential and magnetic induction was carried out using 5 iterations of the WSIRT algorithm (cf. Section 3.6). A separate reconstruction of only the exterior field was carried out employing five conjugate gradient LSQR iterations (Paige & Saunders, 1982). Similar to the previously discussed GaAs–$Al_{0.33}Ga_{0.67}As$ nanowire, the number of iterations was determined by visually inspecting the reconstructed data for an acceptable balance between spatial resolution and noise. The thereby obtained spatial resolution of the reconstructed electric potential is well below 10 nm.

6.3.2 Results

The main results of the tomographic reconstruction of the interior fields within the Co nanowire are summarized in Fig. 6.11. Accordingly, a Mean

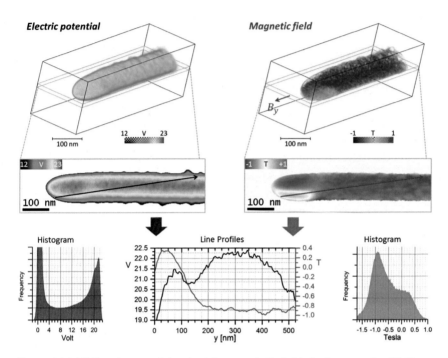

Figure 6.11 (A) Electric potential and axial magnetic B_y-field of a Co nanowire. (B) 15 nm thick 2D cross-sections as indicated by orange boxes. (C) Histograms of 3D volumes and 1D line profiles along the arrows in (B). The peak in the histogram of the electric potential at 21.5 V corresponds to the Mean Inner Potential of this Co NW. The most frequent value in the histogram of the magnetic induction is −0.9 T.

Inner Potential of $\Phi_0 = 21.5$ V is obtained, which is in reasonable agreement with previously reported values for nanoparticles ($\Phi_0 = 18.7$; Gao, Shindo, Bao, & Krishnan, 2007). An inspection of the magnetic field exhibits a large domain (blue) with an induction of about −0.9 T, where the spins are aligned along the negative y-direction. Noting the bulk induction of cobalt $B_0 = 1.76$ T (Dunin-Borkowski, McCartney, Smith, & Parkin, 1998), we ascribe the comparatively low value of $B = 0.9$ T to the reduced Co content of the FEBID grown nanowire (Takeguchi, Shimojo, & Furuya, 2005). Furthermore, a small magnetic inversion domain (red) with a reduced field of 0.3 T pointing in opposite direction is formed at the apex of the nanowire. Moreover, an electric potential drop at the position, where the magnetic induction changes its direction, indicates a structural modulation of material resulting in a local reduction of the magnetization at that area, which could have triggered the formation of the observed magnetic inversion.

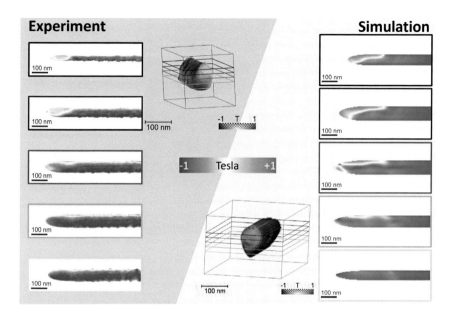

Figure 6.12 Comparison between experimentally reconstructed and simulated *B*-field in axial (*y*-)direction.

The observed remanent C state separated by a 180° domain wall is common in soft magnetic nanowires (Usov, Zhukov, & Gonzalez, 2006), where the strong demagnetizing field twists the magnetization at the apex. Fig. 6.12 exhibits a micromagnetic simulation taking into account the tomographically determined shape of the nanowire. Accordingly, an excellent agreement with the experimental tomographic reconstruction with some slight deviations in the arrangement of the inversion domain is visible. Namely, the inversion domain is formed in the thicker region in the simulation, whereas it is formed in the thinner region of the nanowire cross-section in the experiment (Fig. 6.12). These differences could be caused by effects not considered in the simulation, such as roughness, presence of defects, local change of composition along the wire or influence of local magnetocrystalline anisotropy.

Finally, we turn our attention to the reconstruction of the weak magnetic stray field around the nanowire. Fig. 6.13A shows that the standard reconstruction fails to recover the weak stray field emerging predominantly from the tip, mainly because of non-local reconstruction artifacts resulting from dynamical scattering, alignment issues and problems with the separation of the magnetic phase shift. As stated previously, the support theorem

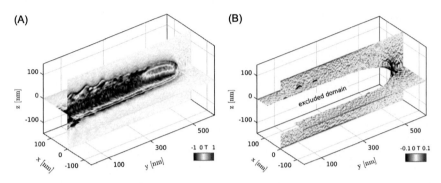

Figure 6.13 Comparison between full reconstruction (A) and reconstruction of the exterior (stray) field only (B). Note the different color scale ranging being reduced by one order of magnitude in the exterior reconstruction. Accordingly, the (weak) stray field is only recovered in the exterior reconstruction. The characteristic shape of the 3D stray field resembles that of a nanobar magnet.

(cf. Section 3.2.4) also ensures a unique reconstruction from a set of line integrals outside a convex region (Fig. 3.9). It is obvious that all object-related issues such as dynamical scattering, phase unwrapping as well as subtraction of two flipped tilt series, do not inflict such a reconstruction (Lubk, Javon, et al., 2014). The reconstruction of the exterior (stray) field, shown in Fig. 6.13B, agrees very well with that of a homogeneously magnetized nanobar magnet, where the magnetic field emerges from the tip region (see Lubk, Javon, et al., 2014 for magnetostatic simulations of a similar nanowire). This result corroborates the monodomain structure of the nanowire only slightly perturbed by a small inversion domain in the tip region.

6.4. ELASTIC AND INELASTIC ATTENUATION COEFFICIENTS

Mean free path lengths (MFPL) between successive scattering events of beam electrons may be divided into elastic ones, due to elastic scattering at screened Coulomb potentials of the atoms, and inelastic ones, largely determined by low-loss excitations, such as plasmons. The latter are characteristic for the electronic behavior of materials, while the former directly reflect the chemical composition (cf. Sections 2.3 and 5.1). Consequently, the measurement of MFPLs is frequently used to characterize materials in the TEM (Reimer, 1989; Egerton, 1996). The tomographic determination of the corresponding elastic and inelastic attenuation coefficients is referred to as bright field TEM tomography and energy filtered TEM tomography, respectively. It remains difficult, however, to perform a combined (i.e., correlated) measurement of elastic and inelastic attenuation coefficients as well as electromagnetic potentials at nanometer resolution in 3D, yielding a more comprehensive picture of the materials' properties. Subsequently, we elaborate on off-axis EHT to achieve this goal.

A comprehensive discussion of EHT applied to recover electrostatic potentials and magnetic fields has been given in the previous sections. For several reasons, however, holographic measurements of projected attenuation coefficients have been rare (McCartney & Smith, 1994; Harscher & Lichte, 1997; Chung, Smith, & McCartney, 2007) (cf. Section 5.1) and tomographic reconstructions have been reported only recently (Lubk, Wolf, Kern, et al., 2014). First, electromagnetic potentials typically exhibit a close relationship to the properties and functionality of the material under investigation, inducing a certain disregard toward the attenuation data. Second, attenuation signal-to-noise ratios (SNRs) are typically small (Lenz, 1988; Harscher & Lichte, 1996; Röder & Lubk, 2014), complicating their tomographic reconstruction. Third, the attenuation coefficients were difficult to relate to calculated ones or those obtained from other techniques such as EELS.

In Section 5.1, the last point has been successfully resolved by noting the accurate expression for the influence of elastic and inelastic scattering on an off-axis hologram. There, we discussed the necessary steps to accurately separate elastic and inelastic attenuation coefficients and established their connection to the chemical composition and dielectric losses in the sample. In the following, we combine these findings with the tomographic reconstruction techniques developed above, thereby facilitating the simultaneous reconstruction of electromagnetic potentials, elastic and inelastic MFPLs in 3D from one off-axial holographic tilt series. We demonstrate the current state of the technique at the example of the GaAs–$Al_{0.33}Ga_{0.67}As$ core–shell nanowire from Section 6.1, where the combined holographic data allow separating chemical composition variations (visible in both attenuation and electrostatic potential) and purely electronic effects such as space charges (visible in the reconstructed potential only).

As a short reminder we first recapitulate the pertaining expressions facilitating a separate reconstruction of inelastic and elastic attenuation coefficients from the holographically measured side- and centerband attenuation in the limit of a sufficiently large objective aperture semiangle α containing all inelastically scattered electrons. Following Section 5.1, the centerband attenuation is given by elastic scattering into angles larger than the aperture semiangle α

$$\ln\left(2\frac{I_{CB}(\mathbf{r})}{I_{CB,0}} - 1\right) = -\int_{-\infty}^{\infty} \mu_{el}^{(\alpha)}(\mathbf{r}, z)\, dz. \tag{6.21}$$

In the sideband, additionally all inelastically scattered electrons are missing, which facilitates the following separation of inelastic attenuation coefficients:

$$\ln\left(2\frac{I_{CB}(\mathbf{r})}{I_{CB,0}} - 1\right) - 2\ln\frac{I_{SB}(\mathbf{r})}{I_{SB,0}} = \int_{-\infty}^{\infty} \mu_{in}(\mathbf{r}, z)\, dz. \tag{6.22}$$

Expressions (6.21) and (6.22) serve as basis for the tomographic determination of elastic and inelastic attenuation in the previously considered GaAs–AlGaAs core–shell nanowire below.

It is furthermore important to emphasize that the linear relationships (6.21) and (6.22), and hence the separation of the inelastic damping according to (6.22), are only valid up to a certain thickness, because multiple scattering eventually increases the angular distribution of the beam, thereby increasing the probability of scattering absorption in a nonlinear manner. The linear thickness range increases toward materials with low scattering power, large acceleration voltages and large aperture angles (Smith & Burge, 1963; Wang, Zhang, Cao, Nishi, & Takaoka, 2010; Zhang, Egerton, & Malac, 2012). In order to achieve linear damping as well as separability of elastic and inelastic damping, we therefore employed a large acceleration voltage (300 kV) and used no objective aperture in the following experiments.

6.4.1 Experimental Implementation

Most of the experimental details pertaining to the recording and reconstruction of the tomographic tilt series of the GaAs–Al$_{0.33}$Ga$_{0.67}$As core–shell nanowire have been already noted in Section 6.1. Here, we only reiterate that the TEM was operated at 300 kV acceleration voltage in aberration corrected Lorentz mode without inserting an objective aperture. Thus, we had to gauge the effective aperture semi-angle from a reference measurement (cf. Section 5.1), which gave $\alpha \approx 22$ mrad (Lubk, Wolf, Kern, et al., 2014). The latter is well above the critical angle (Bethe ridge), below which inelastic plasmon scattering occurs. We normalized the center and sideband intensity in each tilt with the vacuum intensities

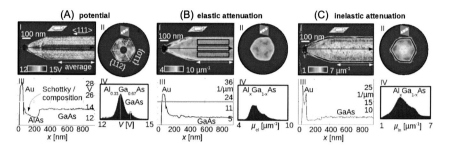

Figure 6.14 3D reconstruction of potential (A), elastic (B) and inelastic (C) attenuation coefficients. (I) shows longitudinal sections and (II) cross sections averaged over the interval indicated in (I), respectively. (III) displays linescans along the black arrow in (I), and (IV) the histogram of the volume data, respectively. The boxes in the elastic longitudinal section indicate the interval used for determining the average and error of the attenuation coefficients in Table C.1.

according to (6.21) and (6.22), which has the beneficial side effect of removing overall intensity and coherency fluctuations in the beam without division by a noisy reference hologram. Dynamical scattering effects significantly modulated the intensity only in the thicker part of the tapered region (Lubk, Wolf, Kern, et al., 2014). Before starting the actual tomographic reconstruction, we also checked the projection data for a violation of the linear projection laws (6.22) and (6.21), and found no significant deviations (Lubk, Wolf, Kern, et al., 2014).

The tomographic reconstruction of the elastic and inelastic attenuation coefficients according to (6.21) and (6.22) is almost identical to the one applied for reconstructing the 3D potential described in Section 6.1. The main difficulty here is the inherently small SNR of the intensity data. We therefore align the intensity tilt series by correcting the displacements that are determined with a center of mass method from the unwrapped phase tilt series with a higher SNR. Furthermore, a significantly larger regularization, imposed by employing only four (compared to ten for the phase) conjugate gradient iterations of the LSQR algorithm (Paige & Saunders, 1982), proved adequate to stabilize the values of the attenuation coefficients before amplifying noise and other non-projective artifacts.

6.4.2 Results

Fig. 6.14 displays a complete EHT reconstruction dataset consisting of histograms, cross-sections and linescans from the reconstructed potential, elastic and inelastic attenuation coefficients of the nanowire. Based on these images, we will first discuss the different discernible features of the nanowire before proceeding with a quantitative analysis of the determined MFPLs from different homogeneous regions.

With respect to the NW properties, the most important findings from the potential reconstruction (Fig. 6.14A, cf. Section 6.1 for details) have been Al-accumulation at nanofacets (faint radial lines in Fig. 6.14A(II)), AlAs alloying at the tapered region (Fig. 6.14A(I), A(III)), core facet modulations (Fig. 6.5), and a potential slope of either electronic (Schottky barrier) or compositional origin at the Au–AlGaAs interface (Fig. 6.14A(III)).

In spite of the lower signal resolution in the reconstructed elastic attenuation coefficient μ_{el} (Fig. 6.14B), one can clearly discern the compositional changes between core and shell (Fig. 6.14B(I), B(II)). The fluctuations of the attenuation in the shell (Fig. 6.14B(II)) may not yet be interpreted in terms of real composition changes, since they change with the iteration number used in the reconstruction. For the same reason, it is difficult to relate the attenuation at the thick part of the tapered region, which was affected by dynamical scattering, to the AlAs region found in the potential. The slope toward the Au nanoparticle in both the attenuation coefficient and potential reconstruction (compare Figs. 6.14A(III) and 6.14B(III)), on the other hand, indicates a chemical composition change close to the Au interface. The particularly large values of the elastic attenuation suggest an accumulation of Au atoms from the catalyst here, which has been previously discussed in similar systems (Perea, Lensch, May, Wessels, & Lauhon, 2006).

The inelastic attenuation coefficients μ_{in} (Fig. 6.14C) exhibit a smeared core–shell contrast, which is not surprising considering that bulk plasmon excitations are delocalized over several tens of nm (Verbeeck, Van Dyck, Lichte, Potapov, & Schattschneider, 2005). Furthermore, one observes a clear drop in the tapered region, abruptly ending at the Au interface. That could be explained by a negligible influence of the composition change, discussed previously, on plasmonic excitations. The remaining sharp features (mainly the abrupt changes at the NW boundary) are ascribed to Begrenzungs effects due to surface excitations (Egerton, 1996). We now proceed with a quantitative discussion of average MFPLs of the GaAs core, the $Al_{0.33}Ga_{0.67}As$ shell and the Au tip obtained from the radial intervals indicated in Fig. 6.14B(I).

We first note that the aperture semiangle $\alpha \approx 22$ mrad is sufficiently large to validate the approximation (5.31) at 300 kV. However, α is not well defined in our case, since the aperture is not situated in the back focal plane of the Lorentz lens due to instrumental limitations at our FEI Titan TEM. With this aperture angle the average elastic λ_{el}^{hol} values of the core and shell agree well with a theoretical approximation for kinematic elastic scattering (2.53), whereas the Au tip deviates significantly (see Table C.1). The latter discrepancy may be explained by the shortcomings of the kinematic model for a strong dynamical scatterer such as Au.

The inelastic MFPLs λ_{in}^{hol} for core and shell, on the other hand, are slightly larger than those obtained from EELS or theoretical considerations (see Table C.1). However, the total inelastic MFPL obtained from holography should be slightly smaller because it considers the effect of all losses including very low ones, e.g., due to phonons buried in the zero loss peak of our EELS. This contradiction could be due to measurement errors (see Table C.1) and small nonlinear effects, noting that the elastic MFPLs are roughly two times larger than the specimen thickness.

6.5. STRAIN FIELDS

Dark-Field Electron Holography (DFEH) is an Off-Axis Holography variant dedicated to the measurement of strain in nanostructures (Hÿtch, Houdellier, Hue, & Snoeck, 2008; Hÿtch, Houdellier, Hüe, & Snoeck, 2011). It typically employs an off-axis setup, where a certain diffracted beam of the strained crystal lattice is superimposed with the same of an unstrained reference region (Fig. 6.15). Alternatively, the unperturbed reference beam may be generated by employing a biprism in the upper part of the condenser system,[4] allowing a deliberate tilt of the reference beam (Röder & Lubk, 2015; Röder, Houdellier, Denneulin, Snoeck, & Hÿtch, 2016). Moreover, the need for an unperturbed reference can be dispensed with, if performing DFEH in an inline (cf. Sections 5.2, 5.3) (Koch, Özdöl, & van Aken, 2010; Song, Shin, Kim, Oh, & Koch, 2013) or Differential Phase Contrast setup (cf. Section 5.4) (Denneulin, Houdellier, & Hÿtch, 2016). The basic principle behind DFEH is that a slightly changing diffraction angle within the strained region translates into a phase shift in the reconstructed wave, which may be interpreted in terms of a displacement field $\mathbf{u}(\mathbf{R})$.[5] From the displacement field one usually derives the compo-

---

[4] This is necessary to facilitate the required amplification of the tilt angle by the following condenser lenses.

[5] Here, the gauge freedom for the phase corresponds to that of the displacement field, depending on a deliberate choice of origin for the reference.

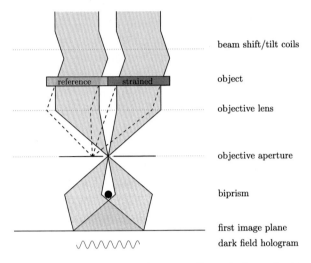

reference strained

beam shift/tilt coils

object

objective lens

objective aperture

biprism

first image plane
dark field hologram

Figure 6.15 Optical setup of dark field electron Off-Axis Holography. By deliberately tilting the incoming beam, a systematically scattered beam is aligned with the optical axis, permitting its selection through a subsequent objective aperture in the back-focal plane.

nents of the gauge invariant (infinitesimal) strain tensor

$$e_{ij} = \frac{1}{2}\left(\frac{\partial u_i}{\partial r_j} + \frac{\partial u_j}{\partial r_i}\right) \tag{6.23}$$

as well as the rigid lattice rotation

$$\omega_{ij} = \frac{1}{2}\left(\frac{\partial u_i}{\partial r_j} - \frac{\partial u_j}{\partial r_i}\right). \tag{6.24}$$

Contrary to other strain measurement techniques, this setup allows large spatial resolution over large fields of view, which can be further enhanced by dedicated lens configurations (Cooper, Rouvière, et al., 2011; Sickmann, Formánek, Linck, Muehle, & Lichte, 2011; Wang, Li, Domenicucci, & Bruley, 2013). The precision of reconstructed strain may attain $\mathcal{O}\left(10^{-4}\right)$ depending on experimental parameters such as exposure time, biprism voltage and sample thickness (Cooper et al., 2009; Béché, Rouvière, Barnes, & Cooper, 2011). Furthermore, several advances in the methodology allow removing other sources of phase shifts such as thickness variations (Hÿtch et al., 2011). A considerable number of DFEH studies, ranging from strained–silicon devices (Hüe, Hÿtch, Bender, Houdellier, & Claverie, 2008; Hüe, Hÿtch, Houdellier, Bender, & Claverie, 2009;

Cooper, Béché, Hartmann, Carron, & Rouvière, 2010) to multilayers (Hartmann et al., 2010; Denneulin et al., 2011) and quantum dots (Cooper, Rouvière, et al., 2011), has been reported so far.

In this section, we will discuss the possibilities to extend DFEH to map strain in 3D. This would provide a unique characterization method to study long range strain fields, such as present in quantum dot structures (Cooper, Barnes, Hartmann, Béché, & Rouvière, 2009) or modern 3D microelectronic devices such as FinFETs (Conzatti et al., 2011), where 3D strain engineering plays a crucial role in enhancing the carrier mobility and hence the performance of the device (Sun, Thompson, & Nishida, 2007; Baykan, Thompson, & Nishida, 2010). Moreover, 3D strain mapping would allow to reveal and take into account the ubiquitous surface relax-ation effect afflicting strain studies in thin TEM lamellas (Treacy, Gibson, & Howie, 1985; Hÿtch, et al., 2011). However, 3D strain mapping in the TEM has been elusive to date of this work, mainly because there is no suit-able tomographic reconstruction principle. For instance, convergent-beam electron diffraction (CBED), the first technique used to study strained-silicon devices (Zhang & Liu, 2006), obeys a very complicated projec-tion law (Clement, Pantel, Kwakman, & Rouvière, 2004), complicating a tomographic reconstruction of the underlying strain field. The situ-ation gets even worse for zone-axis techniques such as high–resolution electron microscopy (HRTEM) (Hüe et al., 2008) or nano-beam elec-tron diffraction (NBED) (Usuda, Numata, Irisawa, Hirashita, & Takagi, 2005), mainly because the number of beams involved is prodigious. The only solution here is to model the strain field, perform scattering and imaging simulations and compare with the experimental data (Houdellier, Roucau, Clément, Rouvière, & Casanove, 2006). To avoid brute-force atomistic multislice calculations (Tillmann, Lentzen, & Rosenfeld, 2000; Chuvilin, Kaiser, de Robillard, & Engelmann, 2005), a Feynman dia-gram technique applied to dynamical theory has been devised (Houdellier, Altibelli, Roucau, & Casanove, 2008) (cf. Section 2.2). Surprisingly, high-angle annular dark-field imaging (HAADF) has seen the most progress toward an analytical approach (Grillo, 2009), following on the earlier anal-ysis in terms of strain-induced inter-band scattering (Perovic, Rossouw, & Howie, 1993).

Subsequently, it is shown that the 2-beam scattering conditions (Howie & Whelan, 1961, 1962), employed in DFEH, facilitate a particularly straightforward analytical projection law of general strain fields. Based on this projection law, inversion strategies, eventually retrieving the underlying

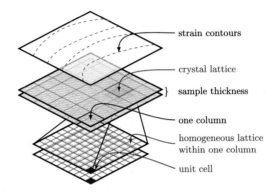

Figure 6.16 Column approximation.

strain distribution, are discussed. A considerable portion of the subsequent results has been published in Lubk, Javon, et al. (2014) and Javon et al. (2014).

6.5.1 Dynamical Scattering and Geometric Phase

To describe electron scattering on strained crystal lattices, we shall assume that the corresponding deformation of the crystal varies slowly on the length scale of the lattice constant. Thus, the lattice periodicity is preserved locally. Only on larger length scales a deviation, which may be described by a spatially varying lattice constant, may be observed. Such an almost or nearly periodic behavior also plays a large role in speech processing, audio signal processing, and music synthesis, and the corresponding theory is that of quasiperiodic functions or signals. The quasiperiodic crystal lattice permits to subdivide the crystal within the field of view of the TEM into smaller periodic patches, for which the scattering problem may be solved independently (Fig. 6.16). This approach, referred to as column approximation, allows the use of Bloch wave techniques (cf. Section 2.2), where the lattice deformation can be incorporated via geometric phase factors in the following straightforward manner.

We begin our considerations with the paraxial equation noted in lateral Fourier representation

$$\frac{\partial \tilde{\Psi}(\mathbf{g}, z)}{\partial z} = -i \left(\underbrace{\frac{\mathbf{g}^2 + 2\mathbf{k}_0 \mathbf{g}}{2k_{0z}}}_{s_G} - \frac{e}{v} \tilde{\Phi}(\mathbf{g}, z) \otimes \right) \tilde{\Psi}(\mathbf{g}, z). \qquad (6.25)$$

In a second step, the z-dependency is dropped, which corresponds to working in the 2D Bloch wave approximation (cf. Section 2.2), and the number of excited beams is restricted to two. This so-called two beam condition is adjusted by orienting the crystal such that the Ewald sphere crosses only two reciprocal lattice vectors, i.e., $s_g \approx 0$. Note that dynamical scattering inevitably violates exact two-beam conditions to some extent, i.e., a diffuse background of other beams is also excited and has to be taken into account in the interpretation of the experimental results. For instance, additional beams may lead to modified phase shifts due to dynamical scattering in analogy to the axial case (cf. Section 2.3). When plugging in the geometric phase of the potential, we have the following coupled equations governing the behavior of the diffracted beam in the presence of strain

$$\frac{\partial}{\partial z}\begin{pmatrix} \tilde{\Psi}_0 \\ \tilde{\Psi}_g \end{pmatrix} = i \underbrace{\begin{pmatrix} 0 & \frac{e}{v}\tilde{\Phi}_{-g}e^{i\mathbf{g}\cdot\mathbf{u}(z)} \\ \frac{e}{v}\tilde{\Phi}_g e^{-i\mathbf{g}\cdot\mathbf{u}(z)} & s_g \end{pmatrix}}_{-\hat{H}} \begin{pmatrix} \tilde{\Psi}_0 \\ \tilde{\Psi}_g \end{pmatrix}. \qquad (6.26)$$

Accordingly, the geometric phase term does not violate the Hermitian nature of the Hamiltonian $\hat{H}$. Thus, the total norm is conserved upon propagation of the beam. Furthermore, we note a gauge freedom

$$\mathbf{u} \to \mathbf{u} + \mathbf{a} \quad \text{and} \quad \tilde{\Psi}_g \to e^{-i\mathbf{g}\cdot\mathbf{a}}\tilde{\Psi}_g \qquad (6.27)$$

reflecting the fact that the displacement is gauge dependent (i.e., depends on the deliberate choice of reference). In order to solve (6.26) we transform the above system of first order differential equations into a second order differential equation

$$\frac{d^2\tilde{\Psi}_g}{dz^2} - is_g\frac{d\tilde{\Psi}_g}{dz} + \frac{e^2}{v^2}\left|\tilde{\Phi}_g\right|^2 \tilde{\Psi}_g = -i\frac{\partial(\mathbf{g}\cdot\mathbf{u})}{\partial z}\frac{d\tilde{\Psi}_g}{dz} - \frac{\partial(\mathbf{g}\cdot\mathbf{u})}{\partial z}s_g\tilde{\Psi}_g \qquad (6.28)$$

amenable to a perturbation treatment.

An analytic solution to the paraxial scattering equations (6.26) or (6.28), in the presence of a z-dependent strain field, may only be obtained for special z-dependencies. The situation is greatly simplified if the right hand side of (6.28) is small, i.e.,

$$\frac{\partial(\mathbf{g}\cdot\mathbf{u})}{\partial z} \ll \frac{e}{v}\left|\tilde{\Phi}_g\right| \quad \text{and} \quad s_g \ll \frac{e}{v}\left|\tilde{\Phi}_g\right|. \qquad (6.29)$$

In that case a perturbation expansion is possible. The zeroth order (unperturbed) solution is obtained by solving

$$\frac{d^2\tilde{\Psi}_{1,2}}{dz^2} - is_g\frac{d\tilde{\Psi}_{1,2}}{dz} + \frac{e}{v^2}\left|\tilde{\Phi}_g\right|^2 \tilde{\Psi}_{1,2} = 0 \tag{6.30}$$

under appropriate boundary conditions. With

$$k_{1,2} = \frac{s_g}{2} \pm \kappa_g \tag{6.31}$$

and

$$\kappa_g = \sqrt{\frac{s_g^2}{4} + \frac{e^2}{v^2}\left|\tilde{\Phi}_g\right|^2} \tag{6.32}$$

the transmitted beam ($\tilde{\Psi}_1(0) = 1$ and $\tilde{\Psi}_1'(0) = 0$) reads

$$\tilde{\Psi}_1(z) = \frac{1}{k_1 - k_2}\left(k_1 \exp\left(ik_2 z\right) - k_2 \exp\left(ik_1 z\right)\right) \tag{6.33}$$

$$= \exp\left(i\frac{s_g}{2}z\right)\left(\cos\left(\kappa_g z\right) - \frac{is_g}{2\kappa_g}\sin\left(\kappa_g z\right)\right).$$

Similarly, the diffracted beam ($\tilde{\Psi}_2(0) = 0$ and $\tilde{\Psi}_2'(0) = -\frac{i\tilde{\Phi}_g}{v}$) is computed to

$$\tilde{\Psi}_2(z) = -\frac{k_1 k_2 v}{e\tilde{\Phi}_{-g}(k_1 - k_2)}\left(\exp\left(ik_1 z\right) - \exp\left(ik_2 z\right)\right) \tag{6.34}$$

$$= -\frac{ie\tilde{\Phi}_g}{v\kappa_g}\exp\left(i\frac{s_g}{2}z\right)\sin\left(\kappa_g z\right),$$

where we note the familiar contraction of the extinction length

$$\xi_g \equiv \pi/\sqrt{\frac{s_g^2}{4} + \frac{e^2}{v^2}\tilde{\Phi}_g\tilde{\Phi}_{-g}}, \tag{6.35}$$

if the excitation error deviates from zero (deviation from two-beam conditions).

We now proceed by noting that the two solutions of the second order differential equation form a fundamental system. The corresponding Wronskian is obtained from Abel's identity

$$W(\tilde{\Psi}_1, \tilde{\Psi}_2)(z) = \begin{vmatrix} \tilde{\Psi}_1(z) & \tilde{\Psi}_2(z) \\ \tilde{\Psi}_1'(z) & \tilde{\Psi}_2'(z) \end{vmatrix} \tag{6.36}$$

$$= W(0)\exp\left(-\int_0^z p(z')\,dz'\right)$$

$$= -\frac{ie\tilde{\Phi}_g}{v}\exp(is_g z).$$

Upon variation of parameters, a perturbed solution may be obtained by expanding the diffracted beam into the fundamental solutions

$$\tilde{\Psi}_g(z) \approx \underbrace{\tilde{\Psi}_2(z)}_{\text{0th order}} + \underbrace{A(z)\tilde{\Psi}_1(z) + B(z)\tilde{\Psi}_2(z)}_{\text{1st order}}, \tag{6.37}$$

where the z-dependent expansion parameters $A(z)$ and $B(z)$ may be integrated from the fundamental solutions respecting the boundary conditions $A(0) = B(0) = 0$. In the following, we only note the crucial steps of that computation. The first parameter reads

$$A(z) = \frac{v\kappa_g}{ie\tilde{\Phi}_g}\int_0^z \exp\left(-is_g z'\right)\tilde{\Psi}_2 \frac{\partial\,(\mathbf{g}\cdot\mathbf{u})}{\partial z'}\left(i\frac{d\tilde{\Psi}_2}{dz'} + s_g\tilde{\Psi}_2\right)dz' \tag{6.38}$$

$$\overset{\text{p.I.}}{=} \frac{ie\tilde{\Phi}_g}{v\kappa_g}\int_0^z (\mathbf{g}\cdot\mathbf{u})\left(\frac{s_g}{2}\sin\left(2\kappa_g z'\right) + i\kappa_G\cos\left(2\kappa_g z'\right)\right)dz'$$

$$- \frac{ie\tilde{\Phi}_g}{v\kappa_g^2}(\mathbf{g}\cdot\mathbf{u})\left(\frac{s_g}{2}\sin^2\left(\kappa_g z\right) + \frac{i}{2}\kappa_g\sin\left(2\kappa_g z\right)\right),$$

where a partial integration has been employed in the second step yielding a non-zero boundary term at the upper integration boundary. Note that $A(0) = 0$ as required. The second parameter $B(z)$ is obtained by a similar computation

$$B(z) = -\frac{v\kappa_g}{ie\tilde{\Phi}_g}\int_0^z \exp\left(-is_g z'\right)\tilde{\Psi}_1(z')\frac{\partial\,(\mathbf{g}\cdot\mathbf{u})}{\partial z'}\left(-i\frac{d\tilde{\Psi}_2}{dz} - s_g\tilde{\Psi}_2\right)dz' \tag{6.39}$$

$$\overset{\text{p.I.}}{=} \int_0^z (\mathbf{g}\cdot\mathbf{u})\left(-i\left(\kappa_g + \frac{s_g^2}{4\kappa_g}\right)\sin\left(2\kappa_g z'\right) + s_g\cos\left(2\kappa_g z'\right)\right)dz'$$

$$- \frac{1}{\kappa_g}(\mathbf{g}\cdot\mathbf{u})\left(i\kappa_g\cos^2\left(\kappa_g z\right) - s_g\sin\left(\kappa_g z\right)\cos\left(\kappa_g z\right) - \frac{is_g^2}{4\kappa_g}\sin^2\left(\kappa_g z\right)\right).$$

In this case the boundary term is nonzero at $z = 0$, requiring an additional integration constant to ensure $B(0) = 0$.

We proceed by restricting the discussion to small deviation parameters $s_g \ll \kappa_g$, neglecting terms with s_g as prefactor. We furthermore neglect the boundary term $\sim \sin(2\kappa_g z)$ in $A(z)$ by noting that the thickness range suitable for DFEH is restricted to some interval around half integers of the extinction length. Using $\cos(\alpha - \beta) = \cos(\alpha)\cos(\beta) - \sin(\alpha)\sin(\beta)$, we finally obtain

$$
\tilde{\Psi}_g(t) = -\frac{e\tilde{\Phi}_g}{v} \exp\left(i\frac{s_g}{2}z\right) \left(\frac{i}{\kappa_g}\sin(\kappa_g t)\right.
$$
$$
\left. + \cos(\kappa_g t) \int (\mathbf{g}\cdot\mathbf{u})\cos(2\kappa_g z)\mathrm{d}z + \sin(\kappa_g t)\int (\mathbf{g}\cdot\mathbf{u})\left(\sin(2\kappa_g z)\right)\mathrm{d}z\right)
$$
$$
= \tilde{\Psi}_2(t) \underbrace{\left(1 - i\kappa_G \frac{\int(\mathbf{g}\cdot\mathbf{u})\cos(\kappa_g(t-2z))\mathrm{d}z}{\sin(\kappa_g t)}\right)}_{\approx \exp(i\phi_g)}. \tag{6.40}
$$

The last expression may be expanded to a phase modulation

$$
\phi_g = -\int_0^t f_u^g(z)\,\mathbf{g}\cdot\mathbf{u}(z)\,\mathrm{d}z, \tag{6.41}
$$

where we define the weighting function

$$
f_u^g(z, t) \equiv \frac{\kappa_g \cos(\kappa_g(t-2z))}{\sin(\kappa_g t)}. \tag{6.42}
$$

Expressions (6.41) and (6.42) have been derived in Lubk, Javon, et al. (2014) by employing a slightly different Green's function approach. The variation of parameters used here has the advantage to clearly reveal the employed boundary conditions and to exhibit a simple explanation of the obtained result (Fig. 6.17). A similar projection law may be derived for the transmitted beam, with the crucial difference that the amplitude is modulated in the lowest order perturbation (not shown).

Alternatively, it is possible to derive the weighting function in a less formal way based on a z-dependent Master equation for the two-beam case. Here, one writes the intensity of the diffracted beam in the exit plane as a sum of partial beams, created at a certain depth in the crystal and phase-shifted with the corresponding geometric phase. Both the original transmitted (6.33) and diffracted beam (6.34) contribute to the final diffracted beam by locally exciting partial waves carrying the geometric phase information. The respective weight in the final result is determined

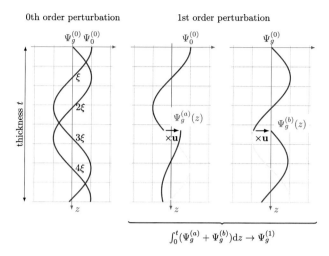

0th order perturbation   1st order perturbation

$$\int_0^t (\Psi_g^{(a)} + \Psi_g^{(b)}) \mathrm{d}z \rightarrow \Psi_g^{(1)}$$

Figure 6.17 Phenomenological derivation of the sinusoidal weighting function $f_u^g(z,t)$.

by the value of the transmitted (diffracted) beam at z multiplied with the value in the exit plane t of the diffracted (transmitted) beam created at z, i.e.,

$$f_{u,1}^g \sim \cos\left(-\kappa_g z\right) \cos\left(-\kappa_g (t-z)\right) \tag{6.43}$$

and

$$f_{u,2}^g \sim \sin\left(-\kappa_g z\right) \sin\left(-\kappa_g (t-z)\right). \tag{6.44}$$

Using again

$$\cos\left(-\kappa_g (t-2z)\right) = \cos\left(-\kappa_g (t-z)\right) \cos\left(-\kappa_g z\right) \\ + \sin\left(-\kappa_g (t-z)\right) \sin\left(-\kappa_g z\right) \tag{6.45}$$

and normalizing, the sum of the two contributions yields the weighting function (6.41). Similar to the more stringent perturbation approach, it was assumed that the original transmitted and diffracted beams are not modified by the scattering, i.e., the perturbation condition was used implicitly.

We will now discuss the general shape of the weighting function f_u^g of the diffracted beam and its influence on the measured phase in more detail. f_u^g depends on two variables, the thickness and the integration variable z (see Fig. 6.18). It is observed that f_u^g depends in an oscillating and symmetric manner on the distance from the specimen center. Moreover, $\int_0^t f_u^g(z)\,\mathrm{d}z = 1$, i.e., in the case of a constant displacement field in

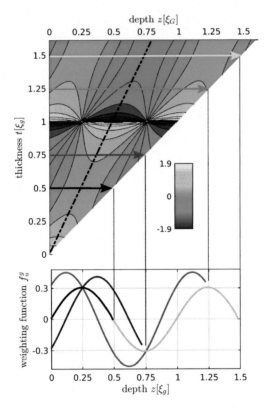

Figure 6.18 Weighting function $f_u^g(z, t)$ (color) with four 1D cuts at special thicknesses $t = 0.5, 0.75, 1.25, 1.5\,\xi_g$. Note that $f_u^g(0 < z < 0.5\,\xi_G, 0.5\,\xi_g) = f_u^g(0 < z < 0.5\,\xi_g, 1.5\,\xi_g)$. The black dashed/dotted line indicates the point of symmetry at $z = t/2$.

z-direction ($\mathbf{u}(z) = \mathbf{u}$) the reconstructed geometric phase corresponds exactly to the 2D displacement field of the crystal lattice, i.e., $\phi_g = -2\pi\,\mathbf{g}\cdot\mathbf{u}$.

We finally note the weighting function for the reconstructed phase of the diffracted beam with non-vanishing excitation error s_g:

$$f_u^g(z, t) = 2\pi\,\frac{\kappa_g \cos\left(\kappa_g(t - 2z)\right) - i\frac{s_g}{2}\sin\left(\kappa_g(t - 2z)\right)}{\sin\left(\kappa_g t\right)}. \tag{6.46}$$

Note that the additional antisymmetric weighting is imaginary, hence it introduces an amplitude (and no phase) modulation proportional to the antisymmetric part of the distortion. Consequently, a non-zero excitation error mainly affects the reconstructed phase by reducing the extinction length ξ_g (6.35). One can therefore fine-tune the extinction length in order to maximize the diffraction amplitude and thus the signal-to-noise ratio in the reconstructed phase (Lenz, 1988), which is common practice in DFEH experiments.

It has been verified experimentally that the above sinusoidal projection of the strain agrees better with experimentally reconstructed phases than the simple projection used before (Cherkashin, Lubk, Hÿtch, & Claverie, 2013; Javon et al., 2014). However, it is also noted that deviations between the theoretical and experimental behavior persist, e.g., because of a systematic violation of two-beam conditions in the vicinity of low-index zone axes (Javon et al., 2014). In preparation for the next section, Fig. 6.19 exhibits the level of agreement between the sinusoidal projection law and experiment at the example of a strain profile in a Si–SiGe multilayer obtained by DFEH (for the details of the experiment see Javon et al., 2014). Accordingly, a small difference in the adjusted deviation from exact two-beam conditions

$$s_g = \frac{g^2 + 2\left(\theta_{\mathrm{Bragg}} + \Delta\theta\right) k_{0z} g}{2k_{0z}} \tag{6.47}$$

as small as $\Delta\theta = 0.1$ mrad may produce a significant (and predictable) change of the reconstructed DFEH phase map.

6.5.2 Toward Strain Tomography

As noted in Javon et al. (2014), the weighted projection formalism suggests several tomographic reconstruction schemes eventually facilitating the recover of 3D strain distributions in strained crystals. In the following, we will shortly elaborate on the conventional tilt series approach before discussing the more promising variation of the incident beam tilt method. Conventional Radon transformation techniques discussed in previous sections allow inverting a set of line integrals under different tilt angles with a constant weighting function $f = 1$. Here, we face the problem of a non-constant and even nonlinear weighting function. Under such circumstances, an inversion of the projection, i.e., a unique tomographic reconstruction, may not exist anymore. Moreover, even if a solution exists, the problem may be ill-conditioned, rendering a solution experimentally unfeasible. The necessary and sufficient condition for the existence of the solution is that the nonlinear projections are bijective, i.e., different strain profiles yield different phases reconstructed from a dark-field holographic tomographic tilt series. It is currently not known, whether the above sinusoidal projection

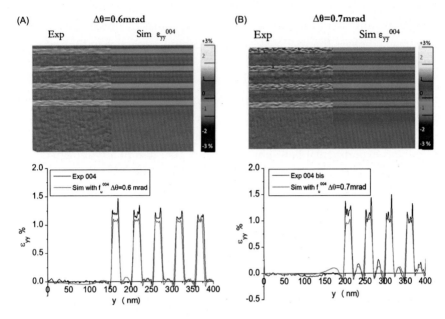

Figure 6.19 Comparison between the experimental strain map obtained from the (004) diffracted beam on a 145-nm-thick Si–Si$_{0.8}$Ge$_{0.2}$ multilayer specimen and the weighted projection of strain profiles simulated with finite element elastic modeling, taking into account (A) $\Delta\theta = 0.6$ mrad deviation from the exact Bragg orientation and (B) $\Delta\theta = 0.7$ mrad (Javon et al., 2014). The difference between experiment and simulation in the first Si$_{0.8}$Ge$_{0.2}$ layer may be attributed to a deviation of the strain relaxation in the theoretic model rather than to a failure of the weighting function formalism.

law is bijective and we shall not discuss this topic within this volume for the following reasons. First, the experimental challenges to be surmounted to record a dark-field holographic tilt series, are numerous and unresolved to date of this work. For instance, the inevitable variation of the thickness in projection direction modifies the diffracted beam intensity and the weighting function in an object dependent manner while tilting the sample. In order to guarantee well-defined scattering conditions (i.e., with zero excitation error) throughout the tilt series, the relative orientation of the sample and the beam have to be controlled with an accuracy in the range of 0.1 mrad, which is at the very edge of current goniometer technology. Second, the weighting function suggests an alternative tomographic reconstruction principle based on slightly varying the beam tilt with more optimistic prospects to date of this work. Thus, we elaborate on this scheme in the following.

We first note that the weighted projection of the displacement field into the reconstructed dark field phase resembles a cosine transform

$$\phi_g\left(\kappa_g\right) = -\frac{2\pi g \kappa_g}{\sin\left(\kappa_g t\right)} \int_0^t \cos\left(\kappa_g\left(t - 2z\right)\right) u\left(z\right) \mathrm{d}z, \qquad (6.48)$$

if the wave number κ_g of the weighting function may be drawn freely from the set $\{-N/2, -N/2 + 1, \ldots, N/2 - 1\} \pi/t$. The notable difference is the normalization of the projection integral, which, however, can be removed if the thickness of the sample is well-known (e.g., determined in advance). Consequently, the symmetric part of the displacement may be reconstructed from a sufficiently complete cosine series of phases

$$u\left(z\right) \sim \sum_{\kappa_g} \phi\left(\kappa_g\right) \cos\left(2\kappa_g z\right), \qquad (6.49)$$

which represents an alternative route toward the reconstruction of strain fields in 3D, therefore.

In contrast to the conventional tilt series tomography, the experimental effort is less demanding, since the beam tilt may be controlled with a significantly higher precision than the tilt of the specimen. However, there are also several challenges in the realization of this reconstruction principle. The most severe consists of the limited range of κ_g values accessible in the TEM. Only a small number of tilts corresponding to integer multiples of reciprocal lattice vectors $\mathbf{g}$ may be realized even at the most advanced TEM instruments to date of this work.[6] That effectively imposes a lower bound

$$\kappa_g > \kappa_{g\mathrm{min}} = \sqrt{\frac{e^2}{v^2}\left|\tilde{\Phi}_{g=\mathrm{max}}\right|^2} \qquad (6.50)$$

and an upper bound

$$\kappa_g < \kappa_{g\mathrm{max}} = \sqrt{\frac{s_g^2}{4} + \frac{e^2}{v^2}\left|\tilde{\Phi}_{g=\mathrm{min}}\right|^2} \qquad (6.51)$$

given by the scattering potential and the maximal excitation error s_g (see Fig. 6.20). The attainable z-resolution is then given by the maximal κ_g from

---

[6] The HITACHI I2TEM TEM equipped with a double stage for Lorentz microscopy permits the recording of dark-field holograms of {008}-diffracted beams in Si (Hÿtch, Gatel, Houdellier, Snoeck, & Ishizuka, 2012).

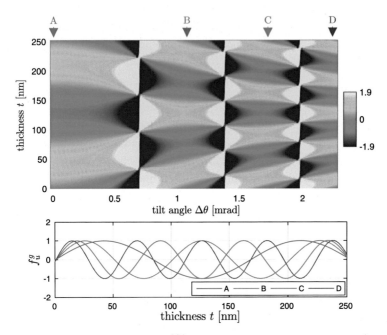

Figure 6.20 Weighting function $f_u^{g=004}$ in silicon in dependence of the deviation from Bragg conditions $\Delta\theta$. 1D line scans are plotted for $(2n+1)\xi_g = 253$ nm with $n = 1, 2, 3, 4$ corresponding to two-beam conditions with maximally excited diffracted beam.

the set $\{-N/2, -N/2+1, \ldots, N/2-1\}\pi/t$. Moreover, the reconstruction will suffer from aliasing artifacts, if that maximal κ_g is below the band limit of the underlying strain variation.

To improve the quality of the reconstruction in that case, additional regularization becomes necessary. In case of step-like deformation fields with sharp interfaces, for instance, compressive sensing in the space of strain, i.e., total variation minimization (cf. Section 3.5), has been proven highly successful in recovering functions from incomplete Fourier series (Candes, Romberg, & Tao, 2006). Similarly, regularity (smoothness) constraints may be applicable. Moreover, the reconstruction may be stabilized, if the strain fields obey some analytic model depending only on a limited number of parameters, which can then be fitted to the experimental results.

We summarize that the reconstruction of sufficiently well-behaved (e.g., slowly varying) strain fields may be possible with the above Fourier synthesis method, provided sufficient experimental control over the sinusoidal weighting function. To date of this volume, a proof-of-concept tomographic reconstruction of 3D strain fields, e.g., in strained transistor channels, is still pending.

REFERENCES

Appenzeller, J., Knoch, J., Bjork, M., Riel, H., Schmid, H., & Riess, W. (2008). Toward nanowire electronics. *IEEE Transactions on Electron Devices, 55*(11), 2827–2845.

Bals, S., Van Tendeloo, G., & Kisielowski, C. (2006). A new approach for electron tomography: Annular dark-field transmission electron microscopy. *Advanced Materials, 18*, 892–895.

Barnard, J. S., Sharp, J., Tong, J. R., & Midgley, P. A. (2006). High-resolution three-dimensional imaging of dislocations. *Science, 313*(5785), 319.

Barton, B., Joos, F., & Schröder, R. R. (2008). Improved specimen reconstruction by Hilbert phase contrast tomography. *Journal of Structural Biology, 164*(2), 210–220.

Batenburg, K. J., Bals, S., Sijbers, J., Kübel, C., Midgley, P., Hernandez, J., ... Tendeloo, G. V. (2009). 3D imaging of nanomaterials by discrete tomography. *Ultramicroscopy, 109*(6), 730–740.

Baykan, M. O., Thompson, S. E., & Nishida, T. (2010). Strain effects on three-dimensional, two-dimensional, and one-dimensional silicon logic devices: predicting the future of strained silicon. *Journal of Applied Physics, 108*(9), 093716.

Béché, A., Rouvière, J., Barnes, J., & Cooper, D. (2011). Dark field electron holography for strain measurement. *Ultramicroscopy, 111*(3), 227–238.

Beleggia, M., Kasama, T., & Dunin-Borkowski, R. E. (2010). The quantitative measurement of magnetic moments from phase images of nanoparticles and nanostructures – I. Fundamentals. *Ultramicroscopy, 110*(5), 425–432.

Biermans, E., Molina, L., Batenburg, K. J., Bals, S., & Van Tendeloo, G. (2010). Measuring porosity at the nanoscale by quantitative electron tomography. *Nano Letters, 10*(12), 5014–5019.

Blanes, S., Casas, F., Oteo, J., & Ros, J. (2009). The Magnus expansion and some of its applications. *Physics Reports, 470*(5–6), 151–238.

Börrnert, F., Müller, H., Linck, M., Horst, A., Kirkland, A. I., Büchner, B., & Lichte, H. (2015). Approaching the "Lab in the Gap": first results from a versatile in-situ (S)TEM. In *Proceedings of microscopy & microanalysis 2015: Vol. 21* (pp. 99–100).

Brandt, S., Heikkonen, J., & Engelhardt, P. (2001). Automatic alignment of transmission electron microscope tilt series without fiducial markers. *Journal of Structural Biology, 136*(3), 201–213.

Brandt, S. S. (2006). Markerless alignment in electron tomography. In J. Frank (Ed.), *Electron tomography* (pp. 187–215). Berlin: Springer.

Bryllert, T., Wernersson, L.-E., Löwgren, T., & Samuelson, L. (2006). Vertical wrap-gated nanowire transistors. *Nanotechnology, 17*(11), S227–S230.

Candes, E. J., Romberg, J., & Tao, T. (2006). Robust uncertainty principles: Exact signal reconstruction from highly incomplete frequency information. *IEEE Transactions on Information Theory, 52*(2), 489–509.

Chen, B., & Stamnes, J. J. (1998). Validity of diffraction tomography based on the first Born and the first Rytov approximations. *Applied Optics, 37*(14), 2996–3006.

Chen, C.-C., Zhu, C., White, E. R., Chiu, C.-Y., Scott, M. C., Regan, B. C., ... Miao, J. (2013). Three-dimensional imaging of dislocations in a nanoparticle at atomic resolution. *Nature, 496*(7443), 74–77.

Chen, F. R., Kisielowski, C., & Van Dyck, D. (2015). 3D reconstruction of nanocrystalline particles from a single projection. *Micron*, *68*, 59–65.

Cherkashin, N., Lubk, S. R., Hÿtch, M. J., & Claverie, A. (2013). Strain in hydrogen-implanted Si investigated using dark-field electron holography. *Applied Physics Express*, *6*(9), 091301.

Chung, S., Smith, D. J., & McCartney, M. R. (2007). Determination of the inelastic mean-free-path and mean inner potential for AlAs and GaAs using off-axis electron holography and convergent beam electron diffraction. *Microscopy and Microanalysis*, *13*(05), 329–335.

Chuvilin, A., Kaiser, U., de Robillard, Q., & Engelmann, H.-J. (2005). On the origin of HOLZ lines splitting near interfaces: Multislice simulation of CBED patterns. *Journal of Electron Microscopy*, *54*, 515–517.

Clement, L., Pantel, R., Kwakman, L. F. T., & Rouvière, J. L. (2004). Strain measurements by convergent-beam electron diffraction: The importance of stress relaxation in lamella preparations. In *Proceedings of EMC 2004: Vol. 85* (p. 651).

Conrad, J. (2015). Reproducibility: Don't cry wolf. *Nature*, *523*, 27–28.

Conzatti, F., Serra, N., Esseni, D., De Michielis, M., Paussa, A., Palestri, P., . . . Lander, R. (2011). Investigation of strain engineering in FinFETs comprising experimental analysis and numerical simulations. *IEEE Transactions on Electron Devices*, *58*(6), 1583–1593.

Cooper, D., Barnes, J. P., Hartmann, J. M., Béché, A., & Rouvière, J. (2009). Dark field electron holography for quantitative strain measurements with nanometer-scale spatial resolution. *Applied Physics Letters*, *95*, 053501.

Cooper, D., Béché, A., Hartmann, J., Carron, V., & Rouvière, J. (2010). Strain evolution during the silicidation of nanometer-scale SiGe semiconductor devices studied by dark field electron holography. *Applied Physics Letters*, *96*, 113508.

Cooper, D., de la Peña, F., Béché, A., Rouvière, J.-L., Servanton, G., Pantel, R., & Morin, P. (2011). Field mapping with nanometer-scale resolution for the next generation of electronic devices. *Nano Letters*, *11*(11), 4585–4590.

Cooper, D., Rouvière, J., Béché, A., Kadkhodazadeh, S., Semenova, E., Yvind, K., & Dunin-Borkowski, R. (2011). Quantitative strain mapping of InAs/InP quantum dots with 1 nm spatial resolution using dark field electron holography. *Applied Physics Letters*, *99*, 261911.

Danev, R., Kanamaru, S., Marko, M., & Nagayama, K. (2010). Zernike phase contrast cryo-electron tomography. *Journal of Structural Biology*, *171*(2), 174–181.

Dengler, J. (1989). A multi-resolution approach to the 3D reconstruction from an electron microscope tilt series solving the alignment problem without gold particles. *Ultramicroscopy*, *30*(3), 337–348.

Denneulin, T., Houdellier, F., & Hÿtch, M. (2016). Differential phase-contrast dark-field electron holography for strain mapping. *Ultramicroscopy*, *160*, 98–109.

Denneulin, T., Rouvière, J.-L., Béché, A., Py, M., Barnes, J. P., Rochat, N., . . . Cooper, D. (2011). The reduction of the substitutional C content in annealed Si/SiGeC super-lattices studied by dark-field electron holography. *Semiconductor Science and Technology*, *26*, 125010.

Devaney, A. J. (1981). Inverse-scattering theory within the Rytov approximation. *Optics Letters*, *6*(8), 374–376.

Dunin-Borkowski, R. E., McCartney, M. R., Smith, D. J., & Parkin, S. S. P. (1998). Towards quantitative electron holography of magnetic thin films using in situ magnetization reversal. *Ultramicroscopy*, *74*(1–2), 61–73.

Egerton, R. F. (1996). *Electron energy-loss spectroscopy in the electron microscope.* Plenum Press.

Ferrier, R. P., Liu, Y., Martin, J., & Arnoldussen, T. (1995). Electron beam tomography of magnetic recording head fields. *Journal of Magnetism and Magnetic Materials, 149*(3), 387–397.

Frank, J. (Ed.). (2008). *Electron tomography: Methods for three-dimensional visualization of structures in the cell.* Springer.

Frank, J., & McEwen, B. F. (1992). Alignment by cross-correlation. In J. Frank (Ed.), *Electron tomography* (pp. 205–214). New York: Plenum.

Friedrich, D., Schmidt, B., Heinig, K. H., Liedke, B., Mücklich, A., Hübner, R., ... Mikolajick, T. (2013). Sponge-like Si–SiO$_2$ nanocomposite – morphology studies of spinodally decomposed silicon-rich oxide. *Applied Physics Letters, 103*(13), 131911.

Friedrich, H., de Jongh, P. E., Verkleij, A. J., & de Jong, K. P. (2009). Electron tomography for heterogeneous catalysts and related nanostructured materials. *Chemical Reviews, 109,* 1613.

Friedrich, H., McCartney, M. R., & Buseck, P. R. (2005). Comparison of intensity distributions in tomograms from BF TEM, ADF STEM, HAADF STEM, and calculated tilt series. *Ultramicroscopy, 106*(1), 18–27.

Fujita, T., & Chen, M. (2009). Quantitative electron holographic tomography for a spherical object. *Journal of Electron Microscopy, 58*(5), 301–304.

Gallo, E. M., Chen, G., Currie, M., McGuckin, T., Prete, P., Lovergine, N., ... Spanier, J. E. (2011). Picosecond response times in GaAs/AlGaAs core/shell nanowire-based photodetectors. *Applied Physics Letters, 98*(24), 241113.

Gao, Y., Shindo, D., Bao, Y., & Krishnan, K. (2007). Electron holography of core–shell Co/CoO spherical nanocrystals. *Applied Physics Letters, 90*(23), 233105.

Gatel, C., Lubk, A., Pozzi, G., Snoeck, E., & Hÿtch, M. (2013). Counting elementary charges on nanoparticles by electron holography. *Physical Review Letters, 111*(2), 025501.

Ghiglia, D. C., & Pritt, M. D. (1998). *Two-dimensional phase unwrapping.* Wiley.

Glaeser, R. M. (1985). Electron crystallography of biological macromolecules. *Annual Review of Physical Chemistry, 36*(1), 243–275.

Goris, B., Bals, S., Van den Broek, W., Carbó-Argibay, E., Gómez-Graña, S., Liz-Marzán, L. M., & Van Tendeloo, G. (2012). Atomic-scale determination of surface facets in gold nanorods. *Nature Materials, 11*(11), 930–935.

Goris, B., Bals, S., Van den Broek, W., Verbeeck, J., & Van Tendeloo, G. (2011). Exploring different inelastic projection mechanisms for electron tomography. *Ultramicroscopy, 111*(8), 1262–1267.

Goris, B., Van den Broek, W., Batenburg, K., Heidari Mezerji, H., & Bals, S. (2012). Electron tomography based on a total variation minimization reconstruction technique. *Ultramicroscopy, 113,* 120–130.

Grillo, V. (2009). The effect of surface strain relaxation on HAADF imaging. *Ultramicroscopy, 109,* 1453–1464.

Harscher, A., & Lichte, H. (1996). Experimental study of amplitude and phase detection limits in electron holography. *Ultramicroscopy, 64,* 57–66.

Harscher, A., & Lichte, H. (1997). Determination of mean internal potential and mean free wavelength for inelastic scattering of vitrified iron by electron holography. *European Journal of Cell Biology, 74,* 7.

Hartmann, J. M., Sanchez, L., Daele, W. V. D., Abbadie, A., Baud, L., Truche, R., ... Cristoloveanu, S. (2010). Fabrication, structural and electrical properties of compressively strained Ge-on-insulator substrates. *Semiconductor Science and Technology, 25*(7), 075010.

Hata, S., Kimura, K., Gao, H., Matsumura, S., Doi, M., Moritani, T., ... Midgley, P. A. (2008). Electron tomography imaging and analysis of γ' and γ domains in Ni-based superalloys. *Advanced Materials, 20*(10), 1905–1909.

Heiss, M., Fontana, Y., Gustafsson, A., Wüst, G., Magen, C., O'Regan, D. D., ... Fontcuberta i Morral, A. (2013). Self-assembled quantum dots in a nanowire system for quantum photonics. *Nature Materials, 12*(5), 439–444.

Hindson, J. C., Saghi, Z., Hernandez-Garrido, J. C., Midgley, P. A., & Greenham, N.C. (2011). Morphological study of nanoparticle-polymer solar cells using high-angle annular dark-field electron tomography. *Nano Letters, 11*(2), 904.

Houben, L., & Bar Sadan, M. (2011). Refinement procedure for the image alignment in high-resolution electron tomography. *Ultramicroscopy, 111*, 1512–1520.

Houdellier, F., Altibelli, A., Roucau, C., & Casanove, M.-J. (2008). New approach for the dynamical simulation of CBED patterns in heavily strained specimens. *Ultramicroscopy, 108*, 426–432.

Houdellier, F., Roucau, C., Clément, L., Rouvière, J., & Casanove, M. (2006). Quantitative analysis of HOLZ line splitting in CBED patterns of epitaxially strained layers. *Ultramicroscopy, 106*(10), 951–959.

Howie, A. (1979). Image contrast and localized signal selection techniques. *Journal of Microscopy, 117*(1), 11–23.

Howie, A., & Whelan, M. J. (1961). Dynamical theory of crystal lattice defects. II. The development of a dynamical theory. *Proceedings of the Royal Society of London Series A, 263*, 217–237.

Howie, A., & Whelan, M. J. (1962). Dynamical theory of crystal lattice defects. III. Results and experimental confirmation of dynamical theory of dislocation image contrast. *Proceedings of the Royal Society of London Series A, 267*, 206–230.

Hüe, F., Hÿtch, M., Bender, H., Houdellier, F., & Claverie, A. (2008). Direct mapping of strain in a strained silicon transistor by high-resolution electron microscopy. *Physical Review Letters, 100*(15), 156602.

Hüe, F., Hÿtch, M., Houdellier, F., Bender, H., & Claverie, A. (2009). Strain mapping of tensiley strained silicon transistors with embedded $Si_{1-y}C_y$ source and drain by dark-field holography. *Applied Physics Letters, 95*(7), 073103.

Hÿtch, M., Gatel, C., Houdellier, F., Snoeck, E., & Ishizuka, K. (2012). Darkfield electron holography for strain mapping at the nanoscale. *Microscopy and Analysis, 121*, 6–10.

Hÿtch, M., Houdellier, F., Hue, F., & Snoeck, E. (2008). Nanoscale holographic interferometry for strain measurements in electronic devices. *Nature, 453*(7198), 1086–1089.

Hÿtch, M., Houdellier, F., Hüe, F., & Snoeck, E. (2011). Dark-field electron holography for the measurement of geometric phase. *Ultramicroscopy, 111*(8), 1328–1337.

Javon, E., Lubk, A., Cours, R., Reboh, S., Cherkashin, N., Houdellier, F., ... Hÿtch, M. (2014). Dynamical effects in strain measurements by dark-field electron holography. *Ultramicroscopy, 147*, 70–85.

Jia, C. L., Mi, S. B., Barthel, J., Wang, D. W., Dunin-Borkowski, R. E., Urban, K. W., & Thust, A. (2014). Determination of the 3D shape of a nanoscale crystal with atomic resolution from a single image. *Nature Materials, 13*, 1044–1049.

Jin-Phillipp, N. Y., Koch, C. T., & van Aken, P. A. (2011). Toward quantitative core-loss EFTEM tomography. *Ultramicroscopy, 111*(8), 1255–1261.

Jing, Z., & Sachs, F. (1991). Alignment of tomographic projections using an incomplete set of fiducial markers. *Ultramicroscopy, 35*(1), 37–43.

Jinnai, H., & Spontak, R. J. (2009). Transmission electron microtomography in polymer research. *Polymer*, *50*(5), 1067–1087.

Johansson, J., Karlsson, L. S., Svensson, C. P. T., Mårtensson, T., Wacaser, B. A., Deppert, K., ... Seifert, W. (2007). The structure of <111>B oriented GaP nanowires. *Journal of Crystal Growth*, *298*, 635–639.

Kak, A. C., & Slaney, M. (1988). *Principles of computerized tomographic imaging*. IEEE Press.

Kasama, T., Dunin-Borkowski, R. E., & Beleggia, M. (2011). Electron holography of magnetic materials. In F. Monroy (Ed.), *Holography – different fields of application*. InTech.

Kobayashi, A., Fujigaya, T., Itoh, M., Taguchi, T., & Takano, H. (2009). Technical note: A tool for determining rotational tilt axis with or without fiducial markers. *Ultramicroscopy*, *110*(1), 1–6.

Koch, C. T., Özdöl, V. B., & van Aken, P. A. (2010). An efficient, simple, and precise way to map strain with nanometer resolution in semiconductor devices. *Applied Physics Letters*, *96*(9), 091901.

Koster, A. J., Ziese, U., Verkleij, A. J., Janssen, A. H., & de Jong, K. P. (2000). Three-dimensional transmission electron microscopy: A novel imaging and characterization technique with nanometer scale resolution for materials science. *The Journal of Physical Chemistry B*, *104*(40), 9368–9370.

Krehl, J., & Lubk, A. (2015a). Incorporating Fresnel propagation into electron holographic tomography. In *Microscopy conference*.

Krehl, J., & Lubk, A. (2015b). Prospects of linear reconstruction in atomic resolution electron holographic tomography. *Ultramicroscopy*, *150*, 65–70.

Krogstrup, P., Jorgensen, H. I., Heiss, M., Demichel, O., Holm, J. V., Aagesen, M., ... Fontcuberta i Morral, A. (2013). Single-nanowire solar cells beyond the Shockley–Queisser limit. *Nature Photonics*, *7*(4), 306–310.

Kruse, P., Schowalter, M., Lamoen, D., Rosenauer, A., & Gerthsen, D. (2006). Determination of the mean inner potential in III–V semiconductors, Si and Ge by density functional theory and electron holography. *Ultramicroscopy*, *106*(2), 105–113.

Kuebel, C., Voigt, A., Schoenmakers, R., Otten, M., Su, D., Lee, T.-C., ... Bradley, J. (2005). Recent advances in electron tomography: TEM and HAADF-STEM tomography for materials science and semiconductor applications. *Microscopy and Microanalysis*, *11*(05), 378–400.

Lai, G., Hirayama, T., Fukuhara, A., Ishizuka, K., Tanji, T., & Tonomura, A. (1994). Three-dimensional reconstruction of magnetic vector fields using electron-holographic interferometry. *Journal of Applied Physics*, *75*, 4593–4598.

Lai, G., Hirayama, T., Ishizuka, K., & Tonomura, A. (1994). Three-dimensional reconstruction of electric-potential distribution in electron-holographic interferometry. *Journal of Applied Optics*, *33*, 829–833.

Lavrijsen, R., Lee, J.-H., Fernandez-Pacheco, A., Petit, D. C. M. C., Mansell, R., & Cowburn, R. P. (2013). Magnetic ratchet for three-dimensional spintronic memory and logic. *Nature*, *493*(7434), 647–650.

Leapman, R. D., Kocsis, E., Zhang, G., Talbot, T. L., & Laquerriere, P. (2004). Three-dimensional distributions of elements in biological samples by energy-filtered electron tomography. *Ultramicroscopy*, *100*(1–2), 115–125.

Lehmann, M. (1992). *Eine schnelle alternative Methode zur Rekonstruktion von Bildebenen-off-Axis Elektronenhologrammen* (Diplomarbeit). University of Tübingen.

Lenk, A., Lichte, H., & Muehle, U. (2005). 2D-mapping of dopant distribution in deep sub micron CMOS devices by electron holography using adapted FIB-preparation. *Journal of Electron Microscopy*, *54*, 351–359.

Lenz, F. (1988). Statistics of phase and contrast determination in electron holograms. *Optik*, *79*, 13–14.

Li, S. P., Peyrade, D., Natali, M., Lebib, A., Chen, Y., Ebels, U., . . . Ounadjela, K. (2001). Flux closure structures in cobalt rings. *Physical Review Letters*, *86*(6), 1102–1105.

Li, T., Liang, Z., Singanallur, J. V., Satogata, T. J., Williams, D. C., & Schulte, R. W. (2006). Reconstruction for proton computed tomography by tracing proton trajectories: A Monte Carlo study. *Medical Physics*, *33*(3), 699–706.

Lichte, H., Börrnert, F., Lenk, A., Lubk, A., Röder, F., Sickmann, J., . . . Wolf, D. (2013). Electron holography for fields in solids: Problems and progress. *Ultramicroscopy*, *134*, 126–134.

Lichte, H., & Lehmann, M. (2008). Electron holography – basics and applications. *Reports on Progress in Physics*, *71*(1), 016102.

Liu, E.-S., Nah, J., Varahramyan, K. M., & Tutuc, E. (2010). Lateral spin injection in germanium nanowires. *Nano Letters*, *10*(9), 3297–3301.

Liu, Y., Penczek, P. A., McEwen, B. F., & Frank, J. (1995). A marker-free alignment method for electron tomography. *Ultramicroscopy*, *58*(3–4), 393–402.

Loos, J., Sourty, E., Lu, K., Freitag, B., Tang, D., & Wall, D. (2009). Electron tomography on micrometer-thick specimens with nanometer resolution. *Nano Letters*, *9*(4), 1704–1708.

Lubk, A. (2010). *Quantitative off-axis electron holography and (multi-)ferroic interfaces* (PhD thesis). Technische Universität Dresden.

Lubk, A., Guzzinati, G., Börrnert, F., & Verbeeck, J. (2013). Transport of intensity phase retrieval of arbitrary wave fields including vortices. *Physical Review Letters*, *111*(17), 173902.

Lubk, A., Javon, E., Cherkashin, N., Reboh, S., Gatel, C., & Hÿtch, M. (2014). Dynamic scattering theory for dark-field electron holography of 3D strain fields. *Ultramicroscopy*, *136*, 42–49.

Lubk, A., Wolf, D., Kern, F., Röder, F., Prete, P., Lovergine, N., & Lichte, H. (2014). Nanoscale three-dimensional reconstruction of elastic and inelastic mean free path lengths by electron holographic tomography. *Applied Physics Letters*, *105*(17), 173101.

Lubk, A., Wolf, D., Prete, P., Lovergine, N., Niermann, T., Sturm, S., & Lichte, H. (2014). Nanometer-scale tomographic reconstruction of three-dimensional electrostatic potentials in GaAs/AlGaAs core–shell nanowires. *Physical Review B*, *90*(12), 125404.

Lubk, A., Wolf, D., Simon, P., Wang, C., Sturm, S., & Felser, C. (2014). Nanoscale three-dimensional reconstruction of electric and magnetic stray fields around nanowires. *Applied Physics Letters*, *105*(17), 173110.

Lucic, V., Foerster, F., & Baumeister, W. (2005). Structural studies by electron tomography: From cells to molecules. *Annual Review of Biochemistry*, *74*(1), 833–865.

Mastronarde, D. N. (1997). Dual-axis tomography: An approach with alignment methods that preserve resolution. *Journal of Structural Biology*, *120*(3), 343–352.

Mastronarde, D. N. (2006). Fiducial marker and hybrid alignment methods for single- and double-axis tomography. In J. Frank (Ed.), *Electron tomography* (pp. 163–185). Berlin: Springer.

Matteucci, G., Missiroli, G. F., & Pozzi, G. (2002). Electron holography of long-range electrostatic fields. In P. W. Hawkes (Ed.), *Advances in imaging and electron physics: Vol. 122. Electron microscopy and holography II*. Elsevier.

McCartney, M. R., Agarwal, N., Chung, S., Cullen, D. A., Han, M.-G., He, K., . . . Smith, D. J. (2010). Quantitative phase imaging of nanoscale electrostatic and magnetic fields using off-axis electron holography. *Ultramicroscopy*, *110*(5), 375–382.

McCartney, M. R., & Smith, D. J. (1994). Direct observation of potential distribution across Si/Si p–n junctions using off-axis electron holography. *Applied Physics Letters, 65*(20), 2603.

McIntosh, R., Nicastro, D., & Mastronarde, D. (2005). New views of cells in 3D: An introduction to electron tomography. *Trends in Cell Biology, 15*(1), 43–51.

Midgley, P. A., Ward, E. P. W., Hungría, A. B., & Thomas, J. M. (2007). Nanotomography in the chemical, biological and materials sciences. *Chemical Society Reviews, 36,* 1477–1494.

Midgley, P. A., & Weyland, M. (2003). 3D electron microscopy in the physical sciences: The development of Z-contrast and EFTEM tomography. *Ultramicroscopy, 96*(3–4), 413–431.

Midgley, P. A., Weyland, M., Yates, T. J. V., Arslan, I., Dunin-Borkowski, R. E., & Thomas, J. M. (2006). Nanoscale scanning transmission electron tomography. *Journal of Microscopy, 223*(3), 185–190.

Möbus, G., Doole, R. C., & Inkson, B. J. (2003). Spectroscopic electron tomography. *Ultramicroscopy, 96*(3–4), 433–451.

Mongillo, M., Spathis, P., Katsaros, G., Gentile, P., & De Franceschi, S. (2012). Multifunctional devices and logic gates with undoped silicon nanowires. *Nano Letters, 12*(6), 3074–3079.

Müller, P., Schürmann, M., & Guck, J. (2015). The theory of diffraction tomography. ArXiv e-print.

Nasirpouri, F., & Nogaret, A. (2011). *Nanomagnetism and spintronics – fabrication, materials, characterization and applications.* World Scientific.

Nickell, S. (2001). *Elektronentomographische Abbildung eiseingebetteter prokaryotischer Zellen* (PhD thesis).

Paige, C. C., & Saunders, M. A. (1982). LSQR: An algorithm for sparse linear equations and sparse least squares. *ACM Transactions on Mathematical Software, 8*(1), 43–71.

Parkin, S. S. P., Hayashi, M., & Thomas, L. (2008). Magnetic domain-wall racetrack memory. *Science, 320*(5873), 190–194.

Penczek, P., Marko, M., Buttle, K., & Frank, J. (1995). Double-tilt electron tomography. *Ultramicroscopy, 60*(3), 393–410.

Pennycook, S. J., & Nellist, P. D. (Eds.). (2011). *Scanning transmission electron microscopy.* Springer.

Perassi, E. M., Hernandez-Garrido, J. C., Moreno, M. S., Encina, E. R., Coronado, E.A., & Midgley, P. A. (2010). Using highly accurate 3D nanometrology to model the optical properties of highly irregular nanoparticles: A powerful tool for rational design of plasmonic devices. *Nano Letters, 10*(6), 2097.

Perea, D. E., Lensch, J. L., May, S. J., Wessels, B. W., & Lauhon, L. J. (2006). Composition analysis of single semiconductor nanowires using pulsed-laser atom probe tomography. *Applied Physics A: Materials Science & Processing, 85*(3), 271–275.

Perovic, D. D., Rossouw, C. J., & Howie, A. (1993). Imaging elastic strains in high-angle annular dark-field scanning-transmission electron microscopy. *Ultramicroscopy, 52,* 353–359.

Petford-Long, A. K., & De Graef, M. (2002). Lorentz microscopy. In *Characterization of materials.* John Wiley & Sons, Inc.

Phatak, C., Petford-Long, A. K., & De Graef, M. (2010). Three-dimensional study of the vector potential of magnetic structures. *Physical Review Letters, 104*(25), 253901.

Pozzi, G. (2016). *Particles and waves in electron optics and microscopy. Advances in Imaging and Electron Physics: Vol. 194.* Cambridge: Academic Press.

Pozzi, G., Beleggia, M., Kasama, T., & Dunin-Borkowski, R. E. (2014). Interferometric methods for mapping static electric and magnetic fields. *Comptes Rendus. Physique, 15,* 126–139.

Prete, P., Marzo, F., Paiano, P., Lovergine, N., Salviati, G., Lazzarini, L., & Sekiguchi, T. (2008). Luminescence of GaAs/AlGaAs core–shell nanowires grown by MOVPE using tertiarybutylarsine. *Journal of Crystal Growth, 310*(23), 5114–5118.

Reimer, L. (Ed.). (1989). *Transmission electron microscopy.* Springer Verlag.

Rez, P., & Treacy, M. M. J. (2013). Three-dimensional imaging of dislocations. *Nature, 503*(7476), E1.

Röder, F., Houdellier, F., Denneulin, T., Snoeck, E., & Hÿtch, M. (2016). Realization of a tilted reference wave for electron holography by means of a condenser biprism. *Ultramicroscopy, 161,* 23–40.

Röder, F., & Lubk, A. (2014). Transfer and reconstruction of the density matrix in off-axis electron holography. *Ultramicroscopy, 146,* 103–116.

Röder, F., & Lubk, A. (2015). A proposal for the holographic correction of incoherent aberrations by tilted reference waves. *Ultramicroscopy, 152,* 63–74.

Rudolph, D., Funk, S., Doblinger, M., Morkötter, S., Hertenberger, S., Schweickert, L., ... Koblmöller, G. (2013). Spontaneous alloy composition ordering in GaAs–AlGaAs core/shell nanowires. *Nano Letters, 13*(4), 1522–1527.

Saghi, Z., Xu, X., & Moebus, G. (2009). Model based atomic resolution tomography. *Journal of Applied Physics, 106*(2), 024304.

Schönhense, G. (1999). Imaging of magnetic structures by photoemission electron microscopy. *Journal of Physics: Condensed Matter, 11*(48), 9517.

Schowalter, M., Rosenauer, A., Lamoen, D., Kruse, P., & Gerthsen, D. (2006). Ab initio computation of the mean inner Coulomb potential of wurtzite-type semiconductors and gold. *Applied Physics Letters, 88*(23), 232108.

Scott, M. C., Chen, C.-C., Mecklenburg, M., Zhu, C., Xu, R., Ercius, P., ... Miao, J. (2012). Electron tomography at 2.4-angstrom resolution. *Nature, 483*(7390), 444–447.

Shinjo, T. (2013). *Nanomagnetism and spintronics.* Elsevier Science.

Shinjo, T., Okuno, T., Hassdorf, R., Shigeto, K., & Ono, T. (2000). Magnetic vortex core observation in circular dots of permalloy. *Science, 289*(5481), 930–932.

Sickmann, J., Formánek, P., Linck, M., Muehle, U., & Lichte, H. (2011). Imaging modes for potential mapping in semiconductor devices by electron holography with improved lateral resolution. *Ultramicroscopy, 111*(4), 290–302.

Sköld, N., Karlsson, L. S., Larsson, M. W., Pistol, M.-E., Seifert, W., Trägardh, J., & Samuelson, L. (2005). Growth and optical properties of strained GaAs–$Ga_xIn_{1-x}P$ core–shell nanowires. *Nano Letters, 5*(10), 1943–1947.

Slaney, M., Kak, A. C., & Larsen, L. E. (1984). Limitations of imaging with first-order diffraction tomography. *IEEE Transactions on Microwave Theory and Techniques, 32,* 860–874.

Smith, G. H., & Burge, R. E. (1963). A theoretical investigation of plural and multiple scattering of electrons by amorphous films, with special reference to image contrast in the electron microscope. *Proceedings of the Physical Society, 81*(4), 612.

Song, K., Shin, G.-Y., Kim, J. K., Oh, S. H., & Koch, C. T. (2013). Strain mapping of LED devices by dark-field inline electron holography: Comparison between deterministic and iterative phase retrieval approaches. *Ultramicroscopy, 127,* 119–125.

Spontak, R. J., Williams, M. C., & Agard, D. A. (1988). Three-dimensional study of cylindrical morphology in a styrene–butadiene–styrene block copolymer. *Polymer, 29*(3), 387–395.

Stamps, R. L., Breitkreutz, S., Åkerman, J., Chumak, A. V., Otani, Y., Bauer, G. E. W., ... Hillebrands, B. (2014). The 2014 magnetism roadmap. *Journal of Physics D: Applied Physics, 47*(33), 333001.

Stolojan, V., & Dunin-Borkowski, R. E. (2001). Three-Dimensional Magnetic Fields of Nanoscale Elements Determined by Electron-Holographic Tomography. Technical Report.

Sun, Y., Thompson, S. E., & Nishida, T. (2007). Physics of strain effects in semiconductors and metal-oxide-semiconductor field-effect transistors. *Journal of Applied Physics, 101*(10), 104503.

Takeguchi, M., Shimojo, M., & Furuya, K. (2005). Fabrication of magnetic nanostructures using electron beam induced chemical vapour deposition. *Nanotechnology, 16,* 1321–1325.

Tanigaki, T., Takahashi, Y., Shimakura, T., Akashi, T., Tsuneta, R., Sugawara, A., & Shindo, D. (2015). Three-dimensional observation of magnetic vortex cores in stacked ferromagnetic discs. *Nano Letters, 15*(2), 1309–1314.

Thomas, J. M. (2005). A revolution in electron microscopy. *Angewandte Chemie, International Edition, 44,* 5563.

Tillmann, K., Lentzen, M., & Rosenfeld, R. (2000). Impact of column bending in high resolution transmission electron microscopy on the strain evaluation of GaAs/InAs/GaAs heterostructures. *Ultramicroscopy, 83,* 111–128.

Tonomura, A. (1987). Applications of electron holography. *Reviews of Modern Physics, 59,* 639.

Tonomura, A., Allard, L. F., Pozzi, G., Joy, D. C., & Ono, Y. A. (Eds.). (1995). *Electron holography. Proceedings of the international workshop on electron holography. North-Holland delta series* (pp. 267–276).

Toyoshima, C., & Unwin, N. (1988). Contrast transfer for frozen-hydrated specimens: Determination from pairs of defocused images. *Ultramicroscopy, 25*(4), 279–291.

Treacy, M. M. J., Gibson, J. M., & Howie, A. (1985). On elastic relaxation and long wavelength microstructures in spinodally decomposed $In_xGa_{1-x}As_yP_{1-y}$ epitaxial layers. *Philosophical Magazine, 51,* 389–417.

Twitchett-Harrison, A. C., Yates, T. J. V., Newcomb, S. B., Dunin-Borkowski, R. E., & Midgley, P. A. (2007). High-resolution three-dimensional mapping of semiconductor dopant potentials. *Nano Letters, 7*(7), 2020–2023.

Usov, N. A., Zhukov, A., & Gonzalez, J. (2006). Remanent magnetization states in soft magnetic nanowires. *IEEE Transactions on Magnetics, 42*(10), 3063–3065.

Usuda, K., Numata, T., Irisawa, T., Hirashita, N., & Takagi, S. (2005). Strain characterization in SOI and strained-Si on SGOI MOSFET channel using nano-beam electron diffraction (NBD). *Materials Science and Engineering B, 124,* 143–147.

Van Aert, S., Batenburg, K. J., Rossell, M. D., Erni, R., & Van Tendeloo, G. (2011). Three-dimensional atomic imaging of crystalline nanoparticles. *Nature, 470*(7334).

Van den Broek, W., & Koch, C. T. (2012). Method for retrieval of the three-dimensional object potential by inversion of dynamical electron scattering. *Physical Review Letters, 109*(24).

Van den Broek, W., & Koch, C. T. (2013). General framework for quantitative three-dimensional reconstruction from arbitrary detection geometries in TEM. *Physical Review B, 87*(18).

Van den Broek, W., Rosenauer, A., Goris, B., Martinez, G., Bals, S., Van Aert, S., & Van Dyck, D. (2012). Correction of non-linear thickness effects in HAADF STEM electron tomography. *Ultramicroscopy, 116,* 8–12.

Van den Broek, W., Van Aert, S., & Van Dyck, D. (2009). A model based atomic resolution tomographic algorithm. *Ultramicroscopy, 109*(12).

Van Dyck, D., Jinschek, J. R., & Chen, F.-R. (2012). 'Big Bang' tomography as a new route to atomic-resolution electron tomography. *Nature, 486*(7402), 243–246.

Verbeeck, J., Van Dyck, D., Lichte, H., Potapov, P., & Schattschneider, P. (2005). Plasmon holographic experiments: Theoretical framework. *Ultramicroscopy, 102*, 239–255.

Völkl, E., Allard, L. F., & Joy, D. C. (Eds.). (1999). *Introduction to electron holography*. Kluwer Academic/Plenum Publishers.

Vulovic, M., Voortman, L. M., van Vliet, L. J., & Rieger, B. (2014). When to use the projection assumption and the weak-phase object approximation in phase contrast cryo-EM. *Ultramicroscopy, 136*, 61–66.

Wachowiak, A., Wiebe, J., Bode, M., Pietzsch, O., Morgenstern, M., & Wiesendanger, R. (2002). Direct observation of internal spin structure of magnetic vortex cores. *Science, 298*(5593), 577–580.

Wang, F., Zhang, H.-B., Cao, M., Nishi, R., & Takaoka, A. (2010). Determination of the linear attenuation range of electron transmission through film specimens. *Micron, 41*(7), 769–774.

Wang, G., Yu, H., Verbridge, S. S., & Sun, L. (2014). Local filtering fundamentally against wide spectrum. arXiv:1408.6420.

Wang, Y. Y., Li, J., Domenicucci, A., & Bruley, J. (2013). Variable magnification dual lens electron holography for semiconductor junction profiling and strain mapping. *Ultramicroscopy, 124*, 117–129.

Weickenmeier, A., & Kohl, H. (1991). Computation of absorptive form factors for high energy electron diffraction. *Acta Crystallographica Section A, 47*, 590–597.

Wernsdorfer, W., Hasselbach, K., Benoit, A., Barbara, B., Doudin, B., Meier, J., ... Mailly, D. (1997). Measurements of magnetization switching in individual nickel nanowires. *Physical Review B, 55*(17), 11552–11559.

Weyland, M. (2001). *Two and three dimension nanoscale analysis: New techniques and applications*. University of Cambridge.

Weyland, M., Midgley, P. A., & Thomas, J. M. (2001). Electron tomography of nanoparticle catalysts on porous supports: A new technique based on Rutherford scattering. *The Journal of Physical Chemistry B, 105*(33), 7882–7886.

Williams, D. B., & Carter, C. B. (1996). *Transmission electron microscopy. A text book for material science*. Plenum Press.

Winkler, H., & Taylor, K. A. (2006). Accurate marker-free alignment with simultaneous geometry determination and reconstruction of tilt series in electron tomography. *Ultramicroscopy, 106*(3), 240–254.

Wolf, D., Lichte, H., Pozzi, G., Prete, P., & Lovergine, N. (2011). Electron holographic tomography for mapping the three-dimensional distribution of electrostatic potential in III–V semiconductor nanowires. *Applied Physics Letters, 98*(26), 264103.

Wolf, D., Lubk, A., Lenk, A., Sturm, S., & Lichte, H. (2013). Tomographic investigation of Fermi level pinning at focused ion beam milled semiconductor surfaces. *Applied Physics Letters, 103*(26), 264104.

Wolf, D., Lubk, A., Lichte, H., & Friedrich, H. (2010). Towards automated electron holographic tomography for 3D mapping of electrostatic potentials. *Ultramicroscopy, 110*(5), 390–399.

Wolf, D., Lubk, A., Röder, F., & Lichte, H. (2013). Electron holographic tomography. *Current Opinion in Solid State & Materials Science, 17*(3), 126–134.

Wolf, D., Rodriguez, L. A., Béché, A., Javon, E., Serrano, L., Magen, C., ... Snoeck, E. (2015). 3D magnetic induction maps of nanoscale materials revealed by electron holographic tomography. *Chemistry of Materials, 27,* 6771–6778.

Xu, R., Chen, C.-C., Wu, L., Scott, M. C., Theis, W., Ophus, C., ... Miao, J. (2015). Three-dimensional coordinates of individual atoms in materials revealed by electron tomography. *Nature Materials, 14*(11), 1099–1103.

Yurtsever, A., Weyland, M., & Muller, D. A. (2006). Three-dimensional imaging of non-spherical silicon nanoparticles embedded in silicon oxide by plasmon tomography. *Applied Physics Letters, 89*(15), 151920.

Zhang, H.-R., Egerton, R. F., & Malac, M. (2012). Local thickness measurement through scattering contrast and electron energy-loss spectroscopy. *Micron, 43*(1), 8–15.

Zhang, Y.-F., & Liu, Z.-F. (2006). Pressure induced reactivity change on the side-wall of a carbon nanotube: A case study on the addition of singlet O_2. *Carbon, 44*(5), 928–938.

Zheng, C., Wong-Leung, J., Gao, Q., Tan, H. H., Jagadish, C., & Etheridge, J. (2013). Polarity-driven 3-fold symmetry of GaAs/AlGaAs core multishell nanowires. *Nano Letters, 13*(8), 3742–3748.

Ziese, U., Kübel, C., Verkleij, A. J., & Koster, A. J. (2002). Three-dimensional localization of ultrasmall immuno-gold labels by HAADF-STEM tomography. *Journal of Structural Biology, 138*(1–2), 58–62.

CHAPTER SEVEN

Summary and Outlook

Axel Lubk

Institute for Structure Physics, Physics Department, Faculty of Mathematics and Natural Sciences, Technical University of Dresden, Dresden, Germany

e-mail address: axel.lubk@tu-dresden.de

Contents

> *"The future cannot be predicted, but futures can be invented."*
>
> ***(Dennis Gabor)***

In the previous chapter a comprehensive account of the reconstruction of physical, e.g., electric or magnetic in three dimensions by Electron Holographic Tomography (EHT) has been given. Three aspects have been particularly emphasized. We improved the signal-to-noise ratio and the spatial resolution by revisiting the whole holographic and tomographic reconstruction procedure including various regularization schemes. That involved dedicated preprocessing of the holographic tilt series, adapted sampling schemes and reconstruction algorithms, e.g., taking into account symmetries. We elaborated on strategies suited to overcome the current resolution limit of approximately one nanometer, to eventually facilitate the reconstruction of atomic potentials. Our analysis indicates that diffraction tomography within the framework of the Rytov approximation is suited to take into account the crucial impact of electron diffraction currently preventing higher resolution. The next few years will hopefully witness an experimental proof of this assertion. Last but not least we considerably extended the scope of reconstructable fields, now including not only electric fields but also magnetic fields, (in)elastic cross-sections mirroring the chemical composition, and strain fields. Open problems are the accurate reconstruction of all three Cartesian components of the magnetic field and the reconstruction of inelastic cross-sections pertaining to low-loss excitations such as phonons. In the case of strain fields measured with the help of dark field electron holography, we established the pertaining projection law, from which a tomographic reconstruction from a beam tilt series founded in the principles of Fourier synthesis could be derived. Future work will show how this reconstruction can be realized experimentally, and what spatial resolution may be realized.

To establish EHT as a comprehensive nano-characterization technique, we elaborated on the ramifications of dynamical electron scattering (Chapter 2), the theory of tomographic reconstruction (Chapter 3), the details of the imaging process in the TEM including the impact of partial coherence (Chapter 4), and the various forms of electron holography suited for a combination with tomography (Chapter 5). In all these disciplines, we put special emphasize on the mixed state character of the electrons traveling down the column of a TEM, eventually treating the conventional holographic reconstruction of wave functions as a special case to the more general retrieval of generally mixed quantum states. It turned out that the quantum mechanical phase space and the corresponding distribution of the quantum state, the Wigner function, is particularly useful in providing a general language for the pertaining principles. This allowed to reveal the important characteristics of the conventional wave reconstruction schemes, which are frequently not readily visible in the conventional holographic reconstruction. We vividly summarized the above program, i.e., the combination of holographic wave retrieval and tomographic reconstruction, in the subtitle as "from quantum states to 3D potentials (and back)".

Now, the interested reader might wonder, when do we finally get back. Indeed, this way back, i.e., the reconstruction of a general quantum state, has only partially been realized to date of this work and the pertaining accounts have been given, when elaborating on the various forms of holography in Chapter 5. There, it has been revealed that the reconstruction of the electron's quantum state (scattered from static *and* time-dependent electromagnetic fields) requires a generalization of the various holographic setups, consisting of the acquisition of a quorum of holograms with varying imaging parameters (e.g., biprism voltage) and the application of different reconstruction algorithms. In particular, in the course of elaborating on Focal Series Inline Holography, it has been revealed that the associated mixed quantum state reconstruction is founded in the tomographic principle. The intensities recorded in different focal planes correspond to projections of quantum mechanical phase space densities rotated as a function of the defocus. Hence, the retrieval of the quantum state from a quorum of measurements is referred to as quantum state tomography and we showed that similar reconstruction principles also pertain to other generalized holographies. It is this connection, which provides a second important impetus for jointly elaborating on electron holography and tomography. In Chapter 5, we therefore elaborated on the generalization of the four most prominent holographic setups, namely Off-Axis Holography, Focal Series Inline Holography, Transport-of-Intensity Inline Holography, and Ptychography, particularly making use of the quantum mechanical phase space calculus and the tomographic principle.

Based on these considerations we may now set out to compare the different forms of electron holography and their perspective for quantum state reconstruction. First of all, we have noted that a variety of experimental limitations, imposed by the electron optics employed in current TEMs, prevents a complete reconstruction of a generally four-dimensional electron quantum state, no matter which holographic scheme. Quantum state tomography based on Focal Series Inline Holography requires the ability to set any line defocus between near and far field with a varying orientation, which is currently beyond the instrumental capabilities of a modern TEM. Off-Axis Holography employing rotatable biprisms facilitates the recording of a sufficient number of phase space sections required for a quantum state reconstruction. However, the varying aberrations and the beam shaping influence of the biprism, such as shadowing and Fresnel scattering, seriously complicate the reconstruction. Finally, generalized transport-of-intensity holography only requires a differential change of line defoci, which can be accomplished by modern TEMs. However, the analyzable number of phase space moments is severely limited, here. Ptychography elegantly circumvents all these issues by reconstructing a generalized object transparency instead of the quantum state of the scattered electrons. This requires the single axial scattering approximation to be valid, which seriously limits the applicability of Ptychography in the atomic–resolution domain, where dynamic scattering effects become significant.

Until the above limitations have been overcome, the results of a quantum state reconstruction depend on the optimal choice of the holographic method or combinations of which. For instance, Ptychography is optimal within the scope of single axial scattering. Generalized Transport-of-Intensity quantum state reconstructions may be used, if only the first few generalized currents, e.g., the intensity, the probability current and the momentum variance, are required. Off-Axis Holography is well-suited to study coherence over comparatively large distances in a quantum state and focal series holography may be applied to quantum states of particular symmetry, in particular one-dimensional systems.

Once, the above experimental limitations are sorted out, the reward for the increased experimental complexity of quantum state reconstructions is the increased information obtainable from the scattering process, i.e., the specimen. One intriguing field of application is the characterization of inelastic scattering including the coherence of the inelastically scattered electrons (Röder & Lubk, 2014), which may be referred to as coherence spectroscopy. That information can be linked to correlation phenomena in inelastic excitations of solids, such as surface plasmons. A second application concerns TEM studies of static electric, magnetic and strain fields employing strongly incoherent electron sources. Such electron sources are currently used in the emerging field of ultrafast TEM (UTEM), which, e.g., employ ultra-short LASER pulses for photo-excitation of electrons from extended, i.e., spatial incoherent flat cathodes (Barwick, Flannigan, & Zewail, 2009; Kwon & Zewail, 2010; Flannigan & Zewail, 2012). Last but not least Gabor's original idea of using holography for the a-posteriori correction of coherent aberrations may be extended to incoherent ones (nowadays occasionally referred to as superresolution), provided that the whole mixed quantum state can be retrieved.

The above lines show that in order to remove the persisting limitations in the tomographic reconstruction of physical fields and quantum states, a sustainable development of the instrumentation is of paramount importance. To further improve Electron Holographic Tomography, developments in almost all parts of the TEM are necessary. We will shortly discuss some of the most beneficial ones. Computer-controlled specimen stages permitting the acquisition of 360° tilt series around all three rotation axis would allow the tomographic reconstruction of all components of magnetic vector fields, without additional assumptions. Moreover, the full tilt range is an important prerequisite for the reconstruction of atomic electric fields using the diffraction tomography within the scope of the Rytov approximation. Dedicated off-axis holographic setups incorporating multiple biprisms in combination with freely configurable energy filters would allow to study low-loss excitations such as phonons exploiting the coherence filter implicit to the off-axis setup. More general improvements pertain to the disclosure of optical parameters of commercial TEMs, such as lens distances or focal lengths, as well as the improved computer control of TEMs, which would greatly simplify the setup of advanced holographic setups such long-range focal series. Higher-brightness sources are suited to improve the signal-to-noise ratio as well as the resolution in all discussed holographic techniques. Similarly, the repeated acquisition of micrographs in a holographic tilt series or quorum put particularly high demands on both the microscope and specimen stability. The former may be improved, e.g., in highly stable environments such as provided by dedicated TEM laboratories. Last but not least, the sheer amount of data processed in the reconstruction of an arbitrary four-dimensional phase space distribution strongly benefits from the ever-increasing computational resources.

Summing up, we can expect further progress in a not too distant future, which is suited to strengthen the development of Electron Holography and Tomography into a versatile non-destructive 3D nano-characterization method for solid state physics and materials science. As history has shown in various instances, such advanced characterization methods are crucial for scientific progress, among other things because of fostering the following characteristic feature of scientific perception

"The most important discoveries will provide answers to questions that we do not yet know how to ask and will concern objects we have not yet imagined."

(John N. Bahcall)

REFERENCES

Barwick, B., Flannigan, D. J., & Zewail, A. H. (2009). Photon–induced near-field electron microscopy. *Nature, 462*, 902.

Flannigan, D. J., & Zewail, A. H. (2012). 4D electron microscopy: Principles and applications. *Accounts of Chemical Research, 45*(10), 1828–1839.

Kwon, O.-H., & Zewail, A. H. (2010). 4D electron tomography. *Science, 328*, 1668.

Röder, F., & Lubk, A. (2014). Transfer and reconstruction of the density matrix in off-axis electron holography. *Ultramicroscopy, 146*, 103–116.

APPENDIX A

Dynamical Eigenenergy

In this Appendix, we set out to derive first order correction terms to the axial scattering approximation, forming the basis for the various holographic
and tomographic field mappings discussed throughout this volume. Following Section 2.2 the 2D Boch wave approximation serves as the most simple approach to describe dynamical scattering within the paraxial regime,
hence may be used to compute first-order correction terms most efficiently.

We start off by noting the paraxial equation (2.21) in reciprocal space

$$\frac{\partial \tilde{\Psi}_{\mathbf{g}}(z)}{\partial z} = -i \left(\underbrace{\frac{\mathbf{g}^2 + 2\mathbf{k}_{0\perp}\mathbf{g}}{2k_{0z}}}_{s_{\mathbf{g}}} - \frac{e}{v}\tilde{\Phi}_{\mathbf{g}}\otimes \right) \tilde{\Psi}_{\mathbf{g}}(z), \qquad (A.1)$$

which is obtained by inserting the lateral Fourier expansion of the periodic
potential (2D reciprocal lattice vectors $\mathbf{g}$)

$$\bar{\Phi}(\mathbf{r}, z) = \sum_{\mathbf{g}} \tilde{\Phi}_{\mathbf{g}}(z)e^{i\mathbf{g}\mathbf{r}} \qquad (A.2)$$

and the corresponding wave function with lateral Bloch vector $\mathbf{k}_{0\perp} = (k_{0x}, k_{0y})^T$ (the initial lateral wave vector)

$$\Psi(\mathbf{r}, z) = e^{i\mathbf{k}_{0\perp}\mathbf{r}} \sum \tilde{\Psi}_{\mathbf{g}}(z)e^{i\mathbf{g}\mathbf{r}} \qquad (A.3)$$

into (2.21).

We proceed with setting up a perturbation formalism. When discussing
axial scattering in Section 2.3, it has been noted that medium resolution
imaging techniques employ out-of-zone axis crystal orientations, corresponding to large lateral $\mathbf{k}_{0\perp}$, if z is parallel to some low-index zone axis,
to avoid dynamical scattering. Within the Bloch wave formalism, a large
$\mathbf{k}_{0\perp}$ ensures

$$\forall \mathbf{g} \neq 0 : \left| s_{\mathbf{g}} \right| \gg \frac{e}{v}\tilde{\Phi}_{\mathbf{g}}. \qquad (A.4)$$

Consequently, we may set up a perturbation scheme in the off-diagonal
coupling terms $\frac{e}{v}\tilde{\Phi}_{\mathbf{g}}$, which, in lowest-order, decouples the above equation

Advances in Imaging and Electron Physics, Volume 206
ISSN 1076-5670
https://doi.org/10.1016/S1076-5670(18)30055-7

(A.1) directly, i.e.,

$$\frac{\partial \tilde{\Psi}_{\mathbf{g}}^{(0)}(z)}{\partial z} = -i\left(s_{\mathbf{g}} - \frac{e}{\nu}\tilde{\Phi}_0\right)\tilde{\Psi}_{\mathbf{g}}^{(0)}(z).$$ (A.5)

The solution to this equation is obtained by plugging in the starting conditions, i.e., a plane wave, yielding

$$\tilde{\Psi}_{\mathbf{g}}^{(0)}(z) = \delta_{\mathbf{g},0}e^{i\frac{e}{\nu}\tilde{\Phi}_0 z}.$$ (A.6)

Accordingly, phase of the zero beam is proportional to the projected potential and Glaser identified $\frac{e}{\nu}\tilde{\Phi}_0$ as the index of refraction following the convention used in optics.

To obtain the higher-order terms, we first write down the first order perturbation (in $\tilde{\Phi}_{\mathbf{g}\neq 0}$) to the off-center beams. A particular off-center beam is a solution to the inhomogeneous paraxial equation

$$\frac{\partial \tilde{\Psi}_{\mathbf{g}}(z)}{\partial z} + i\left(s_{\mathbf{g}} - \frac{e}{\nu}\tilde{\Phi}_0\right)\tilde{\Psi}_{\mathbf{g}}(z) = i\frac{e}{\nu}\tilde{\Phi}_{\mathbf{g}}\tilde{\Psi}_0^{(0)}(z)$$ (A.7)

with starting condition $\tilde{\Psi}_{\mathbf{g}}(z) = 0$. Obviously, the inhomogeneous term is first order in $\tilde{\Phi}_{\mathbf{g}\neq 0}$, which is transferred to the solution

$$\tilde{\Psi}_{\mathbf{g}}(z) = \frac{es_{\mathbf{g}}\tilde{\Phi}_{\mathbf{g}}}{\nu}e^{i\frac{e}{\nu}\tilde{\Phi}_0 z}\left(1 - e^{-i\frac{\mathbf{g}^2 + 2\mathbf{k}_{0\perp}\mathbf{g}}{2k_{0z}}z}\right).$$ (A.8)

Consequently, the diffracted beams show an oscillating dependency on z.

Next, we analyze to second order perturbation (in $\tilde{\Phi}_{\mathbf{g}\neq 0}$) for $\mathbf{g} = 0$. This can be done in a self-consistent manner by inserting the formal solution to (A.7) into the right hand side of (A.5), giving[1]

$$\frac{\partial \tilde{\Psi}_0(z)}{\partial z} - i\frac{e}{\nu}\tilde{\Phi}_0\tilde{\Psi}_0(z) =$$

$$-\frac{e^2}{\nu^2}\sum_{\mathbf{g}\neq 0}\left|\tilde{\Phi}_{\mathbf{g}}\right|^2\int_0^z e^{-i\left(s_{\mathbf{g}} - \frac{e}{\nu}\tilde{\Phi}_0\right)(z-z')}\tilde{\Psi}_0(z')\mathrm{d}z'.$$ (A.9)

This differential equation can be solved by substituting

$$\tilde{\Psi}_0(z) \rightarrow e^{i\frac{e}{\nu}\tilde{\Phi}_0 z}\tilde{\Psi}_0(z),$$ (A.10)

---

[1] This self-consistent perturbation approach is closely related to the Random Phase Approximation employed in condensed matter physics.

and noting that (A.9) is a Volterra type integro-differential equation

$$\frac{\partial \tilde{\Psi}_0(z)}{\partial z} = \int_0^z K\left(z - z'\right) \tilde{\Psi}_0(z') dz' \tag{A.11}$$

with a convolution kernel

$$K\left(z - z'\right) = -\frac{e^2}{v^2} \sum_{\mathbf{g} \neq 0} \left|\tilde{\Phi}_{\mathbf{g}}\right|^2 e^{-is_{\mathbf{g}}(z-z')}. \tag{A.12}$$

Such equations can be solved with the help of the Laplace transform (denoted by ˆ) yielding

$$s\hat{\tilde{\Psi}}_0(s) - 1 = \hat{K}(s)\, \hat{\tilde{\Psi}}_0(s) \tag{A.13}$$

with

$$\hat{\tilde{\Psi}}_0(s) = \frac{1}{s - \hat{K}(s)} \tag{A.14}$$

and $\tilde{\Psi}_0(0) = 1$.

The Laplace transform of the convolution kernel reads

$$\hat{K}(s) = -\frac{e^2}{v^2} \sum_{\mathbf{g} \neq 0} \left|\tilde{\Phi}_{\mathbf{g}}\right|^2 \frac{1}{s + is_{\mathbf{g}}} \tag{A.15}$$

yielding

$$\hat{\tilde{\Psi}}_0(s) = \frac{1}{s + \frac{e^2}{v^2} \sum_{\mathbf{g} \neq 0} \left|\tilde{\Phi}_{\mathbf{g}}\right|^2 \frac{1}{s + is_{\mathbf{g}}}}. \tag{A.16}$$

The inverse Laplace transform, yielding the wave function, is now obtained by identifying the poles of the Laplace transform. Each pole transforms into an exponential in position space. To compute the result, we again make use of the perturbation scheme, i.e., we expand the sum in (A.16), containing the perturbation, around the original pole $s = 0$. In the lowest order, one would set $s = 0$ on the right hand side, which results in a purely imaginary shift of that pole

$$\tilde{\Psi}_0(z) = \exp\left(-i\underbrace{\frac{e^2}{v^2} \sum_{\mathbf{g} \neq 0} \left|\tilde{\Phi}_{\mathbf{g}}\right|^2 s_{\mathbf{g}}^{-1}}_{\Xi_{\text{el}}} z\right). \tag{A.17}$$

Consequently, the dynamical scattering corrections act as an additional energy term, which bears some similarity to the Bethe dynamical potential in literature (Reimer, 1989). In a hand–waving manner, it originates from the electrons being scattered into higher-order $\mathbf{g}$-beams and back to the zero beam, which effectively slows them down. For us it is important to note that the dynamical potential depends on the relative orientation of the beam with respect to the sample via $s_{\mathbf{g}}$. Consequently, as a first order approximation, dynamical scattering results in an apparent modulation of the reconstructed potential, which varies throughout the tomographic tilt series. This erroneous effect may be removed by normalizing each projection with the average potential computed from all tilt angles (see Section 6.1.1).

REFERENCES

Reimer, L. (Ed.). (1989). *Transmission electron microscopy*. Springer Verlag.

APPENDIX B

Mean Inner Potentials (Table B.1)

Table B.1 Compilation of Mean Inner Potentials computed from first principles (DFT – Density Functional Theory) or measured with interferometric techniques (EI – Electron Interferometry, EH – Electron Holography, EHT – Electron Holographic Tomography). For the large number of MIP values obtained from diffraction experiments the reader is referred to the literature (e.g., Völkl, Allard, & Joy, 1999)

Material	Method	Φ_0 [V]
H_2O (Harscher, 1999)	EH	3.5 ± 1.2
Be (Jönsson, Hoffmann, & Möllenstedt, 1965)	EI	7.8 ± 0.4
C (Harscher, 1999)	EH	10.74 ± 0.04
(Keller, 1961)	EI	7.8 ± 0.6
Polystyrene (Harscher, 1999)	EH	8.5 ± 0.7
(Wang, Chou, Libera, Voelkl, & Frost, 1998)	EH	8.2 ± 0.9
MgO (Gajdardziska-Josifovska et al., 1993)	EH	13.0 ± 0.8
Al (Keller, 1961)	EI	13 ± 0.4
(Buhl, 1959)	EI	12.4 ± 1
(Hoffmann & Jönsson, 1965)	EI	11.9 ± 0.7
AlAs (Chung, Smith, & McCartney, 2007)	EH	12.1 ± 0.7
(Kruse, Schowalter, Lamoen, Rosenauer, & Gerthsen, 2006)	DFT	12.3
Si (Wolf, Lubk, Lenk, Sturm, & Lichte, 2013)	EHT	13.5 ± 0.3
(Wang, Chou, & Libera, 1997)	EH	12.1 ± 1.3
(Kruse et al., 2006)	DFT	12.6
aSi (Wang et al., 1997)	EH	11.9 ± 0.3
SiO (Wang et al., 1997)	EH	10.1 ± 0.1
Co (Wolf et al., 2015)	EHT	21.5 ± 0.3
Cu (Wang, 2003)	EH	21.2
(Keller, 1961)	EI	23.5 ± 0.6
(Hoffmann & Jönsson, 1965)	EI	20 ± 1

(continued on next page)

© 2018 Elsevier Inc.
All rights reserved.

Table B.1 (*continued*)

Material	Method	Φ_0 [V]
ZnO (Gan, Ahn, Yu, Smith, & McCartney, 2015)	EH	15.3 ± 0.2
(Ding et al., 2015)	EH	14.3 ± 0.3
(Müller et al., 2005)	EH	15.9 ± 1.5
(Schowalter, Rosenauer, Lamoen, Kruse, & Gerthsen, 2006)	DFT	15.75
ZnS (Buhl, 1959)	EI	10.2 ± 0.5
GaAs (Lubk et al., 2014)	EHT	13.9 ± 0.1
(Chung et al., 2007)	EH	14 ± 0.6
(Kruse et al., 2006)	DFT	14.2
Ge (Wolf et al., 2013)	EHT	14.3 ± 0.3
(Li, McCartney, Dunin-Borkowski, & Smith, 1999)	EH	14.3 ± 0.2
(Kruse et al., 2006)	DFT	14.7
InP (Wolf, Lubk, Prete, Lovergine, & Lichte, 2016)	EHT	14.2 ± 0.3
(Kruse et al., 2006)	DFT	13.9
Ag (Keller, 1961)	EI	17.0–21.8
(Buhl, 1959)	EI	20.7 ± 0.7
Au (Lubk et al., 2014)	EHT	27.8 ± 0.2
(Buhl, 1959)	EI	21 ± 2
(Keller, 1961)	EI	22.1–27
(Schowalter et al., 2006)	DFT	30.0
PbS (Gajdardziska-Josifovska et al., 1993)	EH	17.2 ± 0.1

REFERENCES

Buhl, R. (1959). Interferenzmikroskopie mit Elektronenwellen. *Zeitschrift für Physik, 155*, 395.

Chung, S., Smith, D. J., & McCartney, M. R. (2007). Determination of the inelastic mean-free-path and mean inner potential for AlAs and GaAs using off-axis electron holography and convergent beam electron diffraction. *Microscopy and Microanalysis, 13*(05), 329–335.

Ding, Y., Liu, Y., Pradel, K. C., Bando, Y., Fukata, N., & Wang, Z. L. (2015). Quantifying mean inner potential of ZnO nanowires by off-axis electron holography. *Micron, 78*, 67–72.

Gajdardziska-Josifovska, M., McCartney, M. R., Ruijter, W. J., de Smith, D. J., Weiss, J. K., & Zuo, J. M. (1993). Accurate measurements of mean inner potential of crystal wedges using digital electron holograms. *Ultramicroscopy, 50*, 285–299.

Gan, Z., Ahn, S., Yu, H., Smith, D. J., & McCartney, M. R. (2015). Measurement of mean inner potential and inelastic mean free path of ZnO nanowires and nanosheet. *Materials Research Express, 2*(10), 105003.

Harscher, A. (1999). *Elektronenholographie biologischer Objekte: Grundlagen und Anwendungsbeispiele* (PhD thesis).

Hoffmann, H., & Jönsson, C. (1965). *Zeitschrift für Physik, 182*, 360.

Jönsson, C., Hoffmann, H., & Möllenstedt, G. (1965). Messung des mittleren inneren Potentials von Be im Elektronen-Interferometer. *Physik der Kondensierten Materie, 3*, 193.

Keller, M. (1961). Ein Biprisma-Interferometer für Elektronenwellen und seine Anwendung. *Zeitschrift für Physik, 164*, 274–291.

Kruse, P., Schowalter, M., Lamoen, D., Rosenauer, A., & Gerthsen, D. (2006). Determination of the mean inner potential in III–V semiconductors, Si and Ge by density functional theory and electron holography. *Ultramicroscopy, 106*(2), 105–113.

Li, J., McCartney, M. R., Dunin-Borkowski, R. E., & Smith, D. J. (1999). Determination of mean inner potential of germanium using off-axis electron holography. *Acta Crystallographica Section A, 55*, 652–658.

Lubk, A., Wolf, D., Kern, F., Röder, F., Prete, P., Lovergine, N., & Lichte, H. (2014). Nanoscale three-dimensional reconstruction of elastic and inelastic mean free path lengths by electron holographic tomography. *Applied Physics Letters, 105*(17), 173101.

Müller, E., Kruse, P., Gerthsen, D., Schowalter, M., Rosenauer, A., Lamoen, D., . . . Waag, A. (2005). Measurement of the mean inner potential of ZnO nanorods by transmission electron holography. *Applied Physics Letters, 86*(15), 154108.

Schowalter, M., Rosenauer, A., Lamoen, D., Kruse, P., & Gerthsen, D. (2006). Ab initio computation of the mean inner Coulomb potential of wurtzite-type semiconductors and gold. *Applied Physics Letters, 88*(23), 232108.

Völkl, E., Allard, L. F., & Joy, D. C. (Eds.). (1999). *Introduction to electron holography*. Kluwer Academic/Plenum Publishers.

Wang, Y. C., Chou, T. M., & Libera, M. (1997). Transmission electron holography of silicon nanospheres with surface oxide layers. *Applied Physics Letters, 70*, 1296–1298.

Wang, Y. C., Chou, T. M., Libera, M., Voelkl, E., & Frost, B. (1998). Measurement of polystyrene mean inner potential by transmission electron holography of latex spheres. *Microscopy and Microanalysis, 4*, 146–157.

Wang, Y. G. (2003). Determination of inelastic mean free path by electron holography along with electron dynamic calculation. *Chinese Physics Letters, 20*, 888.

Wolf, D., Lubk, A., Lenk, A., Sturm, S., & Lichte, H. (2013). Tomographic investigation of Fermi level pinning at focused ion beam milled semiconductor surfaces. *Applied Physics Letters, 103*(26), 264104.

Wolf, D., Lubk, A., Prete, P., Lovergine, N., & Lichte, H. (2016). 3D mapping of nanoscale electric potentials in semiconductor structures using electron-holographic tomography. *Journal of Physics D: Applied Physics, 49*(36), 364004.

Wolf, D., Rodriguez, L. A., Béché, A., Javon, E., Serrano, L., Magen, C., . . . Snoeck, E. (2015). 3D magnetic induction maps of nanoscale materials revealed by electron holographic tomography. *Chemistry of Materials, 27*, 6771–6778.

APPENDIX C

(In)elastic Mean Free Path Lengths (Table C.1)

Advances in Imaging and Electron Physics, Volume 206
ISSN 1076-5670
https://doi.org/10.1016/S1076-5670(18)30057-0

Table C.1 Compilation of elastic and inelastic MFPLs evaluated from literature (see Kern, Wolf, Pschera, & Lubk, 2016 for details). In those cases, where a high-resolution mode (objective lens switched on) without objective aperture was used, $\alpha = \infty$ is indicated (presumably the actual value is in the range 30–40 mrad). As the aperture sizes in medium resolution modes have not been explicitly noted in the literature, they were estimated from the type of TEM instrument and imaging mode employed. Hence, the aperture values, indicated with a symbol, should be treated with some caution. The theoretical values are computed from (2.53) and (2.61), respectively

Material	U_A (kV)	α (mrad)	λ_{sb}^{hol} (nm)	λ_{el}^{hol} (nm)	λ_{el}^{theo} (nm)	λ_{inel}^{hol} (nm)	λ_{inel}^{EEL} (nm)	λ_{inel}^{theo} (nm)
C (Harscher, 1999)	100	∞	51 (2)				46 (Angert, Burmester, Dinges, Rose, & Schröder, 1996)	
Al (Kern et al., 2016)	200	17		270 (25)	269	175 (17)		161
Si (McCartney & Smith, 1994)	100	∞	92 (7)				91 (4) (Lee, Ikematsu, & Shindo, 2002)	
Si (Wolf, 2010)	200	4		200	160	200	190 (6)[β] (Lee et al., 2002)	
Si (Pantzer et al., 2014)	200	4[γ]	105 (2)		160		190 (6)[β] (Lee et al., 2002)	
GaAs (Lubk et al., 2014)	300	22		161 (8)	169	208 (14)		199
GaAs (Chung, Smith, & McCartney, 2007)	200	4[α]	67 (4)		59			162
AlAs (Chung et al., 2007)	200	4[α]	77 (4)		87			162

(continued on next page)

Table C.1 (*continued*)

Material	U_A (kV)	α (mrad)	λ_{sb}^{hol} (nm)	λ_{el}^{hol} (nm)	λ_{el}^{theo} (nm)	λ_{inel}^{hol} (nm)	λ_{inel}^{EEL} (nm)	λ_{inel}^{theo} (nm)
InP (Kern et al., 2016)	200	4			62			166
ZnO (Gan, Ahn, Yu, Smith, & McCartney, 2015)	200	4[α]	55 (3)	77	58	212	161 (15) (Berta, Ma, & Wang, 2002)	
SiO (Kern et al., 2016)	200	4		145	189	230	234 (9)[β] (Lee et al., 2002)	
MgO (McCartney and Smith, 1994)	100	∞	71 (5)				77 (Iakoubovskii, Mitsuishi, Nakayama, & Furuya, 2008)	
Cu (Wang, 2003)	200	∞	96				100 (Iakoubovskii et al., 2008)	
Au (Lubk et al., 2014)	300	22		37 (7)	27	56 (8)	101 (Iakoubovskii et al., 2008)	

[α] The Philips CM200 FEG used in these studies is identical to the one installed in our lab, hence the same effective aperture in Lorentz mode is assumed.

[β] These values are estimated by a linear interpolation of the mean free path lengths from the values given in Lee et al. (2002) at 4 mrad.

[γ] The JEOL JEM-2100F used here employed the objective mini-lens as Lorentz lens, which is very similar to the Lorentz lens in a Philips CM200 FEG.

REFERENCES

Angert, I., Burmester, C., Dinges, C., Rose, H., & Schröder, R. R. (1996). Elastic and inelastic scattering cross-sections of amorphous layers of carbon and vitrified ice. *Ultramicroscopy*, *63*(3–4), 181–192.

Berta, Y., Ma, C., & Wang, Z. L. (2002). Measuring the aspect ratios of ZnO nanobelts. *Micron*, *33*, 687.

Chung, S., Smith, D. J., & McCartney, M. R. (2007). Determination of the inelastic mean-free-path and mean inner potential for AlAs and GaAs using off-axis electron holography and convergent beam electron diffraction. *Microscopy and Microanalysis*, *13*, 329–335.

Gan, Z., Ahn, S., Yu, H., Smith, D. J., & McCartney, M. R. (2015). Measurement of mean inner potential and inelastic mean free path of ZnO nanowires and nanosheet. *Materials Research Express*, *2*(10), 105003.

Harscher, A. (1999). *Elektronenholographie biologischer Objekte: Grundlagen und Anwendungsbeispiele* (PhD thesis).

Iakoubovskii, K., Mitsuishi, K., Nakayama, Y., & Furuya, K. (2008). Mean free path of inelastic electron scattering in elemental solids and oxides using transmission electron microscopy: Atomic number dependent oscillatory behavior. *Physical Review B*, *77*, 104102.

Kern, F., Wolf, D., Pschera, P., & Lubk, A. (2016). Quantitative determination of elastic and inelastic attenuation coefficients by off-axis electron holography. *Ultramicroscopy*, *171*, 26–33.

Lee, C. H., Ikematsu, Y., & Shindo, D. (2002). Measurement of mean free paths for inelastic electron scattering of Si and SiO_2. *Journal of Electron Microscopy*, *51*, 143–148.

Lubk, A., Wolf, D., Kern, F., Röder, F., Prete, P., Lovergine, N., & Lichte, H. (2014). Nanoscale three-dimensional reconstruction of elastic and inelastic mean free path lengths by electron holographic tomography. *Applied Physics Letters*, *105*(17), 173101.

McCartney, M., & Smith, D. J. (1994). Direct observation of potential distribution across Si/Si p–n junctions using off-axis electron holography. *Applied Physics Letters*, *65*(20), 2603.

Pantzer, A., Vakahy, A., Eliyahou, Z., Levi, G., Horvitz, D., & Kohn, A. (2014). Dopant mapping in thin FIB prepared silicon samples by off-axis electron holography. *Ultramicroscopy*, *138*, 36–45.

Wang, Y. G. (2003). Determination of inelastic mean free path by electron holography along with electron dynamic calculation. *Chinese Physics Letters*, *20*, 888.

Wolf, D. (2010). *Elektronen-Holographische Tomographie zur 3D-Abbildung von elektrostatischen Potentialen in Nanostrukturen* (Dissertation). Technische Universität Dresden.

INDEX

Printed in the United States
By Bookmasters